THE WORLD BOOK ENCYCLOPEDIA OF
PEOPLE AND PLACES

7

PEOPLE ON THE MOVE: SETTLING THE WORLD'S REGIONS —AND INDEX

a Scott Fetzer company
Chicago
www.worldbookonline.com

For information about other World Book publications,
visit our website at http://www.worldbookonline.com
or call 1-800-WORLDBK (1-800-967-5325).

For information about sales to schools and libraries, call
1-800-975-3250 (United States);
1-800-837-5365 (Canada).

Library of Congress Cataloging-in-Publication Data

The World Book encyclopedia of people and places.
   v. cm.
   Summary: "A 7-volume illustrated, alphabetically arranged
set that presents profiles of individual nations and other
political/geographical units, including an overview of history,
geography, economy, people, culture, and government of each.
Includes a history of the settlement of each world region
based on archaeological findings; a cumulative index; and Web
resources"--Provided by publisher.
   Includes index.
   ISBN 978-0-7166-3758-5
   1. Encyclopedias and dictionaries. 2. Geography--
Encyclopedias. I. World Book, Inc. Title: Encyclopedia of
people and places.
   AE5.W563 2011
   030--dc22
                                   2010011919

This edition ISBN: 978-0-7166-3760-8

Printed in Hong Kong by Toppan Printing Co. (H.K.) LTD
3rd printing, revised, August 2012

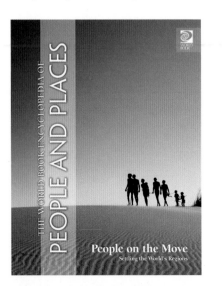

Cover image:
Bushmen near Kalahari
Gemsbok National Park,
South Africa

© Roger De La Harpe,
Gallo Images/Corbis

CONTENTS

INTRODUCTION

Human beings have been on the move for nearly all of history. The earliest humans emerged in the continent now known as Africa about 2 million years ago. Within 300,000 years, groups of primitive humans began migrating out of Africa. By 600,000 years ago, primitive humans were also living in what are now Europe and Asia. Many archaeologists believe the first modern-looking people left Africa about 120,000 years ago. Perhaps as early as 30,000 years ago, people were already spreading into North and South America. Today, there are nearly 7 billion people on Earth, and permanent settlements exist on every continent except Antarctica.

Like many other animal species, early humans migrated to find more plentiful sources of food, to escape threatening changes in climate, or to take advantage of better living conditions. People still migrate for these reasons as well as for political and social purposes. Each year, an estimated 740 million people move within their country and another 200 million people emigrate to a different country, according to the United Nations Development Programme. In the United States alone, nearly 9 million people move annually to a different state or out of the country.

This book tells the story of the journeys people made as they settled the regions of the world and how scientists have traced their migration routes. It's a complicated story, and scientists are still working on—and arguing about—the details.

Human beings are a fairly recent chapter in the 3.5-billion-year story of life on Earth. How did we manage to spread out so far in only about 60,000 years?

OUT OF AFRICA

The earliest human beings emerged about 2 million years ago in what is now east Africa. Fossil evidence suggests that by 1.8 million years ago, there were a number of different *species* (kinds) of humans. Scientists classify all these species using the name *Homo* (Latin for *man* or *human being*). About 1.7 million years ago, members of a species of early human called *Homo erectus* (upright human being) migrated out of Africa. Fossil remains show that *H. erectus* spread as far as modern Java in Southeast Asia before moving into what are now Europe and northern China. Descendants of *H. erectus*, known as Neandertals, spread through much of Europe and the Middle East.

Most scientists believe that *Homo erectus* also gave rise to our own species, *Homo sapiens* (wise human). Some anthropologists have theorized that modern-looking humans evolved independently in several regions inhabited by *H. erectus*. But most anthropologists support another theory called "Out of Africa." According to this theory, early-modern humans appeared first in Africa between 150,000 and 200,000 years ago. Anthropologists often refer to modern humans as *Homo sapiens sapiens*. This name is used for physically modern individuals, as opposed to *archaic* (ancient) members of the species.

Fossil evidence showing significant changes in body shape and brain size has provided important evidence for the transition from archaic to modern forms of human being. For example, a 195,000-year-old fossil skull found at a site in Ethiopia shows the beginnings of modern features, including a large, rounded braincase, and a projecting chin. Earlier *Homo* species had sloping foreheads, large brow ridges, and receding chins.

Anthropologists have also unearthed early-modern fossils from the same time period in North Africa. For instance, a group of researchers in 2007 reported finding the jawbone of a modern-human child at a site in Morocco. Using several advanced dating techniques, the scientists determined that the fossil was from a child who lived about 160,000 years ago. The researchers said this and other recently discovered fossils in the region suggest that North Africa may have been more important in the development of modern humans than most anthropologists had thought.

According to the Out of Africa theory, archaic humans stayed in Africa for tens of thousands of years. Some of them migrated within the continent but not beyond it. Then, about 160,000 years ago, groups of archaic humans migrated northward to lands bordering the eastern end of what is now known as the Mediterranean Sea. Scientists have found no evidence that these early migrants survived for long or spread throughout the region.

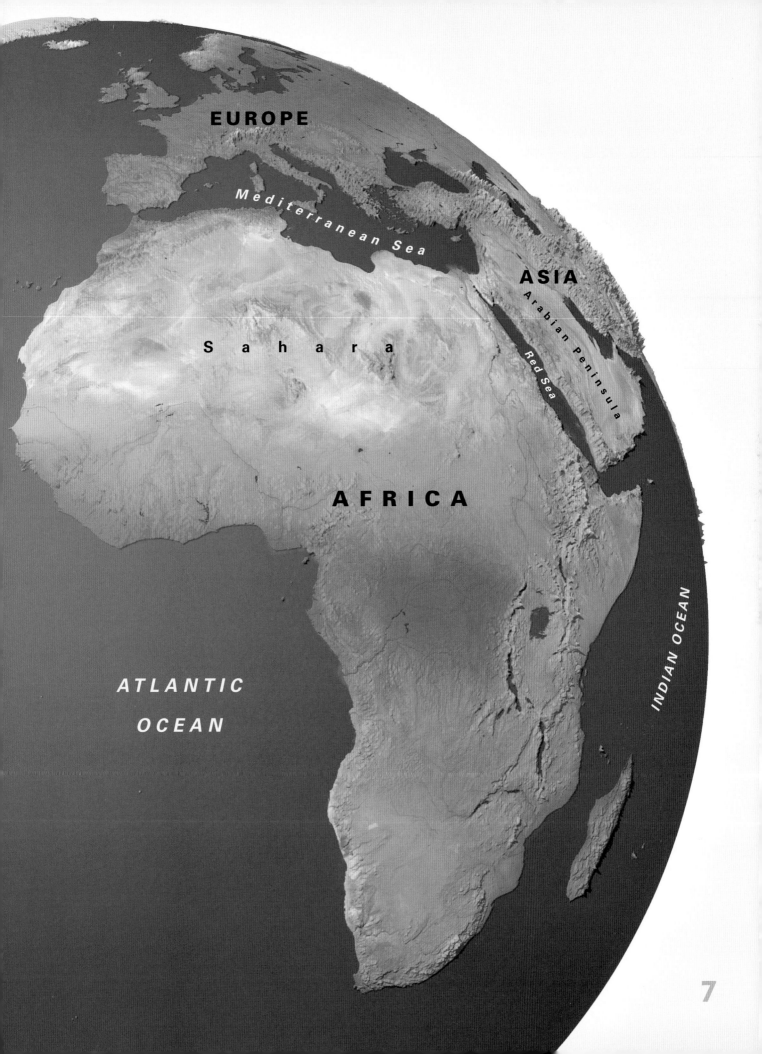

EUROPE

Mediterranean Sea

ASIA

Arabian Peninsula

Red Sea

Sahara

AFRICA

ATLANTIC

OCEAN

INDIAN OCEAN

7

LEAVING AFRICA

Why did early humans leave their African homeland? Some researchers think that climate changes in Africa were the reason. Africa has undergone several major climate shifts during the past 200,000 years. In some periods, the continent was wetter. During those times, what is now the Sahara was covered with trees and lush vegetation. In other times, Africa was gripped by terrible droughts that returned the Sahara to desert conditions and even caused the desert to expand.

The main climate theory for early-human migrations holds that during a wetter period, people were able to move north from sub-Saharan Africa (Africa south of the Sahara). People were able to travel through forests, which offered readily available food, rather than through perilous desert.

In 2009, a team of German and New Zealand scientists reported research that pinpointed the periods of drought and rainfall in Africa over time. In particular, they studied ocean sediments collected from the sea floor off the coast of Guinea in west Africa. The sediments included deposits created by dust blown from the African mainland. Mixed with the dust were waxes from plant leaves. The depth of each deposit could be linked to a particular time period. The deeper a deposit was, the older it was.

The scientists' analysis of the plant waxes in each level of sediment revealed the kinds and amounts of plants growing in Africa during various periods. The scientists concluded that Africa had three wet-climate periods over the past 200,000 years. Those periods were from 120,000 to 110,000 years ago, from 50,000 to 45,000 years ago, and from 10,000 to 8,000 years ago.

The first of those wet periods corresponds to the time of the first migration, so that migration may well have been spurred by wetter conditions. But what about the second migration, which occurred about 65,000 years ago? According to the researchers, that was a time of drought and expanding deserts. There must have been a different reason for the second migration.

Some scientists think the second migrations were spurred by such cultural developments as improved toolmaking and language abilities. In 2008, a group of scientists at the University

Footprints fossilized in volcanic ash are evidence that prehuman ancestors were walking upright on two legs at least 3.6 million years ago. The footprints were discovered in 1978 at a site called Laetoli in Tanzania. Additional footprints discovered since then suggest that upright walking, a feature unique to humans and human ancestors, originated more than 6 million years ago.

of Wollongong in Australia published research supporting this theory. These abilities might have given people the courage to head off into unknown lands in the first migration. And by about 60,000 years ago, humans had learned to make two-sided spear points, making them more effective hunters. For the first time, people had begun gathering shellfish and other seafoods. They had also begun to make decorative personal items, including shell necklaces. The scientists have found such artifacts at nine sites in South Africa.

Fine-grained stone, needed for skillfully made tools and weapons, was transported from miles away. These advances, the scientists said, indicate that humans had evolved a fairly sophisticated culture that included "high-level communication" and complex social and even trade networks.

Ancient rivers in the once-lush Sahara may have provided a route for the first humans who migrated from east Africa to the Middle East 120,000 years ago. Scientists located the river channels, buried under desert sand, in 2008 using satellite-mounted radar. Archaeologists have long believed the first migrants traveled northward through the Nile River Valley.

Homo erectus (foreground) and Australopithecus boisei (background), in an artist's illustration, are among the wide variety of hominids (humans and human ancestors) that have lived on Earth over the past 6 million years. The small-brained A. boisei, which lived in eastern Africa, died out by 1 million years ago. Cultural advantages, such as stone tools, helped Homo erectus survive and spread over much of the world.

THE FIRST MIGRATION

About 120,000 years ago, a group of early-modern humans left Africa. They and their descendants moved northward. Eventually, they settled in the region now known as the Levant—lands bordering the eastern end of the Mediterranean Sea.

Scientists had long thought that the Africans in this migration must have traveled northward through the Nile River Valley. Recent research, however, indicates that they followed several other rivers that existed at the time. This finding was reported in 2008 by a team of British scientists.

The researchers studied satellite images of the Sahara. The images, made with radar, revealed traces of several rivers that once ran north across the Sahara to the Mediterranean Sea. The migrants evidently followed these rivers northward to the coast. When they reached the coast, they turned to the east and eventually arrived in the Levant. The scientists said stone tools from about 120,000 years ago found at scattered sites along Africa's Mediterranean coast lend weight to this theory.

In the Levant, the newcomers established a number of settlements. Archaeologists have excavated several caves with many early-modern remains and tools. A site called Hayonim Cave in present-day Israel, for example, shows that the cave was occupied by early-modern humans until about 100,000 years ago. Findings at the site include flint objects and cooking hearths.

For reasons that archaeologists have not been able to determine, those first migrants from Africa apparently never went any farther than the Levant. They and their descendants lived in that one area for about 30,000 years—until about 90,000 years ago. Then they apparently either died out or moved back to Africa. Advancing desert conditions in the Levant may have been the cause of their disappearance. The human migration that would finally populate the world would not occur for tens of thousands of years.

Stone tools are among the artifacts unearthed at varying levels of Hayonim Cave in what is now Israel. Archaeologists have also found the remains of many animals. Hayonim Cave's occupants almost certainly included descendants of people who left Africa during the first migration, about 120,000 years ago.

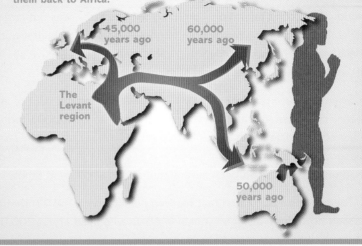

45,000 years ago

60,000 years ago

The Levant region

50,000 years ago

TRACKING HUMAN MIGRATIONS

Scientists have a number of ways of retracing the migration routes of early humans and to determine the relationships between different groups. The study of fossils and stone tools has long been a useful method of tracking human development and living styles. By examining skulls and skeletons, scientists can tell if fossil remains are those of *Homo sapiens* or of an earlier species. By measuring certain radioative materials in the rocks in which fossils were found, scientists can calculate when the fossils were buried.

Technology also changed. Early-modern and modern humans developed several styles of toolmaking, known to archaeologists as "industries." Each new industry was a step forward in design, with stone blades and other tools becoming more efficient.

In recent years, scientists have also been able to track early migrations and the interrelationships of populations with genetics. The genetic material of all organisms, DNA, undergoes changes, called mutations, at a steady rate over time. Because they occur at a known rate, mutations—often referred to as markers—can be used as a "genetic clock." Two kinds of DNA are used as genetic clocks. One is mitochondrial DNA, which is found in cell structures called mitochondria. Mitochondria are inherited only from one's mother, so mitochondrial DNA (mtDNA) can be used to study direct lines of maternal descent. Another genetic clock is based on the Y chromosome, which is inherited only from one's father. Y-chromosome DNA enables scientists to trace lines of paternal descent.

Scientists study the mtDNA and Y-chromosome DNA of present-day populations to learn how peoples are related. By comparing the number of mutations that are shared by two groups, scientists can determine how long ago their ancestors were part of a population that diverged into new groups.

THE SECOND EXODUS

About 65,000 years ago, many anthropologists think, physically modern humans again left Africa. This time, they and their descendants spread around the world. The second migration—probably like the first—was not large. It probably consisted of just a few hundred people who lived as hunter-gatherers. Why these modern humans left Africa is not known. They may have set out in search of lands where the hunting was better.

Scientists have published many studies supporting the "Out of Africa" theory of human migration. Among these studies was a 2005 report by a team of scientists led by geneticist Vincent Macaulay of the University of Glasgow, Scotland. Their conclusion, based on mtDNA analysis, disputed the theories that there had been several migrations at different times.

The scientists calculated that modern humans carry mtDNA that has been passed down from a single woman in Africa who lived about 200,000 to 150,000 years ago. Modern humans also carry Y-chromosome DNA from a single man who lived perhaps 90,000 years ago. Those two individuals are sometimes referred to as Mitochondrial Eve and Y-chromosome Adam.

Workers map the location of stone tools (right in photo) found in India beneath layers of ash ejected by the eruption of Mount Toba on the island of Sumatra 73,000 years ago. The tools, found by a team led by scientists at Oxford University in the United Kingdom, challenge the theory that the eruption caused a steep decline in human populations in Africa before the second wave of migrations 65,000 years ago. Similarities in the tools found above and below the ash layer suggest that the same types of people lived in the area before and after the eruption.

AN INCREDIBLE ERUPTION AND . . . A "BOTTLENECK"?

Human beings are more alike genetically than other living primates, a group that also includes apes and monkeys. Scientists have wondered why. Some have theorized that before leaving Africa the human race experienced a so-called genetic "bottleneck"—that is, the number of people on the planet fell dramatically, leaving populations that, over generations, came to share a limited pool of genetic material. A genetic bottleneck can be caused by a disease, a drastic climate change, or a natural disaster.

Some scientists think that a huge volcanic eruption that occurred in what is now Indonesia before the second migration from Africa may have caused the human population in Africa to plummet. About 73,000 years ago, Mount Toba, a volcano on the island of Sumatra, exploded with tremendous force. The explosion ejected about 190 cubic miles (800 cubic kilometers) of ash into the atmosphere. It was the greatest volcanic eruption of the past 25 million years. In comparison, the 1980 eruption of Mount St. Helens in Washington state ejected just 0.26 cubic miles (1.08 cubic kilometers) of ash.

Advocates of this theory think the ash blown into the atmosphere by Mount Toba created a six-year "volcanic winter." This phenomenon caused global temperatures to decline in some regions by as much as 28 Fahrenheit degrees (16 Celsius degrees) and killed much of the planet's plant life. These effects may have had devastating consequences for the human race. Scientists theorize that the number of modern humans in Africa was reduced to about 10,000. The eruption may have pushed humans to the brink of extinction.

This woman and man were not the only people alive at those times. They are the only people whose genetic lines have been passed down unbroken to the present. Over time, genetic *mutations* (changes) occurred in those two kinds of DNA. The pattern of mtDNA and Y-chromosome mutations in groups that split apart after leaving Africa became more and more different in later generations. Those differing patterns have made it possible for scientists to trace the path of human migrations backward in time.

What did those early-modern humans look like? They undoubtedly had dark skin, and they probably resembled some present-day peoples who have generally lived in isolation for thousands of years, especially people known as Negritos.

CIRCLING THE ARABIAN PENINSULA

Archaeological evidence shows that modern humans settled South Asia and Australia at least 10,000 years before they populated Europe. The reason, scientists concluded, was that the people of the second migration headed east. Only later did a group break away and move north, eventually reaching Europe. Evidence from studies of genetic material called mitochondrial DNA (mtDNA) has supported this theory.

What was the route out of Africa? The migrants may have left the continent by way of the land bridge now called the Sinai Peninsula, which connects North Africa to the Arabian Peninsula. From there, they could have trekked east. Or they could have traveled down the west coast of the peninsula until they reached the southern coastline and then turned east.

Some anthropologists believe they took a shorter route. They think it is possible that those early migrants crossed a small stretch of water now known as the Bab el-Mandeb Strait. The strait, located between east-central Africa and the bottom of the Arabian Peninsula, connects the Red Sea to the Indian Ocean by way of the Gulf of Aden. It is divided into two channels, the widest of which is about 16 miles (26 kilometers). Anthropologists think early humans would have been capable of building simple boats or rafts for this short trip.

Some scientists have found support for an eastern migration from east Africa and across Arabia in archaeological finds in the lower Arabian Peninsula and South Asia. For example, Amanuel Beyin of the State University of New York at Stony Brook compared stone tools from that same period unearthed in east Africa, the Nile River Valley, Arabia, and the Levant. He said the Arabian and African tools shared the most similarities.

Some of the most persuasive research was published in 2006 by Paul Mellars of Cambridge University in the United Kingdom. Mellars reported that certain 60,000-year-old artifacts excavated in India and Sri Lanka are almost identical with ones found in East Africa from the same period. These artifacts included small stone tools that may have been arrowheads or spearheads. They also included decorative beads carefully made from fragments of ostrich eggshell. Another piece of ostrich eggshell had a distinctive crisscross design used in Africa at that time. Mellars contended that the objects found in India and Sri Lanka were made by people whose ancestors had lived in the same part of Africa and shared the same cultural traditions.

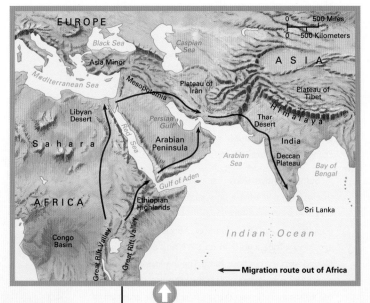

Scientists continue to debate the route taken by modern humans during the second migration from east Africa beginning about 65,000 years ago. Some scientists argue that the migrants followed the Nile River Valley north, then crossed into Asia. Other scientists think that the migrants crossed a strait at the southern edge of what is now known as the Red Sea.

However, other archaeologists have argued that modern humans left Africa from 125,000 to 75,000 years ago by following the Nile Valley in what is now Egypt. They then traveled around the southern border of modern-day Oman.

Evidence from mtDNA studies of modern Arab populations indicates that their early ancestors came from northern regions—they are not descended from early migrants to Arabia. The people who ventured into this harsh region probably had no interest in staying there. They pushed on toward the east in search of better lands.

A scientist searches in the Red Sea for sites that may contain tools and other objects left by migrants traveling from Africa 65,000 years ago. Such ancient sites would have been flooded as the coastline of the Red Sea changed over thousands of years.

ARRIVAL IN SOUTH ASIA

Genetic studies indicate that after leaving Africa, modern humans moved around the Arabian Peninsula fairly rapidly. Migrants seem to have reached the Indian subcontinent by about 63,000 years ago. From there, early-modern humans spread throughout South Asia and Australia.

Some of the evidence supporting the southern migration route out of Africa and the time of India's settlement comes from genetic studies of people known as the Andaman Islanders. The hunter-gatherer inhabitants of these islands, located in the Bay of Bengal off the west coast of India, lived in isolation for many thousands of years.

The Andaman Islanders have long been a mystery to scientists. They have very dark skin and are very short. They look much like African Pygmies, and for many years some people thought there might be a link between them. They are referred to as Negritos. These people are also found in a few other places in Southeast Asia. In the Malay language, Negritos are called "the original people."

The Andaman Islanders were able to live undisturbed for thousands of years because they had a fierce reputation. Anyone who landed on their islands was likely to be killed. In recent years, their population has been dwindling, and they are nearly extinct.

A boy from the Jawara tribe is 1 of fewer than 1,000 members of five tribes that have lived on the Andaman Islands for more than 60,000 years. In a study published in 2008, scientists reported that the Andaman Islanders, also known as Negritos, are the direct descendants of the first people to inhabit much of Southeast Asia.

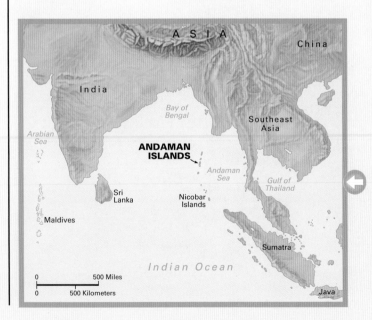

The Andaman and Nicobar islands in the Indian Ocean are a part of India. Most of the islands are uninhabited.

Findings about the origins of the islanders were reported in 2008 by geneticists with the Anthropological Survey of India. When they studied a certain type of genetic material from Andaman Islanders, the researchers found evidence that the Andaman people had settled the islands at least 60,000 years ago and had not interbred with other people since then. The scientists concluded that the Andaman Islanders are the direct descendants of the original settlers of India. They said their findings support the theory that migrants from Africa took the southern route. The newcomers settled in India and other parts of South Asia before some of their kin headed north to central Eurasia. On the way, these migrants may have encountered the Denisovans.

WHO WERE THE DENISOVANS?

In 2010, a team of archaeologists and geneticists reported the discovery of a previously unknown group of early humans. A fossilized finger bone and a single molar tooth had been discovered in a cave called Denisova, in the Altai Mountains of southern Siberia. The fossils were dated from more than 50,000 to 30,000 years ago. Scientists refer to the population represented by these fossils as Denisovans.

Scientists were able to extract DNA from the remains. Their analysis of the genetic material showed that the individual was from a population related to, but separate from, the Neandertals, an extinct early human population known from Europe. The ancestors of Neandertals moved slowly into Europe be-tween about 800,000 and 600,000 years ago. The ancestors of the Denisovans became separate from the Neandertals in Siberia. Because the only fossil remains of these people are a single finger bone and tooth, scientists do not know how they may have differed from Nendertals in appearance.

Surprisingly, scientists found the Denisovans also share about 4 to 6 percent of their genetic material with the present-day inhabitants of New Guinea. The scientists explained that modern humans migrating out of Africa may have encountered Denisovans across southern Asia, acquiring some of their genes through interbreeding as they made their way to Southeast Asia.

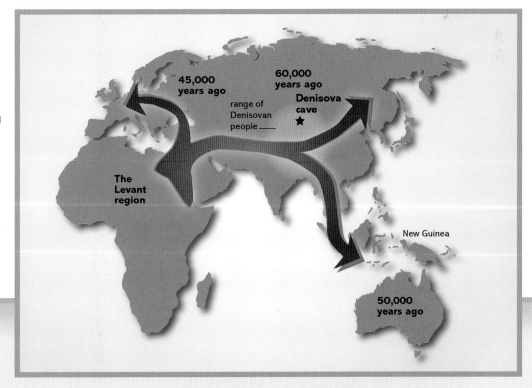

A mysterious group of humans called the Denisovans apparently lived in southern Siberia about 60,000 years ago. Genetic material from these early humans has been found in the present-day inhabitants of New Guinea.

45,000 years ago

60,000 years ago

Denisova cave

range of Denisovan people

The Levant region

New Guinea

50,000 years ago

INTO SOUTH ASIA, NEW GUINEA, AND AUSTRALIA

Who were the first early-modern humans to settle in what is now Southeast Asia more than 60,000 years ago? Genetic studies have revealed that they were the same people known today as Negritos. *Negritos* is a term first used by Spanish explorers to refer to several ethnic groups that still live today in small, isolated populations in Southeast Asia. Negritos are an *aboriginal* people—that is, their ancestors were the first modern-looking human beings to live in the area. They spread from the Andaman Islands of the Indian Ocean to the Malay Peninsula, Indonesia, and the Philippines. Over time, the settlers of the mainland and their descendants mixed with other groups.

The earliest inhabitants of modern-day India, like most early-modern humans, took shelter in caves. Excavations in caves have revealed such evidence of modern human habitation as stone tools from the Middle *Paleolithic* (Stone Age) Period, which ended in this area about 40,000 years ago. Among the tools are a variety of finely made blades and scrapers.

Human remains from this period in India's history are scarce. In fact, the only known human remains from this period are skeletons discovered in the 1970's and 1980's in three caves in Sri Lanka, an island nation off the southern coast of India. The oldest remains were dated to about 36,000 years ago, based on an analysis of charcoal from cooking fires. The technique used to date the Sri Lanka remains is called radiocarbon dating (see sidebar, page 21).

Despite the scarcity of human remains in India, there is abundant evidence of a flourishing culture during the Middle Paleolithic. Colorful painted images of people and animals decorate the walls of many of the caves inhabited by early-modern people. The most famous of these sites is a group of caves called the Bhimbetka Rock Shelters. A rock shelter is a shallow cave at the base of a cliff. The pictures painted on the cave walls depict many animals, including bison, tigers, wild boars, crocodiles, and elephants.

Many of the cave images are painted on top of one another, showing that the same surfaces were used by different artists at different times. The earliest paintings depict mostly animals. But beginning in the Middle Paleolithic Period, the paintings changed. The images include just as humans as animals. Some of the pictures show hunters armed with spears and women and children performing everyday chores. Archaeologists think this development indicates that people were forming complex societies.

19

DEVELOPMENT IN INDIA

Civilization slowly took root in present-day India and Pakistan. By about 6500 B.C., people in India were growing wheat and other crops and raising livestock. India was one of several places around the world where agriculture was invented after the end of the last ice age, about 10,000 years ago. An ice age is a period in Earth's history when ice sheets covered vast regions of land. The most recent ice age occurred mainly during the Pleistocene Epoch, a period that began around 2.6 million years ago and ended about 11,500 years ago.

By 4000 B.C., a culture that scientists call the pre-Harappan had emerged along the Indus River in western India, including what is now Pakistan. The culture is named for the city of Harappa, which became a major center of a great civilization, now known as the Indus Valley civilization. The Indus Valley civilization matured around 2800 B.C. and flourished for about 1,000 years.

At its height, the Indus Valley civilization included hundreds of towns and cities. It may have had a population of more than 5 million. The Indus Valley people developed a system of written symbols, but it is unclear if those symbols represented an actual written language.

The Indus people loved beautiful objects. Archaeologists have found many beautifully crafted pieces of gold jewelry, *terra-cotta* (unglazed earthenware) pottery and figurines, and other works.

Archaeologists who have excavated the remains of Harappa and other cities have concluded that the Indus Valley civilization appears to have been egalitarian. That is, the people were essentially equal, and wealth was evenly distributed. Archaeologists have found no remains of royal palaces, military facilities, slave quarters, or prisons.

Archaeological evidence indicates that the Indus Valley civilization began to decline in about 1900 B.C., with cities and towns losing population. By 1800 B.C., the civilization had collapsed.

Somewhat later, two new civilizations emerged. They were the Dravidian civilization in the south and the Aryan civilization in the north. It is from those two roots that today's Indian civilization arose.

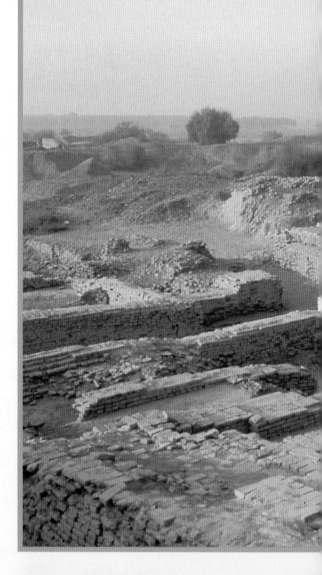

Indus Valley cities were carefully planned. Many buildings were built on mud-brick platforms for protection against seasonal floods. Houses were made of baked or sun-dried brick, and many had two stories. Most homes had a bathing area that was supplied with water from a courtyard well or nearby public well. In larger communities, houses were connected to a citywide drainage system. Other structures include large buildings that may have been used for public purposes.

Images of elephants and a rider decorate the wall of 1 of more than 700 caves at Bhimbetka in central India. The caves, designated as a World Heritage site, contain some of the oldest rock paintings ever found, dating back about 30,000 years. Elephants were first domesticated in India at least 2,300 years ago.

RADIOCARBON DATING

Radioacarbon dating is one of the most valuable scientific tools for determining the age of human living sites. This dating method is based on a phenomenon caused by high-speed particles from space called cosmic rays.

Cosmic rays often smash into atoms of nitrogen, oxygen, and other gases in the atmosphere. The collisions blast the atoms apart into individual particles—protons, neutrons, and electrons. Sometimes one of the neutrons strikes the *nucleus* (center) of a nitrogen atom. The nucleus absorbs the neutron and ejects a proton. The nitrogenatom is thereby transformed into an *isotope* (variation) of carbon, called carbon 14 (C-14).

When an organism—either a plant or an animal—is alive, it constantly absorbs C-14, which is incorporated into its tissues. Some C-14 *decays* (breaks down) into the carbon isotope C-12. After the organism dies, the C-14 in tissues decays slowly to C-12.

C-14 decay occurs at a very precise rate. By measuring the amount of C-14 remaining in a bone or other body remains—or in charcoal from burned wood—scientists can tell how long ago it was part of a living organism. This dating technique is accurate for objects that are up to about 50,000 years old.

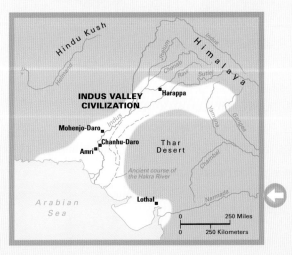

The Indus Valley civilization, also known as the Harappan civilization, flourished from about 2800 B.C. to 1800 B.C., in the vast river plains of what are now Pakistan and northwestern India. Mohenjo-Daro and Harappa were among its largest cities.

MIGRATION INTO MAINLAND SOUTHEAST ASIA

While India was being settled, other early-modern humans moved farther east and south. Archaeological evidence indicates that the first area to be populated after India was mainland Southeast Asia—the region that today consists of Vietnam, Cambodia, Thailand, and several other nations.

Archaeologists have theorized that the expansion into Southeast Asia was fairly rapid. No habitation sites older than about 40,000 B.C. have been found, though people may have begun to spread into this region before that.

One prominent archaeological site in Southeast Asia is Lang Rong Rien Cave in Thailand. Excavations in the cave in the 1980's unearthed human skeletons, pottery, and shells. Radiocarbon dating determined that the bones and artifacts are 37,000 to 40,000 years old, making them the oldest known early-modern remains in Thailand.

Rock shelters from about 33,000 B.C. have been discovered in Vietnam. And in Malaysia, archaeologists found an ancient workshop where early humans made stone tools. This site, named Kota Tampan and dated at 31,000 B.C., has yielded a complete early-human "tool kit." Archaeologists use the term *tool kit* to refer to a collection of tools used for specific tasks. The kit from Kota Tampan included *hammerstones* (stones used as hammers) and stone cores from which blades were flaked. The findings at Kota Tampan showed that toolmaking had reached a significant degree of organization by this time.

Kota Tampan is now set into the side of a hill. But archaeologists have determined that the site was originally on the shore of an ancient lake. Some archaeologists have called Kota Tampan the "missing link" in the route followed by early humans in the path that led them to present-day Australia. The Malaysian Peninsula is a long, narrow strip of land that extends southward almost to the islands of what is now Indonesia. Early humans probably populated the peninsula and then migrated farther south to Indonesia and Australia.

Further evidence for the migration into Southeast Asia comes from the Niah Caves in Borneo, a large island off the Malaysian mainland. This series of vast caves, now a part of a national park, lies in the part of Borneo that belongs to the modern-day country of Malaysia.

In the largest of the caverns, called the Great Cave, an early-modern skull was discovered in 1958. The skull, dubbed the Deep Skull by archaeologists, was radiocarbon dated to about 40,000 years ago. It is the oldest early-human skeletal remains found outside of Africa.

The Niah Caves have also shed light on how people were living and eating in this region some 40,000 years ago. In 2002 and 2003, a group of British and Malaysian archaeologists made a study of plant and animal remnants in the caves. They learned that the cave people ate a varied diet that included freshwater fish, monkeys, and wild pigs, as well as a wide assortment of plants. These findings indicated that the people were well adapted to their rain forest environment and had the skill to acquire all the food they needed.

The dating of the Niah Caves habitation shows that the caves were being used at least 9,000 years before the Kota Tampan tool workshop was in operation. So however long it took early-modern humans to reach Southeast Asia from India, they spread quickly once they got there.

SETTLING INDONESIA AND NEW GUINEA

Between 40,000 to 50,000 years ago, people from the mainland of Southeast Asia began spreading rapidly through the large islands south of the mainland. This migration took them through Indonesia and what is now New Guinea and, finally, to the continent of Australia.

There is little doubt that early-modern humans settled much of Indonesia, though scientists are still arguing about the date. The migrants probably made their way from one island to another on boats or rafts. Very few living sites have been found, though, so archaeologists have largely had to infer migration times and paths.

One site, on the Indonesian island of Sulawesi, consists of rock shelters that have been dated to about 30,000 B.C. Another site is Lene Hara Cave on the island of Timor. Australian archaeologists excavated the cave in the 1990's and found artifacts in a number of layers of the cave floor, indicating that the site was occupied over a long period. The deepest—and thus the oldest—level was dated to perhaps 35,000 years ago. No earlier evidence of modern-human presence in Indonesia has been found.

The migrations of early humans eventually took them to New Guinea. The first settlers in New Guinea are called Papuans. It is from these people that present-day Papua New Guinea—an independent nation on the eastern half of the island—gets its name. Many archaeologists think that the first Papuans arrived in New Guinea by boat or raft about 40,000 years ago. A group of Australian and New Guinean archaeologists reported finding living sites of about that age on the north coast of Papua New Guinea.

A skull from what may be a previously unknown human species (left) that existed until about 13,000 years ago on the Indonesian island of Flores is about half the size of a modern human skull. The skull was among a collection of bones named *Homo floresiensis*. Nicknamed "hobbits," the tiny species stood only about 1 meter (3 feet) tall.

A group of "hobbits," classified by some scientists as a separate species called *Homo floresiensis*, butcher a pygmy elephant on the Indonesian island of Flores, in an artist's reconstruction. Flores's hobbits and elephants may have shrunk because of their long isolation on the small island, a process observed in a number of other mammal species.

Archaeologists have also found *waisted* stone blades—blades that are narrower in the middle than on either end—from this period in both New Guinea and Australia. The scientists think the blades were attached to wooden handles to form powerful axes. The axes may have been used to clear forests to make way for the planting of vegetables and plants with edible fruits and leaves. By 6,000 years ago, the people of New Guinea had cleared massive amounts of forest.

For many years, the Papuans lived in complete isolation from the rest of the world and, often, from one another. When anthropologists began studying the approximately 1 million inhabitants of the New Guinea highlands in the early 1900's, they were astonished by the degree of the people's isolation. They found that groups—divided by mountains and valleys—were often completely unaware of others living just a few miles away.

Because so many groups lived apart from one another, a great number of languages evolved on the island. Today, the *indigenous* (native) people of New Guinea speak an estimated 860 languages. That's more than 1/10 of all the languages in the world!

AN ARCHAEOLOGICAL SURPRISE: THE "HOBBITS"

In 2003, Australian and Indonesian archaeologists working on the present-day Indonesian island of Flores made a surprising discovery. In a cave called Liang Bua, they found the skeleton of a tiny person, about 1 meter (3 feet) tall. At first, the scientists thought the remains were those of a child. But further examination revealed that the skeleton was that of an adult. The skull was very small, with a braincase about the size of a chimpanzee's. (The braincase is the part of the skull enclosing the brain.)

The scientists soon found other remains of the same small size. The researchers believed that they had found a new kind of tiny human. They nicknamed these people "hobbits," a name taken from a series of books called *The Lord of the Rings*.

Radiocarbon dating of charcoal found next to the remains yielded dates of 38,000 to 18,000 years ago. Other evidence indicated that the hobbits occupied the cave from about 95,000 to 13,000 years ago.

In addition to skeletal remains, the scientists found thousands of spear points and other stone tools and the bones of animals the people had hunted and eaten. Those animals included now-extinct pygmy elephants called stegodons.

Scientists disagree on just what species these tiny people represent. Some archaeologists have suggested that the one fairly complete skeleton was an early-modern individual who had suffered from a condition that stunted its growth. The other remains, they said, were probably of children.

But the discovery of another, partial adult skull of the same size indicated strongly that the hobbits are a previously unknown species. Many scientists now refer to the hobbits as *Homo floresiensis*. They believe this species probably descended from *Homo erectus*, a species of early-modern human that may have been an ancestor of modern people.

MIGRATION TO AUSTRALIA

Genetic evidence suggests that Australia was settled at roughly the same time as New Guinea and by the same population of early humans. Migrants from Indonesia apparently followed a single route. Then they split, some of them going to New Guinea, others to Australia.

That conclusion was reported in 2007 by a group of geneticists and archaeologists at Cambridge University in the United Kingdom and elsewhere. The scientists based their finding on studies of genetic material from New Guinea Papuans and Australian Aborigines. They discovered that the two groups share some material dating from ancient times. That genetic similarity, the researchers said, indicates that both regions were settled by members of a single migrating population from Indonesia. The scientists also concluded from their genetic analysis that the migrants arrived in New Guinea and Australia 40,000 years ago. This proposed date was also supported by archaeological findings.

The oldest human remains in Australia have been found at a group of sites called Lake Mungo. This area, in the south-eastern part of the continent, is now dry. But tens of thousands of years ago, it was a lush landscape dotted with lakes.

In 1969, a University of Melbourne geologist, Jim Bowler, was digging at Lake Mungo searching for evidence of the ancient lakes. He found something more significant—a grave containing the poorly preserved partial skeleton of a young early-modern woman, who became known as Mungo Woman. The condition of the bones showed that the body had been cremated after death. Bowler also found the fragmentary remains of another individual.

In 1974, Bowler discovered a third grave, this one holding the nearly complete skeleton of a man, who was dubbed Mungo Man. The people who buried Mungo Man applied red ocher to his body. This substance, a form of iron oxide that yields a red pigment when heated, was often used in ancient burials. The color red may have symbolized death to early peoples. The Mungo Man burial is one of the earliest known burials in which red ocher was used.

In 1999, a team of Australian researchers dated Mungo Man at 62,000 years old. But there were doubts about that date, and it challenged the Out of Africa theory. There was little evidence of such an ancient migration to Australia.

In 2003, teams at four laboratories, including one led by Bowler, ran additional tests on sand taken from the graves. The teams all reached the same conclusion: Both Mungo Man and Mungo Woman were buried about

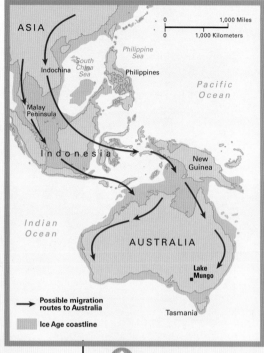

The ancestors of the Aborigines migrated from Asia to Indonesia over a land bridge that existed during the Ice Age. Archaeologists believe these people used boats or rafts to reach Australia.

40,000 years ago. That date seemed to agree with the theory that Australia was settled 40,000 years ago. However, the scientists also dated stone tools from the site to 50,000 years ago.

Scientists continue to debate the timing of the first migration to Australia and Southeast Asia. Was it 40,000 years ago, 50,000 years ago, or even earlier? What is certain is that people were living in Australia by 40,000 years ago. The continent contains more than 150 early-human archaeological sites that have been dated to at least that time. Some scientists think that human beings probably arrived in northern Australia from Asia even earlier but took time to migrate farther into the arid interior regions.

The oldest dated human remains ever found in Australia emerge in 1974 from a burial site at Lake Mungo in the present-day Australian state of New South Wales. The nearly complete male skeleton, known as Mungo III or Mungo Man, is believed to be about 40,000 years old. Mungo Man's skeleton was covered with pigment called red ocher, and the hands were interlocked over the body, indicating some kind of burial ritual.

THE ABORIGINES IN THEIR NEW HOMELAND

The settlers of the continent now known as Australia found themselves masters of a vast new land. The continent covers almost 3 million square miles (7.7 square kilometers), an area more than twice the size of India.

At that time—at least 40,000 years ago—Australia had a mild climate and was populated by many huge animals. These massive creatures—known as megafauna—included giant kangaroos weighing more than 500 pounds (227 kilograms) and ostrich-sized flightless birds. Among other megafauna were 25-foot (7.6-meter) lizards and horned tortoises the size of a Volkswagen.

Sometime after the arrival of humans in Australia, these enormous animals and many others vanished from the fossil record. What happened to them?

Some scientists have claimed that Australia's large animals died out slowly after coexisting with humans for thousands of years. That theory was based largely on a site called Cuddie Springs in southeast Australia. Layers of earth at Cuddie Springs held Aboriginal tools mingled with the bones of large animals. Scientists used radiocarbon dating and other forms of dating to establish a time range of 30,000 to 40,000 years ago for the deposits. Investigators of the site said the megafauna may have died out after that time because of a worsening climate.

The coexistence theory, however, was challenged in 2010 by scientists at the University of Adelaide. After making a thorough study of the Cuddie Springs site and artifacts, they concluded that the tools and bones had gotten mixed up in the deposits. That mingling occurred over many thousands of years, long after the animals had become extinct. Thus, it only *seemed* that the megafauna and early humans had existed together for such a long time. The researchers reported that the megafauna died out soon after people arrived in Australia. In other words, the Aborigines may have been responsible. But how did they drive so many animals to extinction?

One theory is that many animals died as a side effect of massive fires set by the Aborigines. The theory was proposed in 1999 by Gifford Miller, a geologist at the University of Colorado in Boulder.

Scientists have found large deposits of charcoal in soil levels from about 38,000 years ago, indicating the widespread burning of forests. When Europeans began visiting Australia in the 1600's, they observed Aborigines deliberately setting fires. The Aborigines explained that the fires were their way of "cleaning up the land." They also used burning to promote the growth of certain food plants that increase their production after a fire.

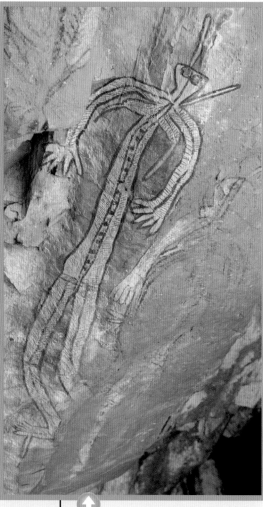

A cave painting in what is now Kakadu National Park in Australia's Northern Territory shows the external shape as well as the bones of a creature. Such images were created in what is called an X-ray style, which is still used by modern Aboriginal artists.

In an Aborigine camp on the Murray River (shown in an artist's illustration), men make tools of stone and wood and fish. A woman prepares and cooks food, such as wild plants and fish speared in the river. The group may be expecting trouble from a neighboring clan, as a warrior stands guard with shield and spear.

An Australian giant marsupial (show in an artist's illustration) was one of many large species, known as megafauna, that became extinct soon after people arrived on the continent. Scientists continue to debate whether early humans were responsible for the animals' disappearance.

Miller said the Aborigines' extensive destruction of plants in ancient Australia would have deprived many plant-eating animals of their food sources. As those animals began to die out, meat-eating animals were deprived of prey. The end result, after hundreds of years, was large-scale extinctions.

Miller said that the killing of animals by Aboriginal hunters probably also played a part in the megafauna extinctions. Other scientists have said that hunting was the main reason for the animals' disappearance. If so, did the Aborigines kill the huge animals primarily for food or was this another way of clearing the land?

Over thousands of years, the Aborigines developed an elaborate culture based on obedience to rules laid down by ancestral spirits. They believed that all time and all human events were linked together in an infinite spiritual cycle called the Dreamtime.

The Aborigines also created one of the oldest art forms in the world. They painted images on rocks, cliff faces, and cave walls throughout Australia. The art depicts animals, animal tracks, religious beliefs, myths, social activities, and other aspects of people's lives. In some places, the Aborigines made prints or stencils of their hands on rock walls by blowing paint from their mouths around an object. The hands may have been a way for people to show that they had participated in a ceremony.

MIGRATIONS TO THE MIDDLE EAST

Many scientists think that the migrants who arrived at the western border of present-day India an estimated 63,000 years ago split into two groups. One group kept going east into India. A second group migrated north. By 40,000 years ago, the descendants of this second group had settled parts of the region now known as the Middle East.

Other scientists think the Middle East was settled by members of a second wave of migration from Africa. Those migrants, according to this theory, moved from North Africa no more than 50,000 years ago. They slowly spread into the Middle East and, later, into the lands we call Europe and central Asia. This conclusion is based on genetic testing of present-day people in those regions.

Scientists who study genetic material have gained a greater understanding of human evolution and human relationships with other groups of living things by studying the rate at which genetic material called DNA *mutates* (changes). Short sections of DNA, known as genes, determine heredity—that is, the passing on of characteristics in living things. Scientists have discovered that, over time, mutations occur in genes at a fairly constant rate. Using this rate, a number of researchers have concluded that all living people evolved from a small group of human ancestors who lived in Africa about 150,000 years ago.

Over time, people descended from this group developed *variations* (differences) in their DNA. Scientists agree that only one group of early-modern descendants—those with a variation known as L3—left Africa. The question is, did people with the L3 variation leave Africa only once, or could there have been at least one additional migration? Some researchers say the evidence for a second migration is persuasive.

One advocate for a second migration is Spencer Wells. Wells is an American geneticist who directed a National Geographic Society/IBM program called the Genographic Project. This project, which ran from 2005 to 2010, was aimed at producing a detailed map of all early-human migrations. Wells's conclusion that there was a second migration is based on additional variations that developed in descendants of the L3 group.

Modern people with one of these genetic variations, known as the M variation, are common in India, Southeast Asia, and Australia. But the M variation is mostly absent in Europe and central Asia. There, a second variation, called the N variation, is more common. Wells and some other scientists say this fact points to a second migration.

But some scientists argue that an inconvenient fact throws this theory into question: The N variation is also found in India and Southeast Asia and is very common in Australian Aborigines. This finding has led these scientists to conclude that there was only one migration out of Africa. They argue that if you trace the DNA of any group of people alive today backward in time, the result is always the same: People's genetic histories always come together about 65,000 years ago in Africa. Therefore, they say, there could not have been more than one migration that led to the population of the world.

THE LEVANT AND THE NILE VALLEY

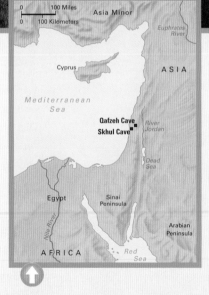

The Middle East became a crossroads that eventually led to the settlement of Europe. And it was a region where Egypt and Mesopotamia, two of the world's great civilizations, arose.

Some of the most interesting early history of the Middle East occurred in the Levant, the eastern Mediterranean region that now consists of Israel, Jordan, Lebanon, Syria, and parts of Iraq. Early-modern humans migrated to this area about 120,000 years ago and stayed for about 30,000 years.

Many early-modern remains and tools have been found in several caves in Israel. One notable site is Qafzeh Cave, a rock shelter in northern Israel. Archaeologists working at Qafzeh from the 1930's to the 1970's unearthed the remains of seven adults and nine children. All of them appeared to have been decorated with red ocher. In a layer of the cave floor dating from about 100,000 years ago, the archaeologists uncovered the bones of what seemed to be a mother and her child. It may be the oldest ritual burial ever found.

The caves in the Levant contained early-modern human remains up to about 90,000 years ago. After that, fossil evidence of modern humans does not reappear

The Nile River, the world's longest river, rises near the equator and flows north in Africa into the Mediterranean Sea. Early-modern people may have followed the Nile River Valley as they migrated from eastern Africa to the Middle East.

until about 47,000 years ago. What happened to modern humans in the Levant during that stretch of almost 45 centuries is not known. Nor do archaeologists know whether the modern humans who arrived in the Levant about 47,000 years ago migrated from Africa in a second migration or came from other parts of Eurasia. Perhaps they came from both places.

What *is* known is that Neandertals also lived in the Levant throughout this long period. Neandertals were prehistoric human beings who lived in Europe and Asia from about 150,000 to 35,000 years ago. At Qafzeh Cave and another site, Skhul Cave, scientists have found the remains of Neandertals in layers dating from about 65,000 to 47,000 years ago. Below and above the Neandertal layers were the remains of modern humans. This finding indicated that modern humans and Neandertals lived in the caves at different times. If the modern humans who resettled the Levant came directly from Africa, they probably traveled through the Nile River Valley. In the river valley, they would have been able to find ample food and water as they trekked northward. Some of the people may have decided to stay along the Nile.

Excavations have revealed many living sites from about 40,000 years ago in the land we now know as Egypt. The climate of Egypt at that time was considerably milder than the harsh conditions there today. The region that is now the upper Sahara was then a grassland, and the regions along the Nile were lush and filled with game animals. At living sites near the Nile, archaeologists have found abundant remains of animals—gazelles, ostriches, antelopes, wild cattle, and others—showing that the hunting was very good.

Around 30,000 years ago, desert conditions began returning to the Nile Valley. People moved closer to the river as the sands advanced.

The climate change was linked to the Ice Age. The Ice Age, a time that scientists call the Late Pleistocene Epoch, had begun some 80,000 years earlier—about 110,000 years ago. Over thousands of years, the glaciers and ice sheets periodically advanced and retreated. Some 30,000 years ago, the ice was again growing. It reached its maximum extent in about 21,000 B.C. When the ice advanced, northern areas, of course, got colder. But regions to the south were also affected. The planet's climate became drier because less water was evaporating from the oceans to produce rain. As a result, deserts expanded. This also occurred in Egypt.

The Ice Age would play a growing role in human migrations as the centuries wore on.

The skeletons of a young woman and a child, which date to about 100,000 years ago, may be the oldest known example of a purposeful burial. Unearthed in Qafzeh Cave in northern Israel, the skeletons may represent populations of early-modern people who migrated to the Middle East from Africa.

ENCOUNTERS WITH NEANDERTALS

In the Middle East and later in Europe, modern humans encountered a different form of human being that we call the Neandertals. Neandertals belonged to the genus *Homo,* and many paleoanthropologists think they evolved from an early form of human being called *Homo erectus* in Europe about 300,000 years ago.

The existence of the Neandertals was discovered in the 1800's. The first widely reported discovery of Neandertal remains was made in 1856 in the Neander Valley of Germany *(Neandertal* means *Neander Valley).* Over time, scientists realized from the appearance of the skull and many bones found there that the remains belonged to a previously unknown type of prehistoric human. Since that discovery, the remains of more than 400 Neandertals have been found.

The Neandertals had heavier bones and were more muscular than modern humans. They had a sloping forehead, a bony ridge across the top of the eye sockets, and a receding chin. Because of these features, the Neandertals have often been depicted as dumb brutes. But that was not the case. The brains of Neandertals were about the same size as those of modern humans. Neandertals also made useful tools, were good hunters, and buried their dead with respect.

What happened when modern humans and Neandertals came face to face? In some cases, they may have learned from one another. At both Neandertal and modern-human levels in Qafzeh Cave, archaeologists found stone tools from a style of toolmaking known as the Mousterian industry. These tools are shaped from flakes struck from a stone core, usually flint. Scientists had thought that only Neandertals made Mousterian tools. Finding these artifacts with modern-human remains suggested to some scientists that the newcomers somehow acquired the knowledge to make Mousterian-type tools from Neandertals.

The use of Mousterian tools by modern humans seems to indicate that they and the Neandertals cooperated. However, it's just as likely that they were competitors. The discovery of alternating human and Neandertal remains in Qafzeh and Skhul caves indicates that the caves changed hands several times. When the first modern humans arrived in the Levant about 120,000 years ago, they may have taken the caves from Neandertals. Or the modern humans may have found the caves empty. Later, some 70,000 years ago, the Neandertals inhabited the caves again, perhaps after migrating south to escape the growing cold of the Ice Age. Then, about 47,000 years ago, modern humans took possession of the caves once more. By about 30,000 years ago, the Neandertals were gone from the Levant forever.

THE NEANDERTAL GENOME PROJECT

Scientists debated for decades whether Neandertals were a variation of our species, called *Homo sapiens Neandertalensis,* or a separate species. By the late 1990's, the issue seemed to be settled: They were a separate species. This conclusion was based in part on computerized comparisons of Neandertal skulls with the skulls of modern humans.

More recently, scientists have turned to genetic studies to unravel the mystery of our relationship with Neandertals. In May 2010, an international team from Germany's Max Planck Institute for Evolutionary Anthropology reported the strongest evidence yet that modern-type humans mated with Neandertals. The group compared

A group of Neandertals tends to daily activities in a cave, in a reconstruction housed at the Neanderthal Museum in Krapina, Croatia. The discovery of the remains of about 75 Neandertals at Krapina in 1899 is one of the most significant archaeological findings in Europe. Evidence from one set of remains suggests that the hand of one man (center in reconstruction) had been surgically amputated.

The skull of a Neandertal (below, right) differed from that of modern people by having a large, projecting face; a low, sloping forehead; and a browridge, a raised strip of bone across the lower forehead. The Neandertal jaw also lacked a chin.

DNA from Neandertal bone fragments found in Croatia's Vindija cave with DNA from five living individuals, from China, France, Papua New Guinea, and southern and western Africa. The researchers concluded that as much as 4 percent of the DNA in modern non-African humans originated in Neandertals—including some genes that function in the immune system. The scientists proposed that a small group of modern-type humans bred with Neandertals sometime from 80,000 to 50,000 years ago in western Asia and that the descendants of these matings later spread through Europe and Asia—but not through Africa.

SETTLEMENT OF TURKEY

To the north of the Levant lies Anatolia—or what we now call the Asian part Turkey. This area was a natural gateway to Europe for early-modern humans from the Levant. A great number of people must have passed through Turkey in their migrations from Eurasia, and many of them stayed on.

Turkey seems to have been widely settled during the later stages of the Paleolithic Period, known as the Upper Paleolithic Period. Because of Turkey's large size and a shortage of scientific resources, however, excavations have been limited. Nonetheless, a few sites have yielded valuable evidence of life in the Upper Paleolithic.

One of the most important sites is Ucagizli *(OOCH au zluh)* Cave, located on Turkey's Mediterranean coast. Archaeologists from Turkey and the University of Arizona have been working in the cave since its discovery in the 1980's. Radiocarbon dating of ashes from ancient fires has revealed that people lived in the cave from about 41,000 to 29,000 years ago.

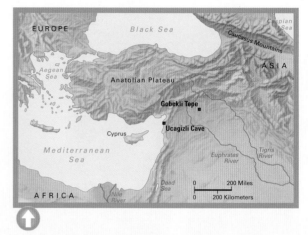

Present-day Turkey was a natural gateway for ancient people migrating from the Middle East to Europe. Many migrants also settled there, attracted by a good climate and plentiful game and water.

T-shaped stones form one of the seven rings at an ancient temple at Gobekli Tepe, whose construction began about 9000 B.C. Gobekli Tepe may be the first large structure ever built by human beings. Some archaeologists think that the temple, built by hunter-gatherers, may have led to the development of an agricultural society.

Images of a boar, the wild ancestor of the domestic pig, and three ostrich-like birds decorate one of the T-shaped stones at Gobekli Tepe. Many of the temple's stones, which are up to 16 feet (5 meters) tall, depict animals, especially dangerous animals.

The only human remains unearthed in the cave have been teeth, which seem to be those of *Homo sapiens*. But the cave contained many tools and animal remains. These remains include the bones of many large hoofed animals, including deer and wild cattle. The abundance of animal bones indicates that Turkey had a good climate in the Late Stone Age, with plenty of water and vegetation to support wildlife.

The tools found in the cave were made of bone as well as stone. Most of the bone artifacts were pointed tools that were apparently used as needles and *awls*—tools used for piercing holes in leather.

THE FIRST LARGE STRUCTURE?

Turkey may be the place where humans laid the first foundations of Western civilization. At least, that is a conclusion suggested by a remarkable discovery made in southeastern Turkey at a site called Gobekli Tepe (*Potbelly Hill* in Turkish). This site contains the ruins of what was evidently an ancient temple. It may be the first temple, or at least the first large structure of any kind built by humans.

The site was first noted by Turkish and University of Chicago researchers in the 1960's. They dismissed its importance, thinking that it was the remains of a medieval cemetery. In 1994, a German archaeologist named Klaus Schmidt visited the site and realized that it held something extraordinary. He and five colleagues began excavations the following year.

Their digging turned up several rings of huge stones. Each ring contained a pair of T-shaped stones up to 16 feet (5 meters) tall surrounded by slightly smaller stones. Many of the stones are carved with abstract images or depictions of lions, foxes, leopards, and other animals. What makes this complex so astonishing is that it was constructed an estimated 11,500 years ago—some 7,000 years before the Great Pyramids of Egypt. And it was built by hunter-gatherers before the beginning of agriculture.

Schmidt thinks the religious center may have been a spur to the rise of civilization. Scholars have long assumed that the construction of temples and other large buildings followed the establishment of agriculture and permanent settlements. Schmidt says the opposite may be true: People's religious awakening came first. An important temple would draw hunter-gatherers for ceremonies and feasts. In so doing, it may have become the focus for the development of a civilized society.

MESOPOTAMIA AND IRAN

East of Turkey lie Mesopotamia and Iran. Mesopotamia, which is Greek for *between the rivers,* consists of present-day Iraq and parts of Turkey and Syria. The rivers are the Tigris and the Euphrates. Mesopotamia is one of the most important areas in history because it was one of the regions where agriculture was invented. Mesopotamia and historic Egypt are part of a continuous region called the Fertile Crescent.

During the Upper Paleolithic Period (Late Stone Age), humans migrated to Mesopotamia and the area now called Iran and occupied many caves. Where most of the people came from is uncertain. Scientists think they probably came from both the south and the east. This part of the world was becoming a central cross-roads of Eurasia.

Although archaeologists are certain that Mesopotamia and Iran had many Upper Paleolithic living sites, relatively few have been excavated. And not many early-human remains have been found. However, archaeologists have un-earthed many tools left by Upper Paleolithic hunter-gatherers.

Much of the Upper Paleolithic evidence in this region has been found in the Zagros Mountains, a range that extends through both Iraq and Iran. At a cave called Eshkaft-e Gavi in Iran, archaeologists in the 1970's excavated a variety of stone tools and human skeletal remains that appear to be from modern-looking humans. The remains included four skull fragments and various other bones. Interestingly, the bones had cut marks on them and they had been burned. Archaeologists who studied the remains said they might be evidence of canni-balism. Such remains have also been found in Upper Paleolithic sites in other regions, so early humans may sometimes have eaten their own kind.

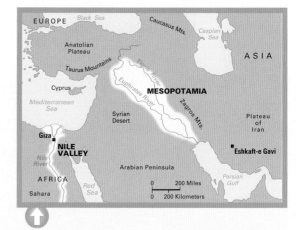

Ancient Mesopotamia. which included the area that is now Iraq, eastern Syria, and southeastern Turkey, was the home of some of the world's first civilizations. The heart of the region was the land between the Tigris and Euphrates rivers.

Archaeologists excavating in the Zagros Mountains have also found several sites containing tools from a style called the Aurignacian (ore ig NAYSH un or ore en YAY shun) industry. These tools, associated mostly with modern humans, probably evolved from the Mousterian tool industry. The Aurignacian tools found at Zagros sites include flint blades that were more finely made than Mousterian blades.

Some archaeologists theorize that the Aurignacian tool industry began in the Zagros region sometime after 40,000 B.C. and then spread to Europe. Other scientists, however, think that it originated in eastern Europe and spread to the Middle East. Moreover, archaeologists have found early Aurignacian tools with some Neandertal remains, so Neandertals may have contributed to the beginnings of this tool industry. The origin of the Aurignacian style is another of the still-unsettled controversies surrounding the migrations and culture of modern humans.

FROM HUNTING AND GATHERING TO FARMING

In humanity's long history, agriculture is a recent development. The first great transition from hunting and gathering to a settled farming life took place in the Middle East, in particular, in Mesopotamia.

The final retreat of the glaciers at the end of the last ice age, about 11,500 years ago, may have contributed to the beginnings of agriculture. Many game animals may have moved north, making it harder to survive by hunting.

Whatever the reason, people became more dependent on such wild grains as wheat and barley for their food supply. They learned to grow these plants and to improve them to obtain a larger harvest. At the same time, people began to domesticate such wild animals as cattle, goats, sheep, and pigs. People probably gave up hunting and gathering slowly. But eventually people began relying on agriculture for most of their food.

The settled life provided a more dependable food supply, but it was not particularly healthful. Farmers got less exercise than hunter-gatherers, and they ate a less varied diet. Skeletal evidence has shown that hunter-gatherers were taller and better nourished than early farmers and suffered from fewer diseases.

People in an early farming community, which flourished about 9,000 to 8,000 years ago in what is now Syria, harvest grain, tend livestock, and perform domestic chores, in an artist's illustration. In the distance, men return from a hunt with a gazelle. Excavations at the site suggest that there was no sudden change from hunting and gathering to a life based on agriculture.

THE LATER DEVELOPMENT OF MESOPOTAMIA AND EGYPT

Mesopotamia and Egypt are part of a region known as the Fertile Crescent, a large arc of land that also includes the greener parts of the Levant on the eastern edge of the Mediterranean Sea.

In the Fertile Crescent, hunting and gathering gave way to settled farming over a span of nearly 5,000 years. This period of change is known as the Mesolithic Period, or Middle Stone Age. During the Neolithic Period, or New Stone Age, agriculture became widespread. Other parts of the world went through these same periods, though not at the same time. In the Middle East, the Mesolithic Period lasted from about 10,000 B.C. to 5500 B.C.

Mesopotamia is often called "the cradle of civilization" because it was one of the first places where people domesticated plants and animals, developed writing, and established cities and states. The first of these civilizations was Sumer, which arose about 3500 B.C. in southeastern Mesopotamia and consisted of 12 city-states. Sumer was followed by a series of other states, including Assyria, Akkad, and Babylonia. The civilizations of Mesopotamia tended to be militaristic, and warfare became a way of life.

Egypt developed roughly in parallel with Mesopotamia but with less conflict. The advances and retreats of the northern ice sheets during the Ice Age had an enormous impact on the Nile Valley. The once-lush expanse of the Sahara was very dry and desert-like during the maximum advance of the ice, about 21,000 years ago. But when the ice sheets began to retreat, the rains did not return to the Sahara. People in the Nile Valley were forced to live close to the river. There, they continued their life of hunting and gathering, as well as fishing.

Over time, several toolmaking styles flourished in the area. A style called the Qadan began about 13,000 B.C. and lasted for 4,000 years. These tools were mainly grinding stones and sickles, a sign that people had begun using grain plants as a food source.

Ancient Egyptian fishers on the Nile River haul in their catch, in an artist's illustration. In addition to providing fish, the Nile River supplied water for living and irrigation and served the ancient Egyptians as their major transportation route.

A complex of temples, courts, and buildings rises along the Euphrates River in the ancient Sumerian city of Uruk, in an artist's illustration. Uruk was the largest of the independent city-states that began to expand rapidly in Sumer, the birthplace of the world's first civilization, in about 3300 B.C. Uruk controlled most of Sumer for a brief time about 2375 B.C.

People seemed to have stopped using these tools for a while, perhaps because of a climate change or a series of natural disasters. People apparently returned to hunting and gathering. This unusual finding demonstrates that the change to agriculture in a region was not always a smooth process.

Archaeological evidence in the Nile Valley indicates that the population increased significantly beginning about 5500 B.C. That finding indicates that farming and the domestication of animals had become firmly established by that time. Agricultural societies support much larger populations than hunting and gathering groups. The Nile Valley culture at this time is called Pre-Dynastic.

In Pre-Dynastic Egypt, two separate civilizations developed on the Nile, each with its own kings. One was in the north, the other to the south. Many historians believe that a southern king named Menes (also known as Narmer) conquered the northern kingdom and unified Egypt in about 3100 B.C. Menes is usually credited with establishing the first of some 30 *dynasties* (series of rulers in the same family) of ancient Egypt. Within just 600 years, Menes's successors were constructing the Great Pyramid of Giza.

The Step Pyramid is the first known pyramid built in ancient Egypt and probably one of the world's first large stone structures. It was constructed in about 2650 B.C. for King Zoser at Saqqarah, near present-day Cairo.

EXPANSION INTO EUROPE

The continent now known as Europe was settled later than other parts of the Eurasian land mass. Who were the first Europeans—often called the Cro-Magnon *(kroh MAG nuhn* or *kroh man YUHN)* people—and where did they come from?

Some Cro-Magnons may have been fairly direct descendants of the original migrants from Africa. The band of migrants taking the so-called southern route out of Africa an estimated 65,000 years ago evidently split when they reached India. Part of the group apparently headed north. Their descendants probably migrated into central Eurasia. From there, some of them turned west. They most likely journeyed through present-day Turkey. Finally, about 40,000 years ago, they reached western Europe.

Some of the evidence for that early parting of the ways on the southern journey comes from genetic studies of Indian and European populations. For example, a group of Indian and American scientists reported in 1999 that they had found a "profound layer of overlap" between western-Eurasian and Indian genetic *lineages* (descendants in a direct line from an ancestor). The scientists concluded the two groups share a genetic variation that appeared 51,000 to 67,000 years ago. Early-modern humans are thought to have reached India and divided into two groups about 63,000 years ago, a time that falls within that range.

Other migrants may have made their way to Europe by way of the Levant (the lands bordering the eastern shore of the Mediterranean Sea). If there was a second migration out of Africa to the Levant, as some researchers believe, a portion of those people's descendants probably moved on to Europe. They, too, would have gone by way of Turkey.

Other migrants to Europe are thought to have come from eastern Asia. By 40,000 years ago, early-modern humans were moving throughout the Eurasian land mass. Some people in eastern Asia were probably moving westward in search of new places to live. Those people would have migrated across the plains of what is now Russia. Some of their descendants are likely to have moved on to western Europe.

The question of where the settlers of Europe originated and what routes they took to get there is still being debated by scientists. What seems clear from archaeological evidence is that the earliest European settlements were in eastern Europe.

Iceland

British Isles

ATLANTIC
OCEAN

AFRICA

ARCTIC OCEAN

EUROPE

ASIA

Ural Mountains

Alps

Black Sea

Caspian Sea

Mediterranean Sea

EARLY SETTLEMENTS IN EASTERN EUROPE

The earliest known modern-human settlements in Europe are in lands that now belong to Russia, Romania, and the Czech Republic. People began moving into these areas about 40,000 years ago from other parts of Eurasia.

The oldest modern-human remains found in Europe are skull fragments from two individuals. They were unearthed in 2002 at a site called Pestera cu Oase (The Cave with Bones) in Romania. Radiocarbon dating established that the bones are about 38,000 years old

Other modern-human remains were discovered at Mladec Cave in the Czech Republic. Bones found in the cave more than 100 years ago were recently dated, using a new form of radiocarbon dating, to 31,000 years ago. The Mladec specimens are a complete set of early-modern fossils from six individuals, including both adults and children. Archaeologists also found at the site a number of tools made in the style known as the Aurignacian *(ore ig NAYSH un* or *ore en YAY shun)* industry. This style of toolmaking is associated mainly with modern humans.

One of the most extensive archaeological sites in central Eurasia is Kostenki, on the Don River in southern Russia. Kostenki actually consists of more than 20 living sites, some established some 40,000 years ago. This area consists largely of unforested grassland called a steppe. When people were living at the Kostenki sites, this area was very cold and dry.

At Kostenki, archaeologists found layers showing that people had inhabited the sites over time. In the lowest levels—therefore, the oldest—archaeologists found tools made in the Mousterian style that they think were made by Neandertal inhabitants. Most of the tools in the upper layers are of the Aurignacian type.

The tools include various kinds of blades, scrapers, and *awls* (tools used for piercing leather). They also include digging tools called mattocks, which resemble pickaxes. Mattocks have been found at other Eurasian sites as well. They were likely used to dig pits for the storage of food, though no pits have yet been discovered at Kostenki.

Animal remains in the modern-human levels include the bones of horses, bison, moose, reindeer, and mammoths. The people of the Don River area were obviously avid hunters. However, they didn't limit themselves to big game. Archaeologists have also found the bones of such small mammals as

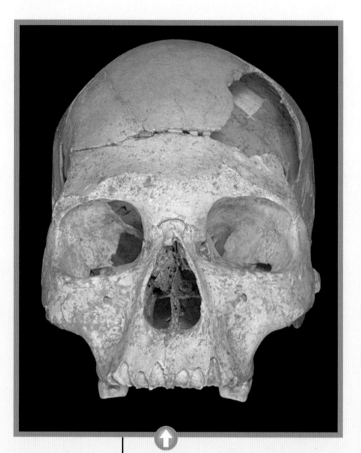

A skull found at an archaeological site called Pestera cu Oase in Romania is one of the oldest modern-human remains discovered in what is now Europe. The skull dates to about 38,000 years ago.

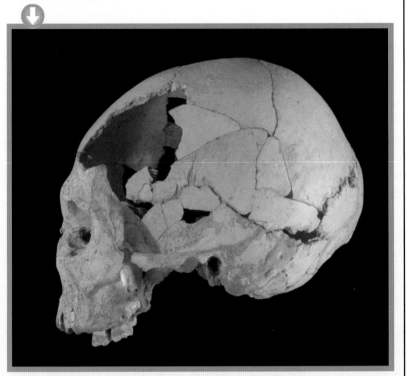

A side view of the Pestera cu Oase skull shows the projecting chin and high, rounded skull characteristic of modern human beings.

foxes and hares, and even of birds. The scientists speculate that the Kostenki people may have invented some sort of dart for killing small, fast-moving animals. They may also have devised traps or snares to capture small animals.

Only a few remains of modern humans have been found at the Kostenki sites. They consist mostly of teeth. One notable exception is the well-preserved skeleton of a young man, aged 20 to 25. It was discovered in 1954 at a site called Markina Gora. The man, who died about 30,000 years ago, had been buried in a crouching position and sprinkled with red ocher.

Scientists have succeeded in extracting genetic material from the skeleton. In 2010, a team of German and Russian researchers at the Max Planck Institute for Evolutionary Anthropology in Leipzig, Germany, reported that the Markina Gora man belonged to a rare group. His form of genetic material is relatively rare in present-day European populations. Even so, the investigators explained, this finding demonstrates a link between Late Paleolithic hunter-gatherers in central Europe and today's European population.

HOW DID PEOPLE BECOME LIGHT-SKINNED?

During their long wanderings through Eurasia, people's appearance slowly changed. They got taller and developed hair with new colors and textures. Perhaps the most striking change was in skin color. How did the black migrants from Africa become the fair-skinned people of Europe?

Many scientists think that the people of Africa and other areas near the equator have dark skin because of the intense tropical sun. Dark skin provides protection from ultraviolet rays in sunlight because it contains relatively high levels of melanin. Melanin, the substance that gives skin its color, is produced in response to sunlight.

Sunlight also causes the skin to produce vitamin D, an essential nutrient. Some researchers think that the need to maintain healthful vitamin-D levels led to lighter skin. As migrants moved north, they got less sunlight and thus produced less of the vitamin. The evolutionary process of natural selection favored people with lighter skin, which would absorb more sunlight.

Physically modern people began moving into Europe about 40,000 years ago. The oldest modern-human remains have been found in eastern Europe in Mladec Cave in the Czech Republic; Pestera cu Oase in Romania; and Kostenki in Russia.

THE EMERGENCE OF THE GRAVETTIAN CULTURE

Sometime around 33,000 B.C., the people of the modern Don River area developed a new style of toolmaking, many archaeologists believe. Called the Gravettian (*gruh VET ee un*) industry, it replaced the Aurignacian style. The Gravettian tool culture reached western Europe by about 29,000 B.C., and it had spread throughout the Eurasian land mass by 26,000 B.C. Gravettian tools included small pointed blades that were used for big-game hunting. Some of these blades were attached to lightweight spears. The blades were made small because larger, heavier blades would have made the spears hard to throw.

An important invention during this period was the atlatl (*AT lat uhl*), or spear thrower. The atlatl was a stick with a small cup or spur at the back end. The user held the thrower with the spear atop it, with the back of the spear up against the spur or cup. The atlatl increased the throwing force of the hunter's arm, enabling him to hurl a spear at a dangerous animal from a safe distance. Some archaeologists think the Gravettian people may also have invented the bow and arrow.

New tool cultures could spread by migrations or by settled populations learning from their neighbors. The Gravettian industry may have spread in both ways. Some archaeologists have suggested that the transition from the Aurignacian to the Gravettian style in western Europe resulted from people carrying the new technology there.

Archaeologists found evidence that the Kostenki people ate a lot of fish from the Don. The researchers said the people may have built *weirs* in the water—barriers made with vertical stakes—to stop fish coming downstream.

The people of the Gravettian culture also learned new ways of coping with the increasing cold of the Ice Age steppe. They built houses with a framework of mammoth bones and covered them with animal hides.

The Gravettian culture is also known for the production of "Venus" figures. These were statuettes of women with exaggerated features, including protruding abdomens and large breasts. These figurines, which were also made by Aurignacian artists, may have been representations of goddesses or statues used in fertility rituals.

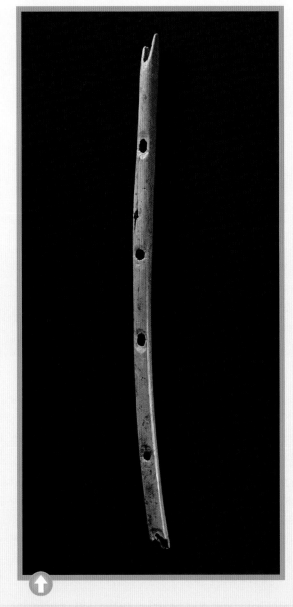

A 35,000-year-old flute, discovered at Hohle Fels Cave in southwestern Germany, is one of the oldest musical instruments ever found. The five-hole flute, which is 8 1/2 inches (21.6 centimeters) long, was made from a hollow bone of a large bird of prey called a griffon vulture. The discovery of this flute and others in the cave suggests that the earliest Europeans had a musical tradition, according to archaeologists who discovered the flute.

An ivory figurine of a female found in the Hohle Fels Cave may be a representation of a goddess or a statue used in fertility ceremonies. The figurine, which stands about 2.4 inches (6 centimeters) high, dates to about 35,000 years ago.

During this period, human communities were getting larger. Better social organization was necessary to enable people to live together peacefully. In other words, rules governing daily life had become a necessity.

Anthropologists theorize that the people of the Gravettian culture lived in groups led by men. This arrangement, which was also noted among American Indians, arises in a society in which a people's survival is mostly in the hands of men. In the Ice Age, the hunting of animals was the main way of obtaining food. The gathering of plant foods by women became less important.

In such bands, all the men of a group relate to one another as brothers. Each band maintains peaceful relations with other bands by "marrying out" the women of the group. That is, female members of a band were required to marry men from another band.

Of course, the structure of Gravettian society can never really be known. Nonetheless, anthropologists think it is likely that the people on the Russian steppe—and elsewhere—were developing new ways of getting along with one another.

The statue of a mammoth found in Volgelherd Cave in Germany is one of the oldest pieces of art ever found. The tiny statue—only 1 1/2 inches (4 centimeters) long—dates to at least 35,000 years ago. It was carved from mammoth ivory.

EARLY MODERN HUMANS IN WESTERN EUROPE

According to fossil and radiocarbon evidence, modern humans migrated to western Europe about 38,000 years ago. Some of them may have come from central Europe and others from the Middle East.

The first modern-human fossils were discovered in 1868 at Cro-Magnon, a rock shelter in southwest France. Excavations at the site turned up the skeletons of five people—three adult males, an adult female, and a child. The people had been ritually buried with stone tools, carved reindeer antlers, ivory pendants, and ornamental shells. The remains were later dated to about 28,000 B.C.

Archaeologists recognized that the skeletons were of individuals who looked the same as present-day humans. They called the people Cro-Magnon, after the name of the cave. In later years, the similar fossils were found elsewhere in France and in Germany, Spain, and central Europe.

For many years, scientists applied the term *Cro-Magnon* to all of the people who came after the Neandertals. Later, it became associated mainly with early-modern Europeans. Scientists no longer refer to all early people as Cro-Magnon. When they speak of prehistoric people who looked like us, they use the terms *anatomically modern humans* or *Homo sapiens sapiens*.

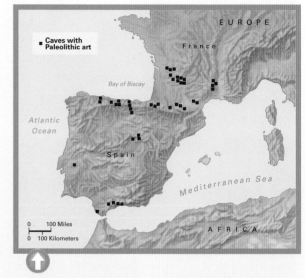

Numerous caves in what are now France and Spain offer examples of the mysterious and magnificent art made by Stone Age artists. Scholars think the art was created for use in rituals.

Stone Age hunters in Europe attack a mammoth with spears, in a reconstruction at Dinosaur Park in Brittany, an area of France. Such hunters often suffered serious injuries while trying to bring down large animals at close range, fossil evidence has revealed.

While the Gravettian culture was spreading in central Europe, the people of western Europe were still part of the Aurignacian culture. The Aurignacian Period in western Europe lasted from about 38,000 to 26,000 B.C.

Living on the steppes, the Gravettians had learned to build houses. In Europe, there were numerous caves and rock shelters that provided homes. Archaeologists have, however, found evidence that some Aurignacians in western Europe also constructed simple huts with rocks, branches, animal fur, and other materials.

Life in prehistoric Europe was hard. Hunters of the Aurignacian culture, like the Neandertals, did not have the atlatl. They killed such large animals as mammoths, bison, and cave bears with heavy spears that they threw at close range or thrust into the animal's side. This was dangerous work.

The first adult skeletons unearthed at Cro-Magnon showed evidence of serious injuries, including damaged neck vertebrae and a skull fracture. The bones showed that some healing had occurred, so the individuals had survived their injuries, at least for a while. Archaeologists believe this indicates that the early Europeans lived in tightly knit communities in which people took care of one another.

By the light of a stone lamp, an Ice Age artist paints the image of a bull on a cave wall with a tamping pad, a sponge-like mat, perhaps made of moss, used to "pat" pigment onto a surface. Another artist uses a stone palette to grind ocher or another naturally occuring pigment used for color.

THE CAVE ARTISTS OF WESTERN EUROPE

The Aurignacian people of western Europe are famed for the many beautiful images they created in caves in France and Spain. Even to the modern eye, many of these pictures are amazing for the artistic skill they display.

The Stone Age artists made hundreds of paintings, engravings and carved *reliefs* (images that stand out from the surrounding surface). These colorful images—almost all of them depicting commonly hunted animals—were executed on the walls, ceilings, and even floors of caves. Archaeologists do not know the purpose of these exquisite pictures. They speculate that early humans may have thought that making images of animals gave hunters power over them.

Cave artists continued this artistic tradition after the Gravettian culture reached western Europe, some 28,000 years ago. It lasted through the final major toolmaking phase of the Upper Paleolithic Period, the Magdalenian Culture, which ended about 10,000 years ago. By that time, Stone Age artists had decorated the insides of nearly 350 caves.

MIGRATION TO AND FROM BRITAIN

Sometime around 30,000 years ago, Aurignacian people from continental Europe made their way to the island now known as Britain. At that time, Britain was connected to mainland Europe by a land bridge. The land bridge—a section of the sea floor between Britain and mainland Europe—was exposed as sea levels dropped, a result of the deepening Ice Age.

The archaeological evidence for Late Stone Age humans in Britain is very sparse. Archaeologists have found a few spear points made of antler or bone and some flints, but not much else in the way of tools or other artifacts. Nor have they found many human remains.

One of the most notable discoveries was made in 1823 in southern Wales. A Welsh minister and geology professor, the Reverend William Buckland, found an ancient human skeleton buried at a site called Goat's Hole Cave (also called Paviland Cave). The person had been covered with red *ocher* (pigment) and buried with an array of grave goods, including decorative items made of ivory. The skeleton was fairly complete, though the skull was missing. The bones had been buried with the skull of a mammoth.

Buckland thought that the skeleton was that of a woman. He thought that the individual might have been a witch who had lived sometime in the first few centuries of the modern era, when the Romans ruled Britain. The skeleton became known as the Red Lady of Paviland.

Later analysis of the skeleton revealed that the remains were those of a man no more than 21 years

STONEHENGE

When humans returned to Britain, new cultures gradually developed over thousands of years. Among them was the culture that constructed Stonehenge, a monument in southwestern England. Stonehenge is composed of enormous rough-cut stones set in circles.

For many years, scholars thought that Stonehenge was built by the Celts *(kehlts or sehlts),* people from western Europe who migrated to Britain about 500 B.C. But scientists have learned that Stonehenge was built much earlier. They have concluded that the monument was constructed in three phases from about 2800 to 1500 B.C. The project required an estimated 20 million hours of labor.

◄ A riverside village from the Middle Stone Age (10,000 to 4,000 years ago) in Great Britain is re-created in a computer simulation of Doggerland by Eugene Ch'ng of the University of Wolverhampton in the United Kingdom. Doggerland was a land bridge that once connected Britain and mainland Europe. Rising sea levels at the end of the Ice Age flooded the area now being explored by underwater archaeologists and submerged evidence of the people who lived there.

of age. Archaeologists theorize that the man may have been a tribal chieftain. Despite their identification as male, the remains continued to be called the Red Lady of Paviland.

Radiocarbon dating of the bones in the mid-1900's gave the remains an age of about 18,000 years. Later, that figure was revised to 25,000 years. However, more precise dating in 2007 by scientists at the British Museum and Oxford University yielded an age of 29,000 years. That finding makes the Red Lady the oldest early-human remains found in Britain. The site also marks the oldest known ceremonial burial in western Europe.

The Aurignacian people stayed in Britain for only about 10,000 years. By 21,000 years ago, the Ice Age had reached its maximum extent. Immense glaciers were grinding relentlessly toward Britain and other parts of northern Europe.

The early Britons retreated back across the land bridge to western Europe and found refuge in Spain and other relatively warm areas. Humans did not return to Britain for another 5,000 years, when the ice was melting.

No one knows for certain why Stonehenge was built. Many people think it was designed to mark astronomical events of the year, such as the summer and winter solstices.

Recently, a new theory has been proposed: that Stonehenge was a holy site where people came to be cured of various afflictions. That theory is supported by the excavation in the area of many skeletons that show signs of physical deformities.

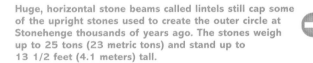
Huge, horizontal stone beams called lintels still cap some of the upright stones used to create the outer circle at Stonehenge thousands of years ago. The stones weigh up to 25 tons (23 metric tons) and stand up to 13 1/2 feet (4.1 meters) tall.

EXTINCTION OF THE NEANDERTALS

For about 12,000 years, modern humans and Neandertals lived together in western Europe. But the Neandertal population steadily declined. Finally, about 28,000 years ago, the last of the Neandertals died out. For more than 200,000 years, before the coming of modern humans, that robust species had ruled supreme. And now they were gone. What happened?

Archaeologists and anthropologists have proposed various explanations for the demise of the Neandertals. A few of the theories have been highly controversial. One scientist has suggested that our modern ancestors hunted and ate Neandertals. Not many of his colleagues agreed with that idea, however.

Some scientists think the Neandertals became extinct because of severe climate change. Others theorize that Neandertals simply could not match the greater hunting skills of modern humans. In fact, both of these factors could have worked against the Neandertals.

The Neandertals perished at a time of increasing cold. Although earlier Neandertals had lived through other ice ages, the latest Ice Age was different. There were wild climate swings, with the glaciers and icecaps repeatedly advancing and retreating.

With each of these shifts, the animals and plants in a region changed, sometimes within a single lifetime. Some scientists speculate that Neandertals were simply unable to cope with these see-sawing conditions. They say the Neandertals had a set way of doing things and were either resistant to change or incapable of it.

Neandertals bury a member of their group, in an artist's illustration based on archaeological findings. At least some groups of Neandertals buried their dead with care.

The advance of glaciers deep into Europe during the Ice Age may have created conditions to which Neandertals were unable to adapt.

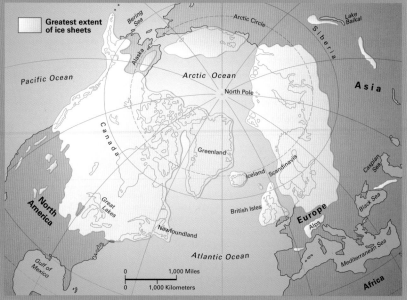

Greatest extent of ice sheets

THE DEPOPULATION AND REPOPULATION OF EUROPE

Britain wasn't the only part of Europe to be abandoned by its inhabitants as the Ice Age grew ever more bitter. The same thing happened on parts of the European mainland.

In about 21,000 B.C., immense glaciers spread southward from the land now known as Scandinavia. Other glaciers moved northward from the Alps. The geological record shows that the area between these two immense ice fields became a barren polar desert that created a virtually impassable barrier between eastern and western Europe. Western Europe became a cultural island.

As northern Europe become uninhabitable, people migrated southward. Archaeologists think that upper Europe became almost completely abandoned at this time. Those fleeing the frigid conditions of the north settled in several relatively mild refuges. Archaeological evidence indicates that the main refuges were in regions known to us as Spain, southern France, Italy, the Balkans, and Ukraine.

The Ice Age began to wane around 16,000 B.C. At that time, people began moving back to the north. They continued their hunter-gatherer traditions until about 7500 B.C. That is when—according to recent genetic studies of present-day populations—migrants from the Middle East brought agriculture to Europe.

In a world undergoing frequent climate changes, it was necessary for human populations to change also, such as by adopting new hunting techniques. The more adaptable modern humans could do that, scientists say. The Neandertals could not. With modern humans taking most of the available game, the Neandertals had less and less to eat. Undernourishment finally killed them.

Archaeological evidence indicates that a group of Neandertals clung to survival in several caves at modern-day Gibraltar, on the southern coast of Spain. They may have been the last of their kind.

A group of British researchers who excavated the caves found the bones of dolphins and seals, along with the remains of shellfish, rabbits, and birds. So the Neandertals, who for tens of thousands of years had lived mostly on the meat of large mammals, had evidently learned to vary their diet.

But apparently it was too late. Perhaps they never really had a chance. With modern humans spreading through the world, the Neandertals' time had passed.

A skull and other Neandertal remains are displayed in Gorham's Cave in Gibraltar, one of the last Neandertal sites inhabited before these prehistoric humans became extinct about 28,000 years ago. Archaeologists have also unearthed knives, spear points, and scrapers.

THE SETTLEMENT OF CHINA AND NORTHERN ASIA

While early-modern humans were moving through what are now the Middle East and Europe, other migrations were occurring on the far eastern side of the Eurasian land mass. These population movements led to the settlement of northern Asia and the area now known as China.

Most scientists think that the modern settlers of China and other parts of the Far East came from the southeast Asian mainland. Genetic evidence supports that theory. In 2009, a team of researchers from 40 institutions in 11 nations reported on a genetic study of more than 2,000 people across Asia. The scientists wanted to test a theory that there had been two early movements from Africa into east Asia. According to that theory, migrants to east Asia came from both central Eurasia and southeast Asia.

Based on their genetic data, the scientists concluded that there had been just one migration that led to the settlement of east Asia. That was the "southern route" migration from Africa perhaps 65,000 years ago across the bottom of the Eurasian land mass. The researchers said that after settling much of southeast Asia, the migrants turned north and moved into China.

The investigators said their data do not point to an exact time for the migration to China. Fossil evidence in China suggests, however, that it occurred about 40,000 years ago. No modern-human remains have been found in China from an earlier time.

The oldest known modern-human remains in China were discovered in 2003 in Tianyuan Cave near Beijing. The remains were later studied by a team of researchers from Washington University in St. Louis, Missouri, and the Chinese Academy of Sciences. Radiocarbon dating showed that the remains—34 bone fragments from a person about 40 to 50 years old at the time of death—are about 40,000 years old. Scientists could not determine the person's sex.

An analysis of the jawbone revealed that the individual ate a lot of freshwater fish. Foods contain various substances that become incorporated into bone, thereby revealing a person's diet. Scientists knew that early hunter-gatherers learned to supplement their meat and plant diets with fish. However, they didn't know how long ago people started fishing. The jawbone provided evidence that fishing began at least 40,000 years ago.

The remains also contained the earliest known evidence for the use of shoes. Two of the scientists who studied the Tianyuan Cave remains, Erik Trinkaus and Hong Shang, studied toe bones from the early human. They noted that though the leg bones were fairly thick, the middle-toe bones were relatively slim. This finding indicated the use of shoes, they explained. When a person walks barefoot, the middle toes curl toward the ground to provide traction. This puts stress on the those toes, causing the bones in them to become thicker. But if a person is wearing shoes, more of the stress of walking is borne by the big toes, so the middle toes are slimmer.

EUROPE

Caspian Sea

Persian Gulf

NORTH AMERICA

ARCTIC
OCEAN

Ural Mountains

S i b e r i a

ASIA

Japan

Tibet

China

PACIFIC
OCEAN

H i m a l a y a

Philippine Islands

THE LATER DEVELOPMENT OF CHINA

Hunter-gatherer cultures lasted in what is now China until about 8,000 years ago. Then, according to archaeological evidence, agriculture emerged at several places almost simultaneously. This finding strongly suggests close communication between people in various regions of China.

The agriculture of China was based on rice in the warm, humid south and millet in the colder, drier north. Millet is a cereal grain that is often used to make flour for bread. Archaeologists have concluded that the people of northern China tried to grow rice, but the climate was too cold. The northern Chinese later began to grow wheat and barley as well.

Archaeologists have learned that the Chinese were the first people in the world to grow rice. By 4000 B.C., rice cultivation had spread from China to southeast Asia and India.

Chinese rice cultivation began in areas around the Yangtze River. Archaeologists think that as early as 14,000 B.C., hunter-gatherers in the Yangtze River Valley were harvesting rice plants from the wild. The oldest rice grains found so far at a living site were discovered in 2005 at a rock shelter called Yuchanyan Cave in southeast China. The grains were radiocarbon dated to 12,000 to 14,000 years ago. By 7000 B.C., rice was being grown throughout the Yangtze Valley.

The Yuchanyan Cave site has also yielded the world's oldest known ceramic pottery. Archaeologists recovered fragments of pots made of clay mixed with small pebbles. Radiocarbon dating of charcoal and bone fragments found in the same level as the pottery showed that the pottery was made about 17,500 to 18,300 years ago. A Japanese people called the Jomon had been credited with creating the first pottery, but the oldest known Jomon pot was made no more than 17,000 years ago.

From the Yangtze Valley, farming spread westward. At a site called Dadiwan in south-central China, archaeologists found, people were growing millet and raising pigs and other animals by about 5850 B.C.

Dadiwan is a remarkable site, covering more than 250 acres (100 hectares). By 2010, archaeologists had uncovered 240 house foundations, 35 pottery *kilns* (small furnaces for firing ceramic objects),

The oldest known modern-human remains in China, dated to about 40,000 years ago, were found at Tianyuan Cave in northeastern China. At Yuchanyan Cave, archaeologists found 14,000-year-old rice grains, the world's oldest from any human-occupied site. People were raising pigs and other animals at a site called Dadiwan by 5850 B.C.

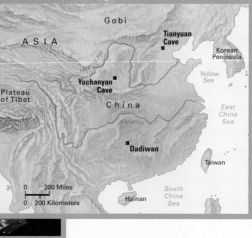

parts of a dozen irrigation ditches, and thousands of artifacts. The artifacts included the oldest known examples of painted pottery.

Agriculture and a shift to living in settlements continued to spread in China. By 3000 B.C., a culture known as the Longshan *(lawng shan)* had developed intensive agriculture, including the domestication of many animals. This culture, centered in the Yangtze River Valley and along the eastern seacoast, is noted for its beautiful black pottery, which was created on a potter's wheel.

Archaeologists have discovered that by about 3000 B.C., many people were living in villages surrounded by strong walls made of rammed earth. This finding suggests that warfare had become a part of Chinese life.

Beginning about 2000 B.C., China entered a period that Chinese historians later called the Age of Kingship. The kings were legendary figures who supposedly laid down the rules for a civilized society. This era was looked back on as a golden age ruled by an enlightened *dynasty* (hereditary ruling family) called the Xia *(SHEE uh)*. For many years, experts doubted that the Xia really existed and thought it was only part of Chinese mythology. But archaeologists found evidence of its existence in what is now Henan province.

In about 1766 B.C., the Shang *(shahng)* dynasty arose from the Longshan and Xia cultures. With this dynasty, the historical period of Chinese civilization began.

Archaeologists explore layers of sediment in Yuchanyan Cave, one of the earliest sites occupied by modern-humans in China. In addition to the oldest known grains of rice, finds at the rock shelter include the world's oldest known pottery.

NEGRITOS AND AUSTRONESIANS

The island of Taiwan lies about 90 miles (140 kilometers) off the Chinese coast in the South China Sea. Migrants from this island played a major role in the peopling of the Pacific Islands.

Genetic evidence from the present-day population of Taiwan shows that the island was first settled by the Negrito people from southeast Asia. Negritos, who settled in what are now the Andaman Islands off the coast of India and New Guinea, represent the original migrants from Africa.

Archaeological findings at caves in Taiwan indicate that the Negritos arrived at least 30,000 years ago and perhaps as early as 50,000 years ago. They were able to walk to Taiwan, which was then connected to the Chinese mainland by a land bridge. Negritos settled in the island chain now known as the Philippines at about the same time.

Evidence of Negrito settlements has been found at a number of sites, including one called Basian Cave. Radiocarbon dating of charcoal from the site showed that the cave was occupied about 25,000 years ago. Among the findings were two sticks that ancient people had rubbed together to make fire.

But archaeological findings are sparse. Not much is known about the Taiwanese Negritos or their culture. Archaeologists believe they played no part in the subsequent history of Taiwan.

The later *indigenous* (native) inhabitants of Taiwan are called Austronesians (South Islanders). Genetic studies indicate that the Austronesians arose from people who migrated to Taiwan from south China and perhaps also from southeast Asia some-

A girl from the Saisiyat ethnic group on Taiwan displays her ceremonial dress at a biannual bonfire and dance held to commemorate the deaths of the last Negritos on the island more than 1,000 years ago. According to Saisiyat tradition, the Negritos died in a Saisiyat attack on their village during a dispute over women.

time after 4,000 B.C. These people, archaeological evidence indicates, were farmers and fishermen. They fished in the Taiwan Strait with both hooks and nets. As they ventured into the sea, they refined their boatbuilding and seafaring skills.

Scientists do not know why people from the mainland migrated to Taiwan. One theory suggests growing populations led some people to seek new territory. Some archaeologists argue that a seafaring culture would find it only natural to cross the 100 miles (160 kilometers) of the strait and settle in a sparsely populated land. Whatever the reason for the move, the migrants settled on Taiwan and transferred their seafaring skills to their new home.

In about 2500 B.C., a group of Austronesian voyagers sailed to the Philippines and settled there. In later centuries, they settled much of what is now known as Indonesia and the coastal areas and offshore islands of New Guinea.

What happened to the Negritos? They may have become part of the newcomers' culture, or they may have been used as slaves on the newcomers' farms. Archaeologists have found evidence that Negritos were used as agricultural slaves in the Philippines, which were colonized by Austronesians when they began to venture away from Taiwan. Many Negritos may have been killed by the Austronesians. According to Taiwanese tradition, the early Austronesians and their descendants wiped out the Negritos over many centuries in a series of tribal disputes.

Scholars think the Negrito people, who closely resembled the first migrants from Africa, became "second-class citizens" in their new homelands. They vanished throughout much of southeast Asia. Negritos continued to exist in just a few isolated places, including the Andaman Islands, the highland areas of New Guinea, and the Philippines, but elsewhere they were replaced by taller, lighter-skinned peoples.

The migrations to Taiwan, New Guinea, and the Philippines were just the beginning. The Austronesians next turned their sights outward to settle all the major islands of the Pacific Ocean.

Archaeological discoveries at Basian Cave on Taiwan have revealed information about the Negritos, the original settlers of the island. Archaeologists believe these dark-skinned people were descended from migrants from Africa who traveled along the coast of India. The Negritos, who may have arrived on Taiwan as early as 50,000 years ago, were apparently wiped out by Austronesian settlers at least 1,000 years ago.

THE SETTLEMENT OF KOREA AND JAPAN

Modern humans migrated onto the peninsula now known as Korea at least 40,000 years ago. From Korea, some people migrated farther east to Japan.

Archaeology in Korea, especially North Korea, has lagged behind that of most other regions. However, several sites containing modern-human remains have been found there.

One notable discovery was reported in 1990 by South Korean archaeologists. They found the nearly complete skeleton of a modern human child about 5 or 6 years old in a site called Hungsu Cave. The body of the child seemed to have been ritually buried. Radiocarbon dating of material at the same level in the cave yielded an age of about 40,000 years.

Korea may have been settled by two waves of migrants. Although migrations into China apparently came mainly from the south, some archaeologists think that Korea was populated from both north and south. The migrants from the south came from China. The migrants from the north apparently came from the region now known as Siberia. By 40,000 years ago, Siberia was being settled by groups from farther west, so some of them may have migrated into Korea.

Some of the evidence for this two-pronged migration comes from a study of tools found at Upper Paleolithic (Later Stone Age) sites in Korea. Archaeologists at Hanyang University in Seoul, South Korea, reported in 2009 that two kinds of tools have been found at Korean living sites. Earlier sites from around 40,000 years ago contain tools associated with the Upper Paleolithic in China. Beginning about 35,000 years ago, the tools are of a type more commonly found in Siberia.

The Chinese-type tools are known as core-and-flake. These tools are made by striking large flakes from a stone core. The Siberian type are called blade tools. Blades are also struck from a core, but they are much longer than flakes.

The settlement of the islands of what is now Japan began at least 30,000 years ago. At that time, the deepening Ice Age resulted in sea levels lower than those today. As a result, the islands of Japan were one continuous land mass joined to the Korean Peninsula by a land bridge. People from Korea would have had no trouble getting to Japan.

Some of the tools found in Japan dating from this period are of the blade type found at various sites in Korea. These findings support the theory that people from Korea migrated to Japan. However, Japanese archaeologists have also discovered evidence that some migrants to Japan came from elsewhere. The scientists studied tools found in parts of the Japanese Ryukyu Islands, southwest of the main

Modern reconstructions of Jomon huts are on display at the Sannai-Maruyama archaeological site in Japan. The site was occupied from about 5,500 to 4,000 years ago. Archaeologists working at Sannai-Maruyama have also unearthed graves, storage pits, roads, pottery, ornaments, tools, and food remains.

Important archaeological sites on the Korean Peninsula include Hungsu Cave in South Korea, where scientists found the 40,000-year-old skeleton of a modern-looking human child. The child, who was about 5 or 6 years old at the time of its death, appeared to have received a ritual burial.

Japanese islands. According to the scientists, the tools resemble ones that were being made 30,000 years ago in Taiwan and southeast Asia. Thus, some migrants might have come from that region.

Modern-human remains from this period discovered in Japan include the bones of a child about 6 years old. They were found in 1968 in a cave on the Japanese island of Okinawa, the largest island in the Ryukyu chain. Radiocarbon dating of charcoal taken from the same level in the cave showed that the bones are about 32,500 years old. They are the oldest modern-human remains that have been found in Japan.

The wandering hunter-gatherer society of Japan eventually gave rise, in about 10,000 B.C., to a more settled one, the Jomon (JOH mon) culture. The Jomon continued their hunting/gathering/fishing lifestyle for a long time, but they also began to cultivate the land.

The name *Jomon* means *cord-patterned* in Japanese. The people of this culture made pottery that was covered with the impressions of ropes or cords. The Jomon culture lasted until about 300 B.C. Then migrants from China arrived in Japan and imposed their culture on the Jomon.

A vase shows the impressions of cords or ropes typical of the style of pottery created by the Joman people, a hunter-gatherer society that first appeared in what is now Japan about 10,000 B.C.

INTO SIBERIA AND NORTHEAST ASIA

Something remarkable happened on the steppes of far-northern Eurasia about 30,000 years ago as the Ice Age tightened its grip: The population increased. By 20,000 B.C., western and central Europe had been largely abandoned, but people were migrating into what are now Siberia and northeast Asia.

When parts of Europe were being abandoned, most of the people headed for more habitable areas to the south. But archaeologists think many people migrated eastward. They first arrived on the Russians steppes at places like the Don River settlements. Some of the migrants may have stayed there, but others pressed on to the northeast into the murderous cold. Why they did so is not known. But these people were involved in the final episode in the settling of Eurasia by modern humans.

The early history of the vast region of Siberia and northeast Asia is poorly understood despite many excavations conducted there. The people of this harsh region were undoubtedly big-game hunters, and mammoths were probably their main prey. Archaeological evidence indicates that mammoth populations began to diminish in much of Eurasia after about 40,000 years ago. Scientists think a major reason for that was overhunting by humans.

Mammoth populations became increasingly concentrated in the colder parts of Eurasia. The pursuit of these animals may have been the reason some people were willing to brave the Ice Age environment of upper Asia. They were able to live in this area because it was not covered by a continuous ice sheet. The climate was too dry to produce the snow that forms such sheets.

One of the earliest modern-human sites found in Siberia is called the Malaya Siya site, which dates from about 35,000 years ago. It is located near Lake Baikal in southeastern Siberia. Bones found at the site suggest the intense hunting of mammoths, horses, reindeer, and other large animals.

By 20,000 B.C., two main cultural traditions had emerged in this region. Archaeologists call them the Malta culture and the Afontova (*AH fahn toh vah*) culture. These sites provide evidence of a society that was becoming more modern in its lifestyle.

The Malta culture is named for a site called Malta located in an area west of Lake Baikal. Archaeologists have learned from their excavations of the area that the people lived in underground houses made of large-animal bones and

Archaeological evidence from the first people to migrate into what is now Siberia 30,000 years ago have been found at a number of sites. These migrants moved northward even as the Ice Age became more severe, perhaps to hunt mammoths and other large animals.

reindeer antlers. The houses were probably covered with earth and animal hides to ward off the cold.

Archaeologists working at Malta have discovered numerous art objects made of expertly carved ivory, bone, and antler. The most commonly found objects are figurines depicting birds and women. The statuettes of women are thought to be fertility figures, such as have been found elsewhere. But their physical features are usually much less exaggerated.

The Afontova culture is named for a site called Afontova Gora in south-central Siberia. The Afontova people evidently lived in portable tents and clothed themselves with tunics, trousers, and fur boots.

In eastern Siberia, a society known as the Dyukhtai *(DEE yook ty)* culture emerged about 18,000 years ago. This culture is named for a site of the same name northeast of Lake Baikal. Archaeologists think the Dyukhtai people may have been the first Asians who colonized North America thousands of years later.

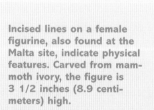

A pattern representing fur overalls covers the figurine of a woman unearthed in Siberia. Dated to about 25,000 years ago, the mammoth-ivory figure stands 1.6 inches (4.2 centimeters) tall. Cord was likely strung through the hole near the figure's feet to create a pendant necklace.

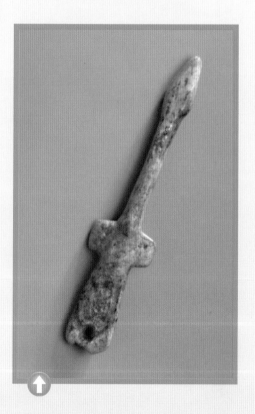

A 25,000-year-old figurine representing a flying bird is one of many Stone Age artifacts found at the Malta archaeological site in Siberia. The hole at the bottom of the bird's body suggests that the carving, made from mammoth ivory, was used as a pendant necklace.

Incised lines on a female figurine, also found at the Malta site, indicate physical features. Carved from mammoth ivory, the figure is 3 1/2 inches (8.9 centimeters) high.

VENTURING INTO THE EURASIAN ARCTIC

Few findings about the history of early-human migrations have surprised archaeologists as much as discoveries by two different scientific teams that early-modern humans may have colonized the Arctic more than 30,000 years ago. Archaeologists had thought that humans did not settle the Arctic until the final stage of the Ice Age, some 13,000 to 14,000 years ago.

The first discovery of ancient far-north people was reported in 2001 by a team of Norwegian and Russian archaeologists. They found the remnants of a human settlement at a site in Russia called Mamontovaya Kurya. The site lies on the Usa River in the Arctic. There archaeologists found an assortment of stone tools and the bones of mammoths, reindeer, wolves, and horses. These animals all seem to have been hunted and eaten.

The most important find was a mammoth tusk marked with a number of small, parallel grooves. Microscopic analysis of the grooves showed that they had been made by a chopping tool. The researchers said the marks may have been made when the animal was butchered. Other possibilities are that the tusk was used as an *anvil* (block used as a base for pounding) or that the grooves had some sort of artistic or symbolic meaning. The tusk and several bones from the site were radiocarbon dated to about 36,000 years ago.

The stone tools found at the site included a large blade and a scraper, both quite sharp. The tools did not fall into any specific toolmaking style. The scientists said they resembled both Mousterian tools and early Aurignacian tools. Mousterian tools are characterized by wedge-shaped flakes struck from stone cores. Tools of this style were made by both Neandertals and modern humans. Aurignacian tools were made after the Mousterian culture died out. These tools include long blades struck from cores. They are associated mostly with modern humans, though Neandertals may have made some of the early Aurignacian tools.

The archaeologists said the humans probably moved into the Arctic during a relatively mild phase of the Ice Age. But even then, the region was extremely cold. A British scientist estimated that the average yearly temperature was probably about 30° F (-1° C), slightly lower than today's average. He said it is unlikely the inhabitants of Mamontovaya Kurya were able to stay in the Arctic once severe Ice Age conditions returned.

Because the tools found at the site could not be definitely identified as those of modern humans, the archaeologists said the inhabitants of Mamontovaya Kurya could have been Neandertals. They added, though, that they felt fairly confident from all the evidence that the Mamontovaya Kurya people were *Homo sapiens,* not Neandertals.

Mammoths were a favorite prey of hunters living in the Arctic 30,000 years ago. However, bones found at sites dating from this period have revealed that the Stone Age people of the region also ate horses, cave lions, and reindeer.

The other surprising discovery was reported in 2004 by a team of Russian archaeologists. They found an ancient hunting camp on Russia's Yana River some 310 miles (500 kilometers) north of the Arctic Circle. Radiocarbon dating of material excavated at the site showed that the area was occupied about 31,000 years ago.

The site contained many pieces of animal bone from the hunters' prey. The people seemed to hunt and eat mostly reindeer, but the researchers also found the bones of mammoths, bears, horses, bison, cave lions, and other animals. The archaeologists also unearthed a large number of tools, including axes and scrapers. The artifacts had been made from stone, mammoth tusks, and animal bones.

The researchers said they were reasonably sure the site had been inhabited by modern humans, not Neandertals. By 31,000 years ago, Neandertals were found mostly in western Europe and were nearing extinction.

Recent archaeological discoveries have shown that people began to settle in the Arctic regions of what is now Europe as early as 30,000 years ago, about 15,000 years earlier than scientists had believed.

ADVANCING INTO UPPER NORTH AMERICA

ATLANTIC OCEAN

Overwhelming archaeological and genetic evidence supports the theory that the first migrants to the region we call North America were people from northern Asia. However, archaeologists have long debated how and when those people reached the Americas.

For many years, the main theory was that humans migrated from Asia less than 12,000 years ago when the Ice Age was winding down. Even though sea levels were rising at that time, they were still lower than they are today. As a result, a wide land bridge, which archaeologists call Beringia, was exposed in what is now known as the Bering Strait between Siberia in Asia and Alaska in North America. Beringia formed at least 25,000 years ago. According to this theory, hunters pursued mammoths and other game in Beringia. But they could not migrate into North America because an enormous glacier blocked their way. Finally, by 12,000 years ago, an open corridor formed between two large melting ice sheets in Canada. People from Beringia could then move into North America.

Some archaeologists believe that while it is almost certain that migrants from Asia came across the Bering Land Bridge, other people may have taken a different route. Some may have voyaged in boats along the Pacific Coast. Some may even have migrated from Europe.

The people who populated Beringia prior to the opening of the ice-free corridor were evidently stalled on the land bridge for about 15,000 years. During that time, they began to change genetically from the north-Asian population they came from, according to a 2007 report from researchers at the University of Illinois at Urbana-Champaign. The scientists studied mitochondrial DNA (mtDNA), a form of genetic material inherited only through the female line, from 20 Native American and 26 Asian populations. The researchers found that the genetic makeup of the Native Americans included several *haplogroups* (collections of mutations) that were not seen in the Asians. This finding, they said, showed that the inhabitants of Beringia had been isolated long enough from their Asian roots to become a genetically separate group.

The research team also found the new haplogroups in all the Native American populations they studied, from Canada to South America. That finding, they said, shows that the peopling of the Americas was carried out by the descendants of a single ancestor group from Asia. After entering North America, those people spread quickly throughout the Americas. Then, the scientists said, Native American populations became isolated from one another and developed new genetic variations. But their "continental" genetic patterns, which marked their origins in Beringia, remain within their mtDNA.

This genetic evidence supports the theory that the Americas were first settled by a single group of Asians migrating across the Bering Land Bridge. The archaeologists said they think the migrants entered North America about 15,000 years ago. Previous genetic studies had also supported the land-bridge theory but suggested the migrants entered from 11,000 to 40,000 years ago. Any date earlier than 12,000 years ago presumes that the migrants were able to find a way into North America before the opening of the ice-free corridor.

Gulf of Mexico

West Indies

Caribbean Sea

THE MYSTERIOUS CLOVIS CULTURE

In 1932, archaeologists began excavating an important new site near Clovis, New Mexico. The site contained the bones of mammoths and other large animals.

The archaeologists also uncovered beautiful spear points of a previously unknown kind. Until then, the oldest known spear points were of a type called Folsom. The Clovis points were found in layers of earth below a Folsom deposit, showing that they were from an earlier time. This discovery was a revelation to the archaeological community. Here was evidence of the first settlers in the Americas. Archaeologists named these settlers the Clovis people and the newly discovered toolmaking style the Clovis culture. Archaeologists have found hundreds of Clovis sites in the United States, Mexico, Central America, and northern South America. The sites have all been dated to more than 12,000 years ago. Archaeologists think the Clovis culture in North America lasted about 500 years—from about 13,300 to 12,800 years ago.

The richest trove of Clovis material found so far was unearthed at a location called the Gault site, in central Texas. Since 1929, archaeologists working at the Gault site have recovered more than 600,000 Clovis artifacts, including projectile points, blades, *adzes* (tools similar to axes), and many others.

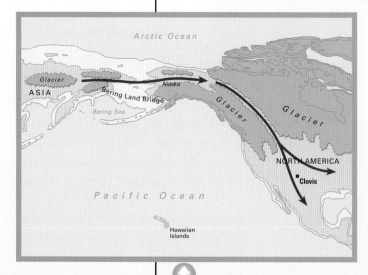

Following a route from Asia across the Bering Land Bridge is the most widely accepted theory of how people first reached the Americas. According to this theory, the first Americans were able to migrate south into North America through a corridor that opened between melting ice sheets about 12,000 years ago.

The projectile points are the artifact that the Clovis people are most noted for today. Clovis points were very carefully made, and some are works of art. The points are fluted, meaning that two shallow grooves—flutes—were cut into the base of the point, one on either side. A notch was cut into the end of the spear shaft, and the two halves of the shaft were slid into the flutes, which were coated with a sticky substance, such as pine tar. This "glue" hardened and held the point tightly in place.

In most Clovis hunts, the spears were probably used with an atlatl, or spear thrower. This device was basically a short stick that increased the leverage of a hunter's arm. The atlatl may have been a simple weapon, but it was deadly. With an atlatl, a light spear could be thrown several hundred feet with great force.

Hunters from Asia travel across the Bering Land Bridge thousands of years ago during the Ice Age, in an artist's illustration. Because so much of Earth's water was locked up in vast ice sheets, sea levels were lower. Hunters were able to walk across land now flooded by the Bering Sea.

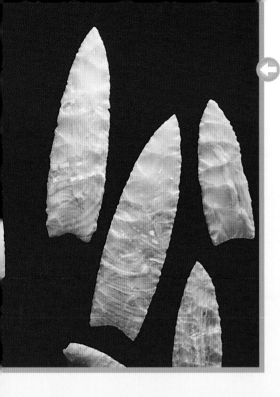

Clovis spearheads, called points, are known for their artistry. They were *flaked* (chipped) into shapes by some of the earliest settlers in the Americas. Clovis points have been found throughout much of North America and parts of Central America.

Although archaeologists have unearthed a great many Clovis points and animal remains, they have found very few human fossils. These include the partial skeletal remains of two women and a child. A genetic analysis of one of the female skeletons, called the Minnesota Woman, produced disappointing results. Researchers had hoped the analysis would help to establish the origin of the woman's ancestors and her tribal relationships. But they could find no link to either the Asian migrants or any modern tribe. This result was a letdown to scientists because the Clovis people are largely a mystery. It has been assumed that they are related to population groups in Asia and came to North America by way of the Bering Land Bridge. But so far there is no proof for that assumption.

THE END OF "CLOVIS FIRST"

For many years, archaeologists agreed that the Clovis people were the first settlers in the Americas. By the 1990's, however, that theory had largely been overturned. Archaeologists kept finding evidence that other people were here before them—perhaps long before. Moreover, genetic and *linguistic* (language) studies indicated that Asian and Native American populations first began to diverge close to 30,000 years ago. In 2007, an article in the journal *Science* declared that the Clovis First theory was officially dead.

Some of the first archaeological evidence that the Clovis culture was not first came from a site in Chile called Monte Verde. Material from that site indicated that people were living there 14,000 years ago. This date was earlier than the widely accepted date for the first Asian migrants entering North America.

A successful hunter in prehistoric America returns to camp in a rock shelter, in an artist's illustration.

In the United States, newly discovered living sites pushed the possible entry of modern humans into the Americas even further back in time. One noteworthy site, called Cactus Hill, is on the coastal plain of Virginia. At this site, discovered in the 1990's, archaeologists found Clovis artifacts. But beneath the Clovis level was evidence of earlier people. Several spear points recovered from those levels were not Clovis points but looked as though they could be an earlier version of the Clovis design. Radiocarbon dating of wood samples yielded dates of about 15,000 to 17,000 years ago. Some archaeologists think the Cactus Hill findings, together with other discoveries, suggest that the Clovis culture may have originated in the southeastern part of what is now the United States.

By 2010, it was clear that the settlement of the Americas was more complex than scientists had thought. If migrations across the Bering Land Bridge were only part of the story, how else might people have arrived in the Americas? And where might they have come from?

Some archaeologists theorized that some people may have migrated from Asia to the Americas in boats, thereby avoiding the huge glaciers on the land. They could have sailed along the west coast of upper North America and stopped at small islands and inlets on their southern journey. Critics of this idea point out that no evidence has been found supporting a coastal migration. In response,

Spearheads found in Clovis, New Mexico, and at the Gault archaeological site in Texas with the remains of extinct prehistoric animals led archaeologists to conclude that the people who made these points were the first migrants to arrive in the Americas. However, the discovery of Kennewick Man and other older artifacts has overturned the Clovis First theory.

At Cactus Hill, a prehistoric living site in what is now Virginia, archaeologists have found spearheads several thousand years older than Clovis points. The spearheads, which could be earlier versions of Clovis-style points, have led some archaeologists to suggest that the Clovis culture originated in the southeastern United States.

advocates of the theory say that any coastal living sites would have been covered by water as sea levels rose.

The most controversial migration theory holds that some people came to the Americas from Europe by boat. Supporters of this theory argue that these migrants would most likely have skirted the southern margin of the Atlantic Ocean ice sheets and then moved down the American east coast. These migrants could have been the ancestors of the Clovis people. Advocates of this theory say that Clovis points are similar to points made in a style called Solutrean. This toolmaking tradition emerged in southern Europe about 21,000 years ago and lasted about 6,000 years. Some archaeologists have cast doubt on this idea by arguing that the Solutrean culture ended at least 3,000 years before the Clovis culture emerged.

There may also have been additional migrations *after* the emergence of the Clovis culture. A skeleton found in Washington state in 1996 and dubbed Kennewick Man was radiocarbon dated to about 9,000 years ago. The skull of Kennewick Man had features that were neither European nor Native American. If anything, archaeologists said, its shape resembled that found among a Japanese people called the Ainu. So did people from Japan also come to the Americas? No one knows.

PALEO-INDIANS AND ANIMAL EXTINCTIONS

As Paleo-Indians—the earliest known inhabitants of North America—spread across the continent, they may have had a devastating impact on native ecosystems. During the Pleistocene Epoch, which lasted from about 2.6 million years ago to about 11,500 years ago, many kinds of giant animals roamed North America. These animals, called megafauna, included mammoths; mastodons; horses; giant ground sloths, bison, and beavers; and relatives of modern camels. Toward the end of the Pleistocene, about 35 kinds of megafauna disappeared. What happened?

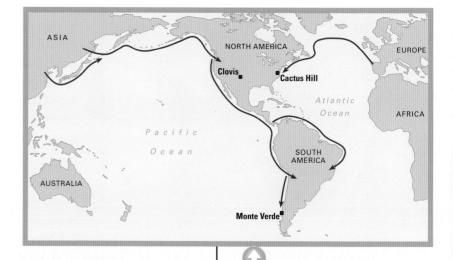

The first inhabitants of North America may have migrated to the continent by a number of routes. Most archaeologists think the migrants crossed a land bridge that once existed between Asia and North America. Some archaeologists have theorized that migrants traveled by boat along the coasts of North and South America. Other archaeologists have suggested that some migrants may have come from Europe.

Many scientists think the disappearance of the Clovis culture was related to the extinction of mammoths and many other prehistoric animal species around the same time. Some experts think that overhunting by Clovis people drove many large prehistoric animals into extinction. Other experts think that climate change, disease, or a combination of factors caused the extinctions. After their main source of food became extinct, Clovis people had to rely more on hunting smaller animals and gathering plant foods. Distinct local cultures began to replace Clovis culture throughout North America. One such culture is recognized by spearheads similar to Clovis points, called Folsom points. These points are also *fluted* (made with long, narrow grooves).

For many years, archaeologists theorized that the Clovis people had hunted America's large animals to extinction. But scientists have now determined that the extinctions took place from about 14,000 to 11,000 years ago—at least in part before the Clovis culture emerged. Some archaeologists have theorized that earlier Paleo-Indians as well as Clovis hunters may have been responsible for the disappearance of North America's megafauna.

Some archaeologists have proposed other reasons for the extinctions. They point out that there is no evidence that Paleo-Indians hunted and ate most of the animals that disappeared. Some scientists have suggested that climate swings near the end of the Ice Age may have put stress on animals and disrupted their habitats. Proponents of this idea cite, in particular, a frigid period of time called the Younger Dryas period, which occurred from about 12,900 to 11,500 years ago. They point out that many animals also went extinct at about this time in South America and northern Eurasia. Many scientists find this theory unlikely, however. They argue that animals that were able

to survive through the depths of the Ice Age would not have been wiped out by a relatively short-lived cold snap.

According to another theory, the megafauna of North America died from germs brought to the New World by the Paleo-Indians. But the examination of animal fossils has found no sign of new diseases. Yet another theory proposes that a comet struck Earth. The collision would presumably have caused widespread fires and a darkening of the atmosphere. Those conditions, in turn, would have killed many plants and the animals that fed on them and brought on the Younger Dryas period. The impact may even have ended the Clovis culture. But this theory also has many critics.

A Paleo-Indian hunter prepares to attack a mammoth using a stone-tipped spear fitted into an atlatl, a shaft with a spur at one end to hold the butt of the spear. This device increased the range and force of a thrown spear.

SETTLEMENT OF THE NORTH AMERICAN ARCTIC

While most Paleo-Indians were settling regions with relatively mild climates, some newcomers to the Americas headed into what is now the Alaskan Arctic. Archaeological evidence indicates that humans were migrating through the Arctic regions of Alaska and regions farther to the east beginning about 12,000 years ago.

The first people to move into Alaska are called Paleo-Eskimos. The Paleo-Eskimos settled in a number of areas. Archaeologists working in Alaska have uncovered more than 50 Paleo-Indian living sites dating from at least 10,000 years ago.

One important site is called Onion Portage, located on the Kobuk River in northwest Alaska. Archaeologists have found evidence that this site was used by groups of people over a period lasting more than 12,000 years. Tools excavated at Onion Portage from the early periods are from a toolmaking tradition called Akmak. They are very similar to tools that archaeologists have found in Siberia from the same era. Archaeologists think this finding shows that the first migrants to the Arctic stayed in close touch with their tribes or clans in Siberia and may have returned to their homeland frequently.

An Inuit harpooner in an *umiak* (open boat) prepares to kill a whale. Whale hunting was a dangerous activity that required many people to work together. The paddlers in the umiak tried to get as close to the animal as possible, to give the harpooner the best chance of success. Because the water was frigid, staying warm meant staying dry. The hunters' outer parkas, mittens, and boots were made of stretched seal intestines, which were waterproof.

A series of Arctic cultures emerged over a period of some 10,000 years. At Matcharak Lake, located in the same general area as Onion Portage, archaeologists have found 4,000-year-old evidence from a group called the Denbigh (*DEN bee*) people. The Denbigh people introduced a new tool-making technique called the Arctic Small-Tool Tradition. Tools produced in this tradition included small, finely made blades, scrapers, spear points, and arrowheads.

In 2008, researchers from several countries analyzed mitochondrial DNA from preserved Paleo-Indian hair found at a site in Greenland. (Mitochondrial DNA is genetic material inherited only through the female line.) The scientists learned that the person from whom the hair came was not genetically related to the people who later inhabited the Arctic—the Inuit and their ancestors. This finding showed that at some time in the past, the ancestors of the Inuit replaced the previous inhabitants of the Arctic.

A similar conclusion resulted from a 2010 analysis of genetic material in the hair of a man dubbed Inuk, also found in Greenland. The researchers believe that Inuk belonged to the now-extinct Saqqaq culture, which inhabited northern Canada and Greenland between about 4,700 and 2,500 years ago. The researchers concluded that the man's closest living relatives would be the Chukchis, people who live at the easternmost tip of Siberia. The genetic analysis suggested to the scientists that a migration from Siberia occurred about 5,500 years ago.

Archaeologists think that another group of Paleo-Indians, seal hunters now known as the Dorset, may have invented the igloo. The Dorset disappeared in about A.D. 1400. They may have been overcome by a stronger group called the Thule (*TOO lee*). The Thule originated in western Alaska about A.D. 1000 and spread eastward across the Arctic. The Thule were apparently the first people in the Arctic to hunt large whales. They pursued whales in large skin-covered boats and killed the animals with harpoons. Archaeologists think the Thule were also the inventors of the dog sled.

The Inuit (formerly called Eskimos) are the direct descendants of the Thule people. The 1800's are considered the beginning of Inuit culture, which is mostly a continuation of Thule traditions.

The portrait of a man, dubbed Inuk, who lived about 4,000 years ago in northwestern Greenland is based on the first reconstruction of the *genome* (set of genetic material) of a human being from an extinct culture. A team led by evolutionary biologists Eske Willerslev of the University of Copenhagen in Denmark and Jun Wang of the Sino-Danish Genomics Center in China analyzed genetic material extracted from tufts of hair from Inuk found in *permafrost* (permanently frozen soil). According to the analysis, reported in February 2010, Inuk had brown eyes, brownish skin, and dark, thick hair. He also had a tendency to baldness and was adapted to living in cold weather.

A number of Paleo-Indian cultures have existed in what is now the Arctic region of North America over the past 10,000 years. Onion Portage is among the oldest human-occupied sites identified by archaeologists.

FROM HUNTER-GATHERERS TO FARMERS

The Paleo-Indians in North America lived strictly as hunter-gatherers for several thousand years after arriving in the Americas. They most likely moved about in bands of 20 to 50 related individuals.

Gradually, as their numbers grew, the Paleo-Indians became grouped into tribes. Many of them settled down and established permanent communities. These people continued their hunting-and-gathering ways, but they also began to grow much of their food. They were becoming farmers.

Archaeologists long thought that Mesoamerica—modern-day Mexico and Central America—was the only place in the Americas where agriculture was independently invented. But by the 1990's, archaeologists had discovered evidence that farming was widespread 4,000 years ago in the eastern part of what is now the United States. The evidence included abundant plant remains and seeds from ancient living sites. Bruce D. Smith, an archaeologist at the Smithsonian Institution in Washington, D.C., was a leader in this work.

Based on these findings, Smith and his colleagues were able to revise the conventional wisdom on early centers of agriculture. They put the Eastern Woodlands region, as it is known, on a par with Mesoamerica. Those two regions are now recognized as two of just five centers in the

Native Americans wearing wolfskins as a disguise hunt buffalo on the Great Plains. Many Indian groups rejected agriculture in favor of hunting.

THE MISSISSIPPIAN CULTURE

Settled agricultural societies often give rise to advanced civilizations, and so it was in North America. Almost.

About A.D. 800, corn-growing communities along what is now the Mississippi River began to join together into a loose federation based on trade. This collection of settlements became known as the Mississippian Culture.

The Mississippian people are often called the mound builders. In their bigger settlements, they constructed large flat-topped mounds of earth. These mounds were used as temple platforms and for the burial of their most important citizens. These status-related structures showed that some people in the society were becoming richer and more powerful than others.

world where people developed farming independently. (The other three are the Middle East, India, and China.)

Evidence from Eastern Woodlands archaeological sites has shown that agriculture got its start more than 4,000 years ago as Native Americans began discovering edible plants that could be cultivated. Within another thousand years, the woodland Indians were growing barley, squash, sunflowers, and several other crop plants.

Eastern Woodlands agriculture continued along the same path for more than 4,000 years. Archaeologists believe the crops remained the same because the Indians could find no other plants worth cultivating. That situation changed dramatically about A.D. 800 when corn made its way north from Mexico. Soon corn was the main crop in the Eastern Woodlands. Corn had arrived via the American Southwest, where Indians had also become corn farmers.

Some Indians rejected agriculture. The Sioux and other inhabitants of the Great Plains looked on farming with contempt, preferring to hunt buffalo. Interestingly, the Sioux had been settled farmers in the Great Lakes region before acquiring horses—brought to the New World by Spaniards—in the 1700's.

Birchbark lodges and cornfields frame a scene of celebration in Secoton, a village of Algonquin-speaking people in what is now North Carolina, in a 1590 engraving based on a watercolor by English illustrator John White.

The most important center of Mississippian culture was Cahokia, in what is now Illinois. At its height around A.D. 1100, it may have had a population of 20,000.

Sometime after A.D. 1200, Cahokia was abandoned, and over the next 200 years the Mississippian culture disintegrated. The reason for the society's demise is not known. Perhaps it was social conflict or a changing climate that caused crop failures.

An earthwork structure now known as Monk's Mound towers above houses at Cahokia. The four-level mound is the largest prehistoric earthwork in the Americas. A large building that once stood at the top of the mound was probably the residence of Cahokia's chief ruler.

Most Indians hunted and fished for their food or gathered wild seeds, nuts, and roots. Farming was the main source of food only in what is now the southwestern United States and Middle America and the Andes region of South America.

MOVING INTO MESOAMERICA AND SOUTH AMERICA

Mesoamerica (also called Central America) have some of the oldest human remains in the Americas. The oldest modern-human remains found in Mexico are four skeletons recovered from underwater caves off the Yucatán Peninsula between 2001 and 2008. Radiocarbon dating of the remains showed they are at least 11,000 years old, with the earliest being about 14,000 years old. The caves in which the skeletons were found were on dry land until about 9,000 years ago. They were flooded by rising seas at the end of the Ice Age. Archaeologists who studied the skeletons' skulls said they resemble those of people in Southeast Asia. This finding supported the contention of many archaeologists that not all the migrants who settled the Americas came from northern Asia.

Finds at several archaeological sites in Mexico have raised the possibility that modern humans were in that region much earlier than 14,000 years ago. At one site, El Cedral in northeastern Mexico, archaeologists excavated what may have been an ancient *hearth* (fireplace), along with with stone tools and the bones of mammoths and horses. Material from the hearth was radiocarbon dated to about 33,000 years ago. Another site, Babisuri Cave on the Baja Peninsula, contained stone tools and clamshells. The shells were dated to 40,000 years ago.

As of 2010, neither of these dates had been widely accepted by archaeologists. Some experts said, for example, that the shells could have been old ones that were collected by the occupants of the cave. And at El Cedral, they pointed out, there was no evidence of human habitation, nor had the animal bones been radiocarbon dated. Dating the bones was not possible because they did not contain enough datable material.

Although scientists are not certain when Mexico was first settled, they have a pretty good idea of when agriculture began. Archaeologists at the Smithsonian Institution in Washington, D.C., and Temple University in Philadelphia reported in 2009 that the ancient Mexicans were cultivating corn (or maize, as it is often called) as early as 8,700 years ago. Corn was domesticated from a wild plant called teosinte *(taa oh SIN tee)*. The researchers based their finding on radiocarbon dates of plant material taken from a Paleolithic (Stone Age) site where tools for grinding grain were also discovered. This date is about 1,200 years earlier than the oldest previously accepted date for the domestication of teosinte in Mexico.

Once agriculture was established in Mexico, civilization followed. A series of civilizations ruled central Mexico over a period of more than 2,000 years. *Indigenous* (native) rule ended with the Aztecs, who were overthrown by Spanish invaders in 1521.

NORTH AMERICA

Gulf of Mexico

West Indies

Caribbean Sea

Central America

ATLANTIC OCEAN

Amazon

PACIFIC OCEAN

SOUTH AMERICA

Andes Mountains

SOUTHERN OCEAN

THE YUCATÁN PENINSULA AND CENTRAL AMERICA

Scholars believe that the Paleo-Indians who settled what is now upper Mexico are the same people who migrated into the Yucatán Peninsula and Central America. When that settlement began is uncertain. Paleolithic (Stone Age) evidence from this area is sparse. Researchers have, however, found evidence that the ancient Indians were growing corn in the region's rain forests more than 7,000 years ago.

The theory that Paleo-Indians farmed the rain forest thousands of years ago was first proposed in the 1950's by an American geographer, Carl Sauer. His theory was largely dismissed because rain forests are not a good environment for growing crops. The soils are thin and acidic.

In the 1980's, researchers from the University of Missouri decided to investigate Sauer's theory. They analyzed rain-forest sediments, focusing on substances called phytoliths. These are microscopic structures that plants form when they absorb *silica* (silicon dioxide) from ground water. The scientists found a way to distinguish phytoliths formed by wild-growing plants from those produced by cultivated plants. They found corn phytoliths in ground sediments from a rock shelter in Panama. Radiocarbon dating of the phytoliths yielded an age of 7,700 years. Some archaeologists, however, question that dating. They contend that the phytoliths might have been contaminated with carbon from other sources.

Nonetheless, at some time in the distant past, Paleo-Indians began growing corn, squash, and other crops in the Central American rain forests. Researchers theorize that for many years, the Indians practiced

The Pyramid of the Sun dominates the ruins of the ancient city of Teotihuacán, one of the largest cities in ancient Mexico. Its population probably reached a peak of between 150,000 and 200,000 during the A.D. 500's. Located near present-day Mexico City, the city was an important religious center for the Aztec, though it was built centuries before the Aztec arrived. The ethnic group that created the city remains unknown.

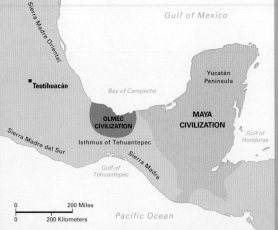

Huge carved-stone heads are the most notable of the artifacts left by the Olmec, whose civilization flourished between about 1200 and 400 B.C. The heads weigh up to 36,000 pounds (16,300 kilograms). Some stand over 9 feet (2.7 meters) high.

WHO WERE THE OLMEC?

Mesoamerica—Mexico and Central America—abounds in mysteries. Teotihuac Teotihuacán is one of them. Another is the civilization known as the Olmec.

Olmec society was the earliest Mesoamerican civilization. It arose about 1500 B.C. in the tropical lowlands of south-central Mexico and lasted until about 400 B.C. The Olmec society is considered one of just a few "pristine" civilizations in the world. A pristine civilization is one that emerges completely on its own.

The Olmec are known for the huge stone heads that they left behind. The heads look African, leading some people to theorize that the Olmec were migrants from Africa. Most archaeologists, however, reject that idea. They say that the heads, which probably represent Olmec kings, were most likely not meant to be realistic depictions. The Olmec were probably the first people in Central America to use numbers and writing.

slash-and-burn agriculture. With this system of farming, people cut down trees in a section of forest and then burn them. The ashes from the trees fertilize the soil, enabling crops to be grown for a few years, after which the soil is depleted. The people then move on to another area and repeat the same process.

Eventually, the inhabitants of the rain forests established a permanent farming system. They learned to create terraces of soil, held in place by stone retaining walls, which they irrigated. The people probably added ashes to the irrigation water, providing a constant flow of nutrients to the soil. They also learned to change the type of crops they grew on a particular section of land from year to year. Although corn was the main crop, the people rotated this crop with beans, squash.

The Maya King Bird Jaguar (right) receives a spear from one of his wives as he prepares for battle in A.D. 755, in a scene depicted on a doorframe at the Maya city of Yaxchilan. The Maya kept records on some buildings as well as on large stone monuments called stelae.

Settled agriculture in the rain forests led to the rise of the Maya in lower Mexico and Central America. The Maya civilization was the most advanced Indian society of the Americas. It reached its height from about A.D. 250 to 900. The Maya civilization then went into decline, but it did not disappear. The Maya culture existed in the 1500's, at the time of the Spanish invasions. Today, many Maya people live in southern Mexico and Central America.

THE SETTLEMENT OF THE ANDES

The early civilizations of the Andes Mountains area of South America were astonishing. The inhabitants of this region were building cities and monumental structures at about the same time that the Egyptians were constructing the Pyramids of Giza. The societies of the Andes were the first civilizations of the Americas.

Archaeologists have not identified the people who first settled the Andes or when they arrived. For many years, archaeologists theorized that the descendants of people who migrated across the Bering Land Bridge some 12,000 years ago eventually settled all of the Americas. If that were true, it would mean that hunter-gatherers migrated from Alaska to the middle of South America—and then established complex societies based on agriculture—within about 7,500 years. Many experts thought that accomplishing so much in that amount of time was unlikely. Recent discoveries have suggested that some people arrived in the Americas much earlier than 12,000 years ago. This finding has widened the period in which civilizations in the Americas may have evolved.

Evidence from early living sites in what is now Peru shows that people in the highlands had begun domesticating plants and animals, including llamas, by about 5000 B.C. At about the same time, other people were establishing settlements along the Pacific Coast.

The first major culture of the Andes was called the Caral-Supe *(KAR al SOO pay)* civilization. It is named for the ancient city of Caral and the Supe River Valley, about 200 miles (322 kilometers) north of present-day Lima. Caral was one of 19 urban settlements in the valley that were established around 2600 B.C. At Caral and the other centers, people constructed large, flat-topped temple pyramids. These monumental structures are now just large mounds of earth and rock. The Caral-Supe civilization came to an end in about 1600 B.C. Archaeologists think it was destroyed by an enormous earthquake or series of earthquakes accompanied by flooding and landslides.

After the demise of the Caral-Supe society, the people of Peru largely turned their backs on the sea. They began to construct settlements in the highlands away from the coast.

As Peruvian communities became bigger, so, too, did their monumental buildings. At a site called Sechin Alto, archaeologists discovered an enormous mound measuring 820 by 980 feet (250 by 300 meters) at its base. The mound was 144 feet (44 meters)

The ruins of Machu Picchu only hint at the grandeur of what was probably an estate for the Inca royal family. Built in the 1400's on a high ridge, Machu Picchu had palaces in addition to houses for the farmers, weavers, and servants who worked for the royal family. Scholars think Machu Picchu was abandoned shortly after the Spanish began their conquest of the Inca in 1532.

The Andes region of South America gave rise to a number of powerful empires. The Inca empire, which lasted from the early A.D. 1400's to 1532, was one of the richest and largest.

high, about the height of a 12-story building. The mound is the remains of an enormous stepped pyramid temple that had been faced with large granite blocks.

In about 800 B.C., Andean civilization was marked by the rise of a society called Chavín. It was named for a large ceremonial center, Chavín de Huantar, in the mountains of central Peru. Over the next three centuries, the Chavín culture spread its influence and art style throughout most of central and northern Peru. Evidence from tombs shows that by this time, significant class divisions had emerged in Andean society. Some tombs from Chavín society are the oldest ones in Peru to contain jewelry and other luxury goods—signs of a rich elite.

The Chavín culture ended about A.D. 200. In the centuries that followed, the Andean region was ruled by a succession of other civilizations. That line ended with the Inca empire, which was overthrown by the Spanish in 1533.

Despite their advances in farming and architecture, none of the Andean societies ever developed a system of writing. So archaeologists must reconstruct their history from the ruins of their civilizations.

A golden plate crafted by the Inca depicts the sun god, Inti *(EEN tee),* their most important god. The Inca's ruling family claimed descent from Inti.

The city of Caral in western Peru was one of about 20 settlements built by the Caral-Supe civilization. Founded in about 2600 B.C., Caral may have been the first city established in the Americas.

SURPRISING FINDINGS IN THE AMAZON

Archaeologists had long assumed that no advanced societies had arisen in the Amazon region of South America. They thought a jungle area with thin soils could have supported only scattered bands of hunter-gatherers.

Yet there was evidence that this view was wrong. In many parts of the Amazon, for instance, there were large expanses of unusual rich soil that the Portuguese called *terra preta* (dark earth). There was also a written record by Spanish explorers who traveled the length of the Amazon River in 1542. A chronicler of the voyage related that the Spaniards observed many Indian settlements with fertile farmlands along the river. This chronicle was dismissed by most archaeologists in later years as a fantasy.

But it wasn't a fantasy. Archaeologists have recently discovered that there was, in fact, an advanced Indian society in the Amazon region. Moreover, that society had ancient roots. From the excavation of human-occupied sites in the Amazon Basin, researchers have found that Paleo-Indians first moved into the region at least 11,000 years ago.

One of the leading researchers on early inhabitants of this region has been the American archaeologist Anna Roosevelt, a great-granddaughter of United States President Theodore Roosevelt. In the 1990's, Roosevelt excavated a cave called Caverna da Pedra Pintada in central Brazil. She dated material in the cave to about 11,000 years ago. Stone artifacts found there appeared to challenge the standard theory of how the Americas were settled by the Paleo-Indians. Because the inhabitants of Caverna de Pedra Pintada lived in the era of the Clovis people, they should have been making Clovis-style tools. Most archaeologists believed that the Paleo-Indians of the Clovis era were big-game hunters using the same kinds of tools. Roosevelt discovered projectile points at the cave that were unlike the Clovis points found at North American sites. Clovis points are narrow and have a shallow groove, called a flute, on either side. The points that Roosevelt unearthed were triangular with a barbed base and unfluted. They were more suitable for spearing fish and small game than for hunting large animals.

American archaeologist Anna Roosevelt (right) and a coworker examines artifacts unearthed in the Amazon. Roosevelt's findings have challenged long-held theories about the settlement of the Amazon and the life of the people who lived there in prehistoric times.

The Amazon region of South America likely once supported a relatively large population of Paleo-Indians. The inhabitants enriched the thin acid soil of the rain forest with a dark, fertile mixture of charcoal, pieces of pottery, manure, and other elements.

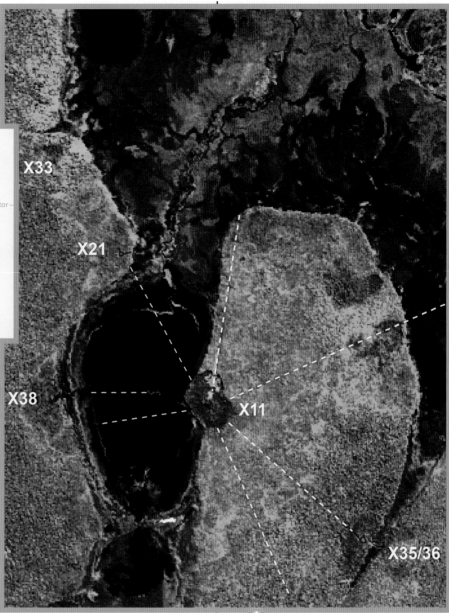

Roosevelt concluded from her research—and from similar findings by other archaeologists working in South America—that the Clovis culture was not universal. It was just one of several regional traditions. People in the Americas adapted to their environments. In the Amazon, she said, the Indians adapted themselves to the tropical rain forest.

It has since become clear that the inhabitants of the Amazon adapted very well. By about A.D. 800, they had turned much of the Amazon region into a garden. This Amazon civilization was rediscovered in the 2000's by several teams of archaeologists. They found the remains of settlements along the Amazon River that had supported populations of up to 5,000 people. The scientists discovered traces of plazas, houses, roads, and other structures. Outside the settlements were large areas covered with terra preta, in which the Amazon people grew their crops. This was specially prepared soil, containing charcoal, manure, and other materials,

When Europeans began spreading through the New World, they transmitted smallpox and other contagious diseases to the Indians. Indian populations plummeted throughout the Americas as people died by the millions. Anthropologists estimate that the population of the Amazon was reduced by at least 90 percent in the 1600's. The survivors drifted away to form small bands, and the jungle covered everything.

Ancient roads (marked with white lines) lead from a central village to smaller surrounding villages in a satellite image of part of the Amazon rain forest in Brazil. In 2008, archaeologists reported finding evidence of an extensive network of clustered settlements linked by wide roads that flourished before the arrival of Europeans in the 1500's.

DISCOVERIES AT MONTE VERDE

In 1975, a mastodon bone sticking out of the ground at a site in Chile led to one of the most important archaeological discoveries ever made in the Americas. The bone was just one of many animal and plant remains and tools found at the site, called Monte Verde (Spanish for *Green Mountain*), when excavation began in 1977. Radiocarbon dating of bones and charcoal showed that the site, which had been a settlement of Paleo-Indian hunter-gatherers, was occupied about 14,000 years ago.

The dating of the site provided persuasive evidence that the so-called Clovis First theory was wrong. This theory held that the first Paleo-Indians to populate the Americas had arrived about 12,000 years ago by way of the Bering Land Bridge. These migrants and their descendants are called the Clovis people because their distinctive tools were first found at a site in Clovis, New Mexico.

The Monte Verde evidence showed that people were living at a place about 9,000 miles (14,500 kilometers) south of the Bering Strait 2,000 years before the Clovis migration. That finding was astonishing enough. But in the following years, archaeologists began compiling even more amazing findings from archaeological sites, genetic studies, and linguistic analyses. They reported a growing body of evidence that the first migrants to the Americas may have arrived more than 30,000 years ago. Some of the earliest archaeological findings pointing to a much earlier migration came from the Monte Verde site. In the 1980's, archaeologists working there discovered a deeper layer of occupation. There, they found stone tools and the remains of several clay-lined hearths. Charcoal from the hearths yielded a radiocarbon date of 33,000 years ago.

Most scientists think that the people of Monte Verde and most other pre-Clovis sites came from northeast Asia, just as the Clovis hunters did. They just got to the Americas earlier. And some of them must have come by a different migration route. Archaeologists think those people must have migrated by boat. They could have sailed south along the western coast of the Americas.

In fact, many archaeologists think that coastal migration by boat was the preferred mode of travel for most of the people who left Asia. It

The archaeological site Monte Verde in modern-day Chile has provided evidence that human beings inhabited the Americas before the Ice Age ended about 11,500 years ago.

would have been faster, and it might even have given the migrants access to a more dependable source of food.

In 2007, a group of American archaeologists proposed an idea they called the "Kelp Highway" theory of migration. They noted that there are large forests of kelp, a type of seaweed, along much of the Asian coast and most of the west coast of the Americas. The kelp forests abound in fish, and the seaweed is edible and nutritious. People may have sailed through the coastal waters, supporting themselves from the bounty of kelp.

Interestingly, archaeologists found extensive remnants of seaweed in the upper level of Monte Verde, even though the site is about 35 miles (55 kilometers) from the ocean. The scientists said seaweed must have been important to the people either as a part of their diet or as medicine.

Prehistoric people who once lived at Monte Verde, only 35 miles (55 kilometers) away from this beach in what is now Chile, gathered various types of seaweed, which they used as food and medicine.

WHO SETTLED TIERRA DEL FUEGO?

At the southern tip of South America lies the *archipelago* (group of islands) of Tierra del Fuego *(tee AIR uh del FWAY go)*. The first European to see this cold, stormy place was the Portuguese explorer Ferdinand Magellan while sailing around the world under the flag of Spain. Observing the islands in 1520, Magellan and his men saw many fires onshore, which the Indians had set to warm themselves. Magellan gave the islands their name, which is Spanish for *Land of Fire*.

Portuguese explorer Ferdinand Magellan (shown in an illustration) uses a compass to plot a course for his voyage around the world, which began in 1519. Magellan and his crew were the first Europeans to visit the southern tip of South America.

The British naturalist Charles Darwin visited Tierra del Fuego in 1832 during his scientific voyage aboard the *Beagle*. In his later account of the journey, *The Voyage of the Beagle*, Darwin told of his meetings with the native Fuegians. He commented that he had never seen people living such a primitive life as these unclothed aborigines. Darwin said they ate mainly fish and shellfish and occasionally caught otters and seals. But he noted that the natives often suffered from famine. Darwin said the members of the four Fuegian tribes on the islands were forced to move constantly in search of food, and that food shortages led to frequent warfare.

Tierra del Fuego is the name of a group of islands lying at the extreme southern tip of South America.

Darwin was both intrigued and appalled by the people of Tierra del Fuego, whom he called "the most abject and miserable creatures I anywhere beheld." He had no idea where the Fuegians had come from. "Whilst beholding these savages," he wrote, "one asks, whence have they come?" Darwin also wondered why any people would choose to live in such a harsh environment. He had no answer to those questions.

So who were those people and where *did* they come from? Scholars had long assumed that Tierra del Fuego was simply the last stop in the southward movement of the people whose ancestors migrated across the Bering Land Bridge about 12,000 years ago. But in recent years, archaeologists have been forced to change that theory.

By the 2000's, the Fuegians were almost extinct. Their numbers had been drastically reduced in the late 1800's by European settlers, who transmitted a variety of diseases to the natives and even hunted them. Fortunately for archaeological research, there were considerable Fuegian remains. Mitochondrial DNA studies of material taken from the bones and teeth of Fuegian skeletons

produced unexpected results. The analyses revealed that the Fuegians were apparently not descended from the Paleo-Indians who came from Siberia and settled most of the Americas.

The most likely explanation for this finding, many archaeologists say, is that the Fuegians are the descendants of an earlier group of north Asian migrants. They speculate that earlier settlers, arriving 30,000 to 40,000 years ago, were displaced—or even exterminated—by the later arrivals. The Fuegians may have been pushed southward into the cold reaches of Tierra del Fuego after losing a territorial battle with new arrivals from Asia. They could have been the last surviving remnants of the earliest migration to the Americas.

Some anthropologists have theorized that the Fuegians weren't from north Asia. They say that the Fuegians looked more like Australian Aborigines or the Negrito people of the Andaman Islands than like the Paleo-Indians from Siberia. These scientists argue that South America may have been populated before North America. Early migrants from southeastern Asia, they argue, could have come across the Pacific Ocean by boat. Most experts do not give this idea much credence. Nonetheless, it emphasizes how much is still not known about early-human migrations.

Native children living on Tierra del Fuego sit outside a hut, in a print based on a drawing made by British naturalist Charles Darwin on his trip to this region in 1832.

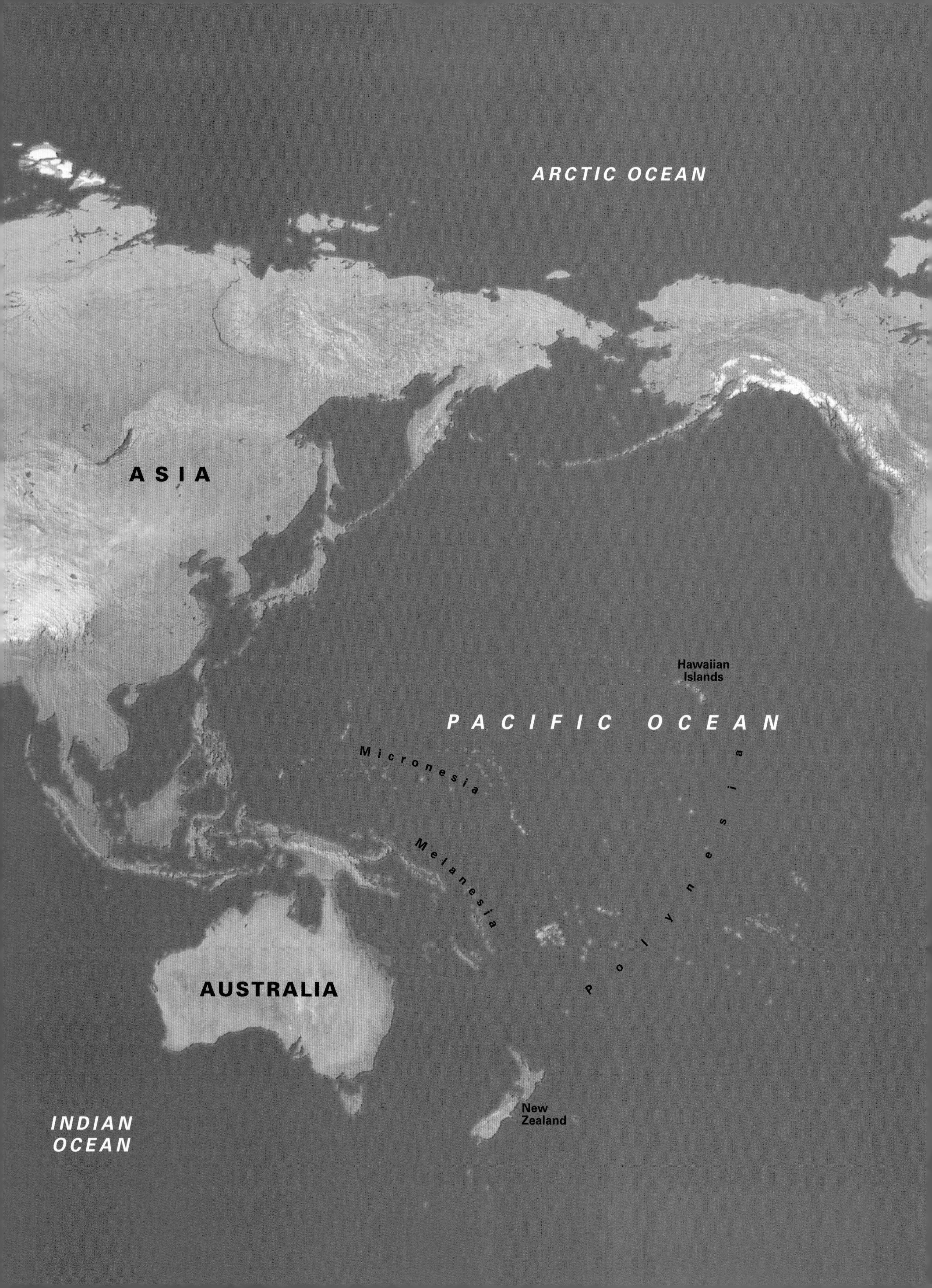
ARCTIC OCEAN
ASIA
Hawaiian Islands
PACIFIC OCEAN
Micronesia
Melanesia
Polynesia
AUSTRALIA
New Zealand
INDIAN OCEAN

POPULATING THE PACIFIC ISLANDS

NORTH AMERICA

SOUTH AMERICA

The islands scattered across the vast expanse of the Pacific Ocean were some of the last places on Earth to be settled. Archaeological findings and genetic and linguistic studies have revealed that these islands were discovered and populated by the Austronesians, who probably originated in what is now Taiwan.

The archaeological evidence includes findings related to a tradition called the Lapita culture that emerged in the western Pacific about 1600 B.C. It is considered the ancestor to later cultures that arose in Polynesia and other parts of the Pacific. Artifacts of this culture include a distinctive type of decorated pottery with horizontal decorative bands. Fragments of pots and bowls made in the Lapita style found on many islands have enabled archaeologists to trace the path of Austronesian migrants as they spread through much of the Pacific.

Genetic studies have established that most of the peoples of the Pacific Islands come from a common ancestral group. Some DNA studies suggest that the Austronesians came from mainland southeastern Asia rather than Taiwan. But most scholars think that Taiwan was the point of origin for the Austronesians, who then spread outward across the Pacific over a period of about 3,000 years.

The Austronesian language group, in particular, shows that all the inhabitants of the Pacific Islands are related. The Austronesian languages are a huge group, comprising some 1,200 languages—about one-fifth of all the languages in the world.

The Austronesians were some of history's most skilled sailors and navigators. Some scientists have credited the oceangoing success of the Austronesians, in part, to their invention of the outrigger sailing canoe. This is a canoe stabilized by one or more horizontal supports, called outriggers, that extend into the water from the side of the vessel. The oceangoing canoes had sails. The largest oceangoing canoes could carry several hundred people.

But canoe design was just part of the picture. The Austronesians were also master navigators. They developed an extensive knowledge of the stars, prevailing winds, ocean currents, and island landmarks that enabled them to find their way across vast expanses of water.

Over the centuries, the Austronesians settled all the major uninhabited islands of the Pacific Ocean. Many archaeologists think the Austronesians advanced through what is called "pause-pulse migration." The Austronesians would arrive at a new island and then pause there for some period of time to settle it. Then, in a new pulse, some of their number would move on to populate a new island.

Heading out in search of new homes was often a response to population pressures. But anthropologists think that "moving on" was also part of the Austronesian culture. Groups who discovered new islands and extended the territory of the Austronesian people achieved high status within the culture.

AUSTRONESIAN MIGRATIONS

Scholars believe that the people now known as the Austronesians first left what is now Taiwan in about 2500 B.C. and settled in the modern-day Philippines. Archaeological evidence for this migration includes pottery like that found in Taiwan from the same period and stone tools. The most significant of these tools was the stone *adze* (a type of ax). In the Philippines, the Austronesians displaced the native hunter-gatherer Negrito people, just as their ancestors did in Taiwan about 1,000 years earlier. The Negritos retreated to the mountains, and the newcomers established themselves on the fertile coastal plains.

The Austronesians also settled in a few parts of mainland Southeast Asia, including what are now Vietnam and Thailand. But these areas already had fairly advanced agricultural societies with large populations, so the Austronesians did not make much of an impact.

Archaeologists think that although the Austronesians had an extensive knowledge of agriculture, especially the growing of rice, not all of them were dedicated to farming. Where game and wild plants were plentiful, they suggest, the wanderers preferred hunting, gathering, and fishing to agriculture. And they remained at heart a seagoing people.

The Austronesians expanded as far west as the island of Madagascar, off the coast of Africa, sometime after 300 B.C., apparently by sailing directly across the Indian Ocean. There is no evidence that they ever ventured onto the African mainland.

Archaeologists think Austronesians began their journey across the Pacific from the Moluccas, a group of islands in eastern Indonesia, in about 1600 B.C. The first leg of this long odyssey was through an island group now known as Melanesia, which lies east of Indonesia and north-northwest of Australia. New Guinea is the largest island of Melanesia. Most of the islands of Melanesia were already occupied when the Austronesians arrived. In New Guinea, the native Papuans resisted being absorbed into Austronesian society. Archaeologists think the people of New Guinea had already begun to cultivate some crops, including a plant called taro with edible roots and leaves, when the Austronesians arrived. Papuans living in the lowlands soon withdrew into the mountains.

Genetic studies have determined that present-day Melanesians are not closely related to the Micronesians and Polynesians. That finding suggests that the Aus-

The Polynesian people trace their ancestry to the Lapita people, Austronesians from modern-day Taiwan who settled in the Solomon Islands and Bismarck Archipelago and mixed with people from New Guinea. Over time, the Lapita people migrated to a number of nearby islands, including Fiji, Samoa, and Tonga.

Elaborate designs charcteristic of the Lapita style cover a large earthenware pot dating from about 3,000 years ago that was found in modern-day Vanuatu. Lapita artists "stamped" the designs on pottery using a comb-like tool with a row of square or rectangular teeth. Sometimes, the patterns represented a human face and eyes.

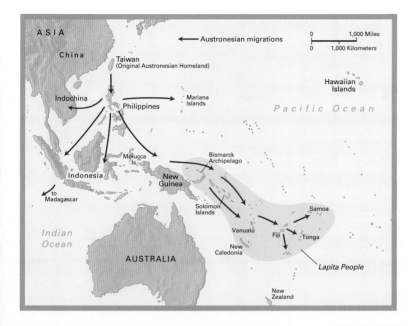

tronesians who came through Melanesia did not have much contact with the native people of the islands and did not stay long.

However, during their stay in Melanesia, the Austronesians may have begun making pottery in the Lapita tradition. Fragments of these distinctive pots and bowls have provided an archaeological trail of Austronesian migrations across the Pacific. Scholars have debated where the Lapita culture originated, but most think it developed in the Bismarck Archipelago, east of New Guinea.

Traveling to the north, the Austronesians entered Micronesia, a name meaning *tiny islands*. Micronesia comprises more than 2,000 islands, most of them low-lying coral isles. They include what are now Guam, the Mariana Islands, and the Marshall Islands. These islands were mostly uninhabited. There, the Austronesians put down roots. In about 1000 B.C., while Micronesia was being settled, other Austronesians set sail from the outer islands of Melanesia in search of new homelands. These people became the Polynesians.

A Tahitian clan prepares an oceangoing canoe for a voyage to colonize a new island home, in an artist's illustration. Provisions for the voyage, as well as the tools, seeds, and farm animals that the clan would need to found a new settlement, were stored in the two hulls of the canoe. The thatched cabin sheltered the passengers from sun and rain. A clay hearth on the deck near the cabin even allowed for a cooking fire while at sea.

THE SETTLEMENT OF POLYNESIA

The islands of Polynesia are some of the most beautiful places in the world. Most of the islands consist chiefly of coral reef material, but some, such as the Hawaiian Islands, were formed by volcanoes. There are more than 1,000 islands in Polynesia, whose name means *many islands*.

Before the coming of the Polynesians, none of the islands—scattered across 16 million square miles (41 million square kilometers) of ocean—were inhabited. Over a period of more than 2,000 years, the Polynesians occupied all the major islands.

On each of the islands they settled, the Polynesians established a society unique to that island. Some developed societies were ruled by family groups. Others established societies ruled by a king. On some islands, there were frequent wars between rival factions. Among the first islands to be discovered were modern-day Samoa and Tonga, by about 900 B.C. The Polynesians then pushed slowly eastward over a period of centuries. By A.D. 700 two major island groups—the Society Islands (which includes Tahiti) and the Marquesas Islands—had been occupied.

King Kamehameha I, wearing a cloak, appears with some of his subjects, in an engraving from the 1800's.

King Kamehameha I of Hawaii (1758?-1819), in the schooner *Fair American* (below, right), uses a raised sword to direct a sea battle against enemy chiefs, in artist's illustration. By 1810, following many years of war, Kamehameha had united all the main islands of Hawaii into the Kingdom of Hawaii.

THE FARTHEST OUTPOST: EASTER ISLAND

One of the most unusual of all the Polynesian societies was that of Easter Island. People today know it for the strange stone heads that stare inland. Easter Island—or Rapa Nui as its people called it—is one of the most remote isolated on Earth. More than 2,000 miles (3,200 kilometers) from the Marquesas Islands, it was the farthest point east in the voyages of the Polynesians.

Scholars are divided as to when the island was first settled. Most think that people from the Marquesas Islands sailed there sometime before A.D. 800. They established a society ruled by a king.

Archaeologists have concluded that everything went well at first. The island was heavily wooded, and the soil was fertile. But then the people began erecting the huge statues, called moai (MOH eye), which represented their ancestors. Erecting moai became the focus of island life. Other activities, such as fishing and farming, were neglected.

But it was an island tradition called the Birdman Cult that may have brought about the society's downfall. Each year, a man who defeated his competitors in retrieving a bird's egg from a nearby isle was named "birdman of the year" and his clan was given control of the island's resources. Over time, the various birdmen and their kin stripped the island of its trees and other resources for their own uses.

The result was starvation, depopulation, and war. When Great Britain's Captain James Cook visited the island in 1774, he found that the earlier population of 15,000 had been reduced to fewer than 1,000. Many of the moai had been toppled by angry islanders.

Giant stone statues called moai stand guard on Easter Island, one of the most remote places in the world. Archaeologists think the statues were erected to represent the ancestors of the island's chiefs.

From Tahiti in the Society Islands, Polynesians traveling in canoes explored and settled islands across a vast section of the central Pacific Ocean.

It was from the Marquesas Islands, scholars think, that Polynesian voyagers set forth on a truly amazing voyage sometime before A.D. 800: a journey across 2,400 miles (3,860 kilometers) of open ocean to the Hawaiian Islands.

The society those settlers established in ancient Hawaii was divided into several rigid social classes. The top tier was the royal class, consisting of various chiefs. The chiefs of different islands often started wars with one another. Below the chiefs was a priestly class that consisted of both priests and people of exceptional abilities, such as navigators. On the next-lowest rung of the social ladder were the commoners: farmers, fishermen, and craftsmen. And below them were slaves, some of whom were captives taken in war.

This system continued until 1795. In that year, a chief named Kamehameha gained control of all the islands after a decade long war and founded the Kingdom of Hawaii.

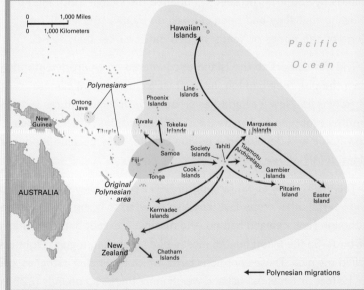

THE FINAL SETTLEMENT: NEW ZEALAND

The settlement of what is now the modern nation of New Zealand was the final chapter in the Austronesian saga of oceanic expansion. It may seem odd that New Zealand—which is made up of two large islands and several small islands southeast of Australia—was not discovered by the Austronesians earlier than it was. Perhaps New Zealand remained unknown because it lies far to the south of the other Polynesian islands. Archaeologists think New Zealand was settled by several waves of voyagers from eastern Polynesia sometime before A.D. 1300.

The Austronesians' arrival date was long debated by experts. Some scholars thought that the first inhabitants arrived as early as A.D. 100. However, archaeological and genetic evidence argues for a later date. The archaeological evidence includes radiocarbon dates for charcoal from burned forests that had no history of fires before the arrival of the Polynesians. The fires were probably set, in part, to clear woodlands for farming. The charcoal studies yielded a date of about A.D. 1280 for the first large fires.

In 2008, a group of New Zealand, Australian, and British scientists arrived at the same date using different evidence. The researchers tested rat bones and rat-gnawed seeds from the islands with radiocarbon dating. The bones were from a species of rat called the Pacific rat. This animal is not native to New Zealand, and so it could only have been brought to the islands by settlers. The seeds contained distinctive tooth marks showing that they had been chewed on by this species of rat. None of the bones or seeds yielded dates earlier than 1280.

The newcomers to the island encountered an environment much different from the one they had left. In contrast to the tropical climate of other parts of Polynesia, New Zealand has seasons, with temperatures in the winter months dropping as low as 35° F (2° C). But the settlers adjusted to their new home and developed a unique culture. They became a people called the Maori *(MAH oh ree or MOW ree)*.

The Maori knew about farming. In New Zealand, they grew sweet potatoes and several other crops. But they also reverted to being hunter-gatherers. Although New Zealand had no land mammals, except bats, it had an abundance of birds. So the Maori hunted mostly birds, including a giant flightless bird called the moa. These enormous birds, which resembled ostriches, could be up to 12 feet (3.7 meters) tall and weigh more than 500 pounds (225 kilograms).

The Maori warrior and chief Tamati Waka Nene (1780?-1871) holds a type of war club called a tewhatewha in an 1890 portrait. Such wooden clubs had feathers and were shaped like an axe. Like other Maori men, Tamati Waka Nene had tattoos covering most of his face.

Maori villagers go about their daily life in their *pa* (fortified village), in an artist's illustration. Because war was a common part of Maori life, most Maori lived in villages protected by fences, watchtowers, ditches, and earthen barriers.

New Zealand, which lies southeast of Australia, consists of two main islands as well as several smaller islands.

In hunting these and other birds, the Maori made things as easy for themselves as they could. Another reason the Maori set forest fires, archaeologists think, was to flush birds out of their hiding places. Within a few hundred years, the Maori had hunted the moas and many other birds to extinction.

The Maori lived in scattered villages. They were skilled woodcarvers, and they decorated their canoes and communal houses with intricate designs. They had a religion that was based on various *taboos* (prohibitions). A prominent feature of Maori culture was tattooing. Men often tattooed much of their face. For women, facial tattoos were restricted to the chin.

The different Maori tribes were often at war with one another. Historians think wars became common after so many types of New Zealand's birds became extinct. Farming became more common, as did intense competition for farmland. Disputes over land and other resources led to feuds—and then to warfare.

MIGRATIONS IN HISTORIC TIMES

In about 500 B.C., the Bantu Migration, one of the most important migrations in African history began. This mass movement transformed sub-Saharan Africa.

Two developments led to this migration—agriculture and *metallurgy* (metal working). About 500 B.C., people in sub-Saharan Africa (Africa south of the Sahara) learned to farm as well as to use iron to make tools and weapons. Although some historians think agriculture may have developed independently in Africa, others think that both agriculture and iron-making spread into lower Africa from Egypt and elsewhere. These advances appeared first in West Africa, in parts of what are now Nigeria and Cameroon.

The inhabitants of West Africa were known as the Bantu, who spoke a number of related languages. Armed with their new technologies of agriculture and metallurgy, the Bantu began to spread to other parts of Africa.

Scholars think population pressures probably spurred the migration. As agriculture replaced hunting and gathering in West Africa, the land could support more people. But more people meant crowding, so some Bantu headed off, apparently in search of less-congested lands. Although the Bantu's iron weapons would have given them an advantage over the Stone Age people they encountered, most historians think the Bantu migrations were mostly peaceful.

The great migration occurred in waves over a period of some 1,500 years, with small groups continually splitting off and moving to new regions. Early on, the Bantu split into two major language families—the Eastern Bantu and the Western Bantu. The Eastern Bantu migrated into areas that are now Mozambique, Zimbabwe, and South Africa. The Western Bantu settled in the modern countries of Angola, Namibia, and Botswana.

As the Bantu populated the lower part of the continent, they merged with the populations they encountered. They established trade networks, states, kingdoms, and empires. The migrations were mostly completed by A.D. 1000. By then, Bantu-speaking populations were spread across most of sub-Saharan Africa.

One of the greatest Bantu empires was that of the Zulus, which arose in the early 1800's in what is now South Africa. The Zulu strongly resisted the Dutch and British colonization of South Africa. In 1879, the Zulus dealt a crushing defeat to the British Army at the Battle of Islandwana. The British then sent a larger force to the area, defeated the Zulus in several battles, and broke up the empire.

Over the centuries, the Bantu language has evolved into many daughter languages. Today, about 200 million Africans speak Bantu languages. The most common Bantu language is Swahili, which is spoken by about 50 million people in the eastern part of the continent. Another well-known group of Bantu speakers is the Kikuyu, the largest group in what is now Kenya.

The Bantu migration, which began in western Africa sometime after 500 B.C., was one of the greatest mass movements in history. Bantu migrants carried the knowledge of ironworking to much of Africa. One Bantu-speaking group, the Karanga, founded the city of Great Zimbabwe in the modern-day country of Zimbabwe in southeastern Africa.

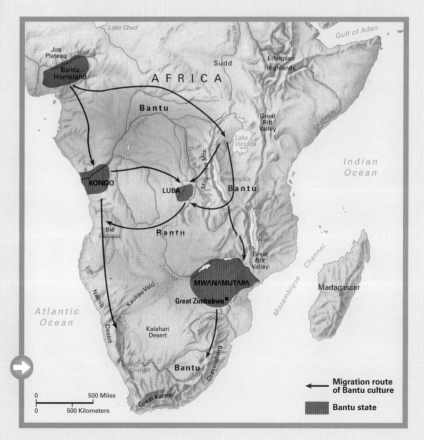

GREAT ZIMBABWE

Sometime after A.D. 1000, the Karanga, a branch of the Bantu-speaking Shona people, established the Mwanamutapa Empire in southern Africa. This empire included most of what is now the country of Zimbabwe. The kingdom is known today for a city and palace complex called Great Zimbabwe, constructed near Masvingo in modern-day southeastern Zimbabwe. The word *zimbabwe* means *House of Stone* in the Shona language.

The Rozwi, a southern Karanga group, rebelled against the Mwanamutapa Empire in the late 1400's and founded the Changamire Empire. The Rozwi eventually took over the city of Zimbabwe, building the city's largest structures. The Changamire Empire was prosperous and peaceful until Nguni people from the south defeated much of the empire in the 1830's. The city was abandoned after the fall of the Changamire Empire.

The chief of the city of Great Zimbabwe looks on as traders display imported cotton cloth, pottery, and other luxury goods offered in exchange for gold, ivory, and copper, in an artist's illustration.

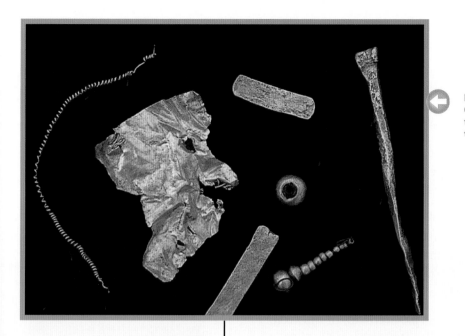

Beads and other gold artifacts found at Great Zimbabwe are evidence of the thriving trade that took place long ago in this ancient African city.

Today, Great Zimbabwe is a complex of stone ruins that includes the Hill Complex, a group of stone buildings at the top of a granite hill. Scholars think that the rulers and other individuals of high social standing lived there. Six large soapstone sculptures of birds found at this location may have served as emblems of royal authority. Scholars estimate that as many as 18,000 people lived in the area around the Hill Complex.

Another feature, the Great Enclosure, includes two thick walls up to 36 feet (11 meters) high. They form a narrow passage leading to a cone-shaped tower 30 feet (9 meters) tall. A wall about 800 feet (240 meters) long surrounds the tower. These structures consist of granite slabs fitted together without mortar. Scholars believe that the Great Enclosure, which was the largest ancient human-made structure south of the Sahara, served as an important ritual center.

Great Zimbabwe's wealth and power came from vast herds of cattle that were kept for meat and milk. Its rulers distributed cattle to the vassal territories in exchange for goods from the interior of Africa. At Great Zimbabwe, rulers taxed gold, ivory, copper, wild animals, and agricultural products, then passed them on to Sofala and other port cities on the southeast coast of Africa. There, the goods were traded for porcelain, cloth, beads, and other articles from China, India, and the Middle East.

Great Zimbabwe declined during the 1400's as Europeans took control of trade along the eastern coast of Africa. Environmental problems—including deforestation and the exhaustion of soils—may have also contributed to the empire's decline. The Kingdom of Zimbabwe lasted until about 1450. By the late 1700's, Great Zimbabwe was abandoned. In 1986, the United Nations Educational, Scientific and Cultural Organization (UNESCO) declared the ruins at Great Zimbabwe a World Heritage Site, an area of unique natural or cultural importance.

SPANISH AND PORTUGUESE MIGRATIONS TO THE AMERICAS

In 1492, Christopher Columbus sailed west from Spain looking for a shorter route to Asia. Instead, he found what would be called the Americas—lands unknown to Europeans. Overnight, the world changed dramatically.

In the years to follow, adventurers from Spain explored much of the Americas and overthrew great Native American civilizations. Later, settlers from Spain moved to the "New World" and built cities and towns. Roman Catholic monks and priests established churches to serve the migrants and missions to convert the Indians to Christianity. During the same period, Portugal made its own claim in the Americas.

The Spanish conquest of the Americas began with the overthrow of the Aztec empire in Mexico. In 1521, Hernán Cortez and a force of soldiers and Indians defeated the Aztecs and took control of the Aztec capital of Tenochtitlán. The Spaniards became the masters of Mexico and rebuilt Tenochtitlán, which was renamed Mexico City, the capital of a Spanish colony called New Spain. The downfall of the Aztecs was followed in 1533 by the conquest of the Inca empire in South America by Spanish explorer Francisco Pizarro.

By 1550, Spain ruled Mexico, Central America, much of western South America, nearly all of the West Indies, and a large portion of what is now the southwestern United States.

Spain grew rich from its New World territories. Gold and silver mines, worked by Indian slaves, were a major source of wealth. With Spain firmly in control of its new possessions, immigration from Spain increased. During the 1500's, about 200,000 Spaniards moved to the New World. In 1565, Spaniards founded the city of St. Augustine in Florida. It was the first permanent settlement established by Europeans in what would become the United States. By 1574, Spaniards had established nearly 200 cities and towns in the Americas.

Although Spain controlled most of the New World territories, Portugal established its own empire in eastern South America. A 1494 agreement between Spain and Portugal, the Treaty of Tordesillas, divided the Americas between the two nations. The treaty gave Portugal the right to possess territory in what is now part of Brazil. A Portuguese naval commander, Pedro Álvares Cabral, landed in Brazil in 1500 and claimed the land for Portugal. In the 1530's, colonists from Portugal began to settle in Brazil. Some of newcomers established sugar plantations, using Indian slave labor. After many Indians died or were killed in uprisings, the Portuguese began importing slaves from Africa.

Montezuma II, the last emperor of the Aztec (atop wall), is surrounded by Spanish conquerors in a painting portraying the ruler's death at the hands of his own people in 1520. According to Aztec tradition, Montezuma was killed by the Spaniards as they retreated from Tenochtitlán.

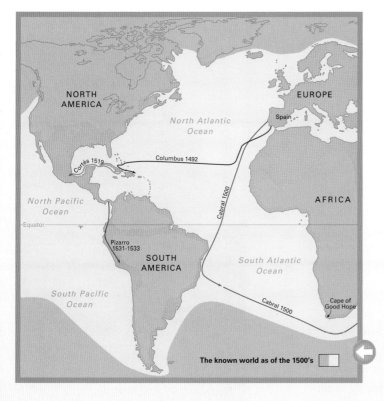

The known world as of the 1500's

Another treaty with Spain in 1750 gave Portugal all of Brazil. In 1763, the capital of Brazil was moved from Salvador to another coastal city, Rio de Janeiro. By about 1800, the population of Brazil was about 3.5 million. More than half of those people were slaves.

Spain and Portugal both began to lose control of their New World possessions in the 1800's. From 1808 to 1814, Spain was involved in a conflict with France called the Peninsular War. During the war, most of Spain's American colonies revolted against Spanish rule and declared their independence. By 1825, Spain had lost all of its major colonies except Cuba and Puerto Rico. Brazil declared its independence from Portugal in 1822.

Christopher Columbus, an Italian who was leading a Spanish expedition to Asia, enountered islands in the Caribbean Sea in 1492. Pedro Álvares Cabral, a Portuguese naval commander, landed in what is now Brazil in 1500 and claimed the land for Portugal.

The Castillo de San Marcos (Fort of Saint Mark), a large, gray stone fortress built by Spaniards in the 1600's and now a national monument, dominates the city of St. Augustine, Florida. Founded in 1565 by Spanish explorer Pedro Menendez de Aviles, St. Augustine is the oldest permanent settlement established in the United States by Europeans.

ENGLISH, FRENCH, AFRICAN, AND ASIAN SETTLEMENT IN THE AMERICAS

Beginning in about 1500, the English and French explorers and fur traders began to venture into the eastern part of North America. In the 1600's, both nations began to establish permanent settlements.

The French concentrated on the region that is now Canada, while the English established settlements farther south on the Eastern Seaboard. The first permanent English settlement was Jamestown, Virginia, founded in 1607. Over the next 150 years, a steady stream of immigrants arrived in English North America, most of them from Great Britain. Eventually, Great Britain established 13 colonies along the East Coast.

The colonists established many industries in their new homeland. Farming was the main economic activity, and it required a lot of labor. In many cases, that labor was supplied by slaves from Africa. Every colony had slaves, but the greatest number were in the Southern Colonies, where settlers established large plantations. By 1750, there were about 200,000 slaves in the colonies, and the number continued to grow rapidly.

During this period, there was constant friction between France and Great Britain. For years, the two countries fought for control of the territory between the Atlantic Ocean and the Mississippi River. Great Britain emerged the victor in 1763, defeating France in the French and Indian War. The Treaty of Paris gave Great Britain almost all of the French territory in Canada. Britain also gained control of all French possessions east of the Mississippi River except the city of New Orleans, which Louis XV of France in 1762 had given to his cousin, Charles III of Spain.

By the 1770's, colonial America had an estimated 2.5 million people, including about 500,000 black slaves. Many Americans by this time wanted to rid themselves of British rule. In the American Revolution (1775-1783), the newly declared United States of America—with the help of France— won its independence. At the end of the Civil War (1861-1865), the United States finally eliminated slavery and granted citizenship to 4 million *freedmen* (freed slaves).

In the late 1800's and early 1900's, the United States became known as "a nation of immigrants." From 1870 to 1916, more than 25 million immigrants poured into the country, the vast majority from Europe. Others

French fur traders paddle along a river in North America, in an illustration from the 1800's. By the early 1820's, fur traders had explored much of what is now Canada from the Atlantic Ocean to the Pacific Ocean and the Arctic coast.

arrived from Asia. Americans were generally open to immigration from Europe, but many were opposed to immigration from Asia, especially from China. Many Chinese came to the United States in the last half of the 1800's. They worked in gold mines, did farm and factory work, and helped to construct railroads in the western United States. Many later started their own businesses.

The growing Chinese presence in the United States started to cause anger and resentment. In 1882, the U.S. government passed the first of several Chinese exclusion acts that drastically reduced Chinese immigration to the United States. The exclusion acts were repealed in 1943, mainly as a gesture of respect to China, an ally of the United States in World War II (1939-1945).

The first African slaves brought to the American Colonies arrive in Jamestown, Virginia, the first permanent English settlement in North America, in August 1619, in an illustration from the 1800's. The slave population of what is now the United States grew from about 200,000 in 1750 to about 4 million in 1860. From the 1500's to the mid-1800's, Europeans shipped more than 12 million slaves from Africa to the Western Hemisphere.

Chinese workers labor on a *trestle* (a structure used as a bridge for railroad tracks) near Sacramento, California, in 1867. By 1860, the United States had a Chinese population of about 35,000, most of whom lived in California. Many newcomers who could not find work in the California gold mines found jobs building railroads.

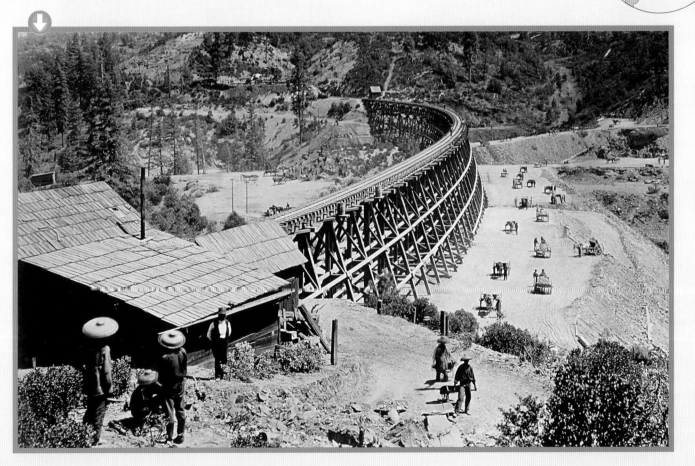

EMIGRATION TO AUSTRALIA AND NEW ZEALAND

British convicts arrive at Van Diemen's Land (now the Australian state of Tasmania) in 1804, in an engraving from 1853. From 1787 to 1868, more than 162,000 convicts were sent from Great Britain (now the United Kingdom) to prison colonies in Australia. Free immigrants began arriving in Australia in 1788.

The islands now known as Australia and New Zealand were unknown to the outside world until the 1600's. Australia was discovered in 1606 by Spanish explorer Luis Vaez de Torres. In 1642, a Dutch ship captain, Abel Janszoon Tasman, became the first European to see New Zealand. Dutch navigators explored much of Australia's coast from 1616 to 1636, but they concluded that the continent was barren and worthless. They missed the fertile east coast. In 1770, Captain James Cook of Great Britain's Royal Navy became the first outsider to visit that part of Australia. He claimed the region for Great Britain and named it New South Wales.

Britain had a policy of shipping convicts to its American colonies to relieve overcrowding in British jails. When the United States won its independence, that outlet was closed. In 1786, Britain decided to use its new Australian possession as a prison colony. The next year, a retired naval officer, Captain Arthur Phillip, transported about 570 male and 160 female convicts to New South Wales. After landing in New South Wales, Captain Phillip directed the establishment of a settlement next to a large harbor about 7 miles (11 kilometers) north of Botany Bay. That settlement became the city of Sydney.

In the 1790's, the British government began making land grants in New South Wales to convicts who had finished their sentences. They also offered

land to British military officers. Immigration from Great Britain, later called the United Kingdom (U.K.), swelled. In 1868, the British stopped sending convicts to Australia. By that time, more than 160,000 convicts had been transported to the continent.

In 1829, a British naval captain, Charles Fremantle, landed on Australia's southwest coast. He claimed the entire western part of the continent for the United Kingdom, solidifying British possession of the entire continent. Britain began dividing the continent into new colonies, which were later renamed territories.

Immigration to Australia soared after gold was discovered in New South Wales and elsewhere, from about 400,000 in 1851 to more than 1.1 million by 1860. During the gold rush, about 50,000 Chinese immigrated to Australia. Growing resistance to any further arrivals from Asia prompted the government to pass restrictive immigration laws. These laws were enacted under what was called the White Australia policy, which was not abolished until 1975.

The settlement of New Zealand began in the early 1800's. Some of the first settlers were seal hunters and traders. Others were Christian missionaries who arrived to convert the native Maori.

At this time, New Zealand was a lawless place with no government. Many of the Maori were abused. The Maori and some British settlers asked the British to step in. In 1840, the British government signed a treaty with the Maori giving the British control of New Zealand in return for government protection. New Zealand thereby became a British colony, and immigrants swarmed in. The settlers clashed with the Maori and war broke out in 1845. Peace did not return until 1872.

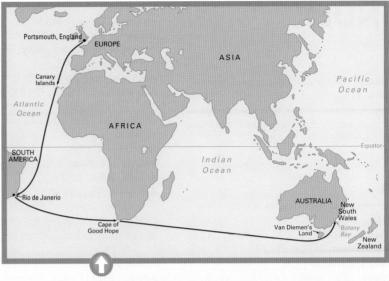

The First Fleet, which sailed from England on May 13, 1787, carried convicts and their guards to establish a prison colony in Australia. The arrival of the 11 ships at Botany Bay on Jan. 18, 1788, resulted in the first permanent European settlement in Australia.

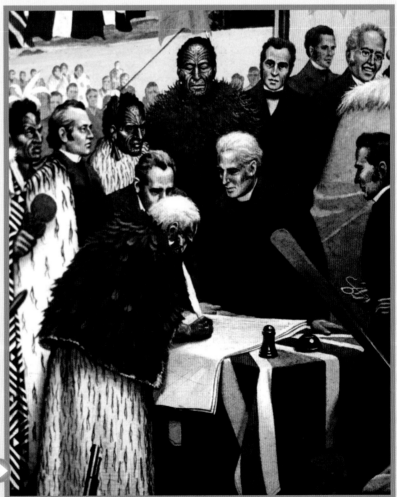

Maori chefs in feather cloaks sign the Treaty of Waitangi as officials representing the United Kingdom look on in February 1840. Under the English version of the treaty, the Maori turned over their authority to govern their territories in what is now New Zealand to the British in return for various rights, including British protection, the rights of British subjects, and recognition of their possession of their lands. Although some important Maori leaders refused to sign the treaty, British officials proclaimed New Zealand a British colony on May 21, 1840.

INDEX

Note: A volume number appears before a slash. A page number appears after a slash. Page numbers in *italic* type are references to illustrations. Page numbers in **boldface** type are references to maps.

A

A Famosa fort (Malaysia), 4/102
A-Ma, 2/86
Aachen (Germany), 3/38, *3/39*
Aalto, Alvar, 2/229, *2/229*
Aaron, Hank, 6/145
Abacá, 5/61, 5/68
Abacha, Sani, 4/239
Abay. *See* Blue Nile River
Abbas, Mahmoud, 3/213
Abbasid dynasty, 3/197
Abbey of Melk (Austria), *1/103*
Abbey Theatre (Dublin), 3/206
Abbottabad (Pakistan), 5/17
Abboud, Ibrahim, 5/229
Abd al-Hafidh, 4/159
Abd al-Krim, 4/159
Abdullah (king of Jordan), 3/282, 3/283
Abdullah (king of Saudi Arabia), 5/145, 5/147
Abdullah II (king of Jordan), 3/283
Abdullah bin Ahmad Badawi, 4/95
Abhisit Vejjajiva, 6/28, 6/29
Abidjan (Côte d'Ivoire), 2/126, *2/127,* 2/129
Abkhazia (Georgia), 3/15, 5/111
Abominable Snowman, 4/190
Aborigines, Australian, 1/71, 1/84, 1/85, 7/89
  early, 1/69, 1/86, **7/26,** 7/26–29, *7/26–29*
  language, 1/91
  today, 1/77
  traditional culture, 1/76–77, *1/76–77, 1/90*
Aborigines, Taiwanese, 6/14, *6/14*
Abortion, 3/223
Abruzzese dialect, 3/233
Absalon of Roskilde, 2/160, 2/161
Abseron Peninsula, 1/110
Abu Bakr, 5/151
Abu Dhabi (United Arab Emirates), 6/84, 6/85, 6/87
Abu Sayyaf Group, 5/65
Abu Simbel, Temple of, *2/193*
Abubakar, Abdulsalami, 4/239
Abuja (Nigeria), 4/242, *4/243*
Abyssinia, 5/228, 6/198
Acacias, 1/74–75
Academy (ancient Greece), 3/65
Acadia National Park (United States), 6/133

Acajutla (El Salvador), 2/197
Acalmán (Mexico), 4/142
Acapulco (Mexico), 4/142, *4/142*
Acción Democrática party, 6/173
Accra (Ghana), 3/43
Aceh province (Indonesia), 3/163
Achaemenid Empire, 3/182, 3/184, 5/14, 6/16
  See also Persian Empire
Acid rain, 2/33
Aconcagua, 1/55
Aconcagua River, 2/52
Acosta, Julio, 2/123
Acropolis (Greece), 3/51, 3/58, *3/58,* 3/64, *3/64*
Acropolis (Zimbabwe), *6/210–211*
Act of Union (United Kingdom), 3/204, 6/112
Adam (first man), 5/227, 7/12–13
Adam's Peak. *See* Sri Panda (Sri Lanka)
Adana (Turkey), 6/65, 6/67
Adapazari (Turkey), 6/66
Addis Ababa (Ethiopia), 2/208
Adelaide (Australia), 1/73
Adelaide, University of, 7/28
Aden (Yemen), 6/194, 6/196, 6/199, 6/201
Aden, Gulf of, 6/194
Aden Protectorate, 6/199
Adja people, 1/147
Administrando Imperio, De (Constantine VII), 5/164
Admiralty Building (St. Petersburg), *5/123*
Adobe houses, 5/49
Adriatic Sea
  Albania on, 1/22, 1/23, *1/23*
  Croatia on, 2/130
  Italy on, 3/224, 3/242
  Montenegro on, 4/156
  Slovenia on, 5/182
Aegean Islands, 3/66–67, *3/66–67,* **3/67**
Aegean Sea
  Greece on, 3/58, 3/60, 3/61, 3/66
  Turkey on, 6/60, 6/67, 6/69, 6/70, 6/72, *6/72–73*
Aeolian zone, 4/191
Aeta people, 5/62
Afar language, 2/163
Afar people, 2/162, 2/163, 2/209, *2/209*
Afghanistan, 1/10–19, **1/13**
  climate, 1/16
  economy, 1/17, **1/17**
  history, 1/10–13, 6/136, 6/137
  Iran, power in, 3/183
  Khyber Pass and, **1/18,** 1/18–19, *1/18–19*
  land, 1/16, *1/16–17*
  people, 1/14–15, *1/14–15*
Afghanistan War (2001–), 1/11, 1/13, 6/137
Afonso I (king of Portugal), 5/90
Afonso Henriques, 5/94
Afontova Culture, 7/62, 7/63

Africa, 1/4
  arts, 3/250, 4/112–113, *4/112–113,* 4/244–245, *4/244–245*
  climatic periods, 7/8
  human origins, 7/5, 7/6
  See also Human migrations; *and specific countries*
African Americans, 6/123, 6/129, 6/138, 6/144–145, *6/144–145*
  history, 6/136
  music, 6/139, 6/159
  Southern states, 6/124
African National Congress, 5/190, 5/193
African Party for the Independence of Cape Verde, 2/40
African Party for the Independence of Guinea and Cape Verde, 2/40, 3/90, 3/91
African religions, 1/166, 2/47
  See also Animism; *and specific African countries*
African Union, 2/108, 5/187
Afrikaans language, 5/194
Afrikaners, 5/189, 5/193, 5/194–195
Afro-Colombians, 2/104
Afro-European people, 3/246
  See also Mulattoes
Afwerki, Issaias, 2/202
Agadez (Niger), 4/232
Agadir (Morocco), 4/168
Agalega (Mauritius), 4/124
Agbades, 1/147
Age of Kingship, 7/57
Age of Liberty (Sweden), 5/243
Age of Reason, 2/252
Agel, Mount, 4/148
Aghas, 3/199
Agora (Athens), 3/58, 3/59, 3/64
Agra (India), 3/125, 3/149
Agrias scardanapalus, 2/125
Agriculture, Early
  Africa, 7/98
  China, 7/56, 7/57
  Japan, 7/61
  Middle East, 7/33–39, 7/39, 7/41
  North America, 7/76–77, **7/77,** *7/77,* 7/78, 7/80–81
  Pacific Islands, 7/92, 7/96
  South America, 7/82, 7/84–85
Agriculture, Modern. *See specific crops and places*
Aguinaldo, Emilio, 5/70, *5/70*
Aguiyi-Ironsi, Johnson, 4/238
Ahaggar Mountains, 1/28, *1/28*
Ahidjo, Ahmadou, 2/19
Ahmad, Muhammad, 5/229, *5/233*
Ahmadabad (India), 3/136, 3/144
Ahmed ibn Tulun Mosque, 2/190, *2/192*
AIDS, 1/159, 2/113, 4/57
Aiga, 5/138
Ailey, Alvin, 6/158
Ainu people, 3/272, 3/274, 7/71
Aïr Mountains, 4/234
Air pollution. *See* Pollution
Aircraft plants, *1/176–177,* 6/99

Ajanta (India), 3/128, *3/140*, 3/149
Ajman (United Arab Emirates), 6/84
Ajvar, 5/165
Ak Orda (Kazakhstan), *4/9*
Akan people, 2/128, *2/129*, 3/42
Akaro language, 5/44
Akayev, Askar, 4/38
Akbar, 3/128, *3/128*
Akershus Castle (Norway), 4/254
Akido, 3/276
Akihabara district (Tokyo), *3/270*
Akihito, 3/254
Akkad, 3/197, 7/40
Akmola (Kazakhstan). *See* Astana
    (Kazakhstan)
Akosombo Dam, 3/40
Aksum (Ethiopia), 2/206
Aktamar Island, *6/60–61*
Alabama (United States), 6/121, 6/124,
    6/145
Alak people, 4/46
Alamannic dialect, 1/100
Aland Islands, 2/223, *2/227*
Alanya (Turkey), *6/65*
Alapanis, 1/122
Al Aqabah (Jordan), 3/285
Alaska (United States), 6/118, 6/130–131,
    6/130–131, **6/131**
    prehistoric settlement, 7/74
Alaska Range, 6/130
Al Askari shrine, *3/194–195*
Al-Assad. *See under* Assad
Alaungpaya, 4/178
Alawi people, 4/54, 4/158
Alay Mountains, 4/38
Al-Azhar Mosque (Egypt), 2/188, *2/189*
Al-Azhar University, 2/188–189, *2/188–
    189*
Al-Aziziyah (Libya), 4/72
Alba, 6/110
Alba Longa (Italy), 3/230
Al-Bakr, Ahmad Hassan, 3/188
Albania, 1/20–23, **1/21,** *1/21–23*
    economy, 1/23
    history, 1/20–21, 4/83
    land and climate, 1/22, *1/22–23*
    people, 1/22, *1/22*
Albanian language, 4/32, 4/33
Albanian people, 3/232, 4/32, 4/33,
    4/83, 5/157, 5/182
    See also Albania
Albanian Riviera, 1/22
Al Basra (Iraq). *See* Basra (Iraq)
Al Batinah (Oman), 5/8
Albert I (prince of Monaco), 4/148
Albert II (prince of Monaco), 4/148
Albert, Lake, 6/80
Albert Cuypmarkt, 4/207
Albert Library (Brussels), 1/141
Albert Market (Gambia), *3/13*
Albert Nile River, 6/80
Alberta (Canada), 2/26, 2/33, 2/37, 2/39
Al bu Said family, 5/9
Albuquerque, Alfonso de, 3/154
Alcazars, 5/199

Alderney (Channel Islands), 6/100
Alderney cattle, 6/100
Aldrin, Buzz, *6/136*
Aleijadinho, 1/172
Aleppo (Syria), *5/269, 5/270–271,* 5/271,
    5/272, *5/272*
Aleutian Islands, 6/131
Aleuts, 6/131
Alexander I (ruler of Yugoslavia), 5/158,
    5/159
Alexander II (Russian czar), *2/228,* 5/113
Alexander III (Russian czar), 5/113, 5/128
Alexander VI (pope), 1/164
Alexander Nevsky Cathedral (Bulgaria),
    1/181
Alexander the Great, 3/62, *3/62,* 4/82
    campaigns and conquests
        Afghanistan, 1/10, 1/15, 1/18
        Cyprus, 2/143
        Egypt, 2/192
        Indus River Valley, 5/14
        Iraq, 3/187
        Lebanon, 4/50
        Mesopotamia, 3/197
        Persia, 3/182
        Tajikistan, 6/16
        Turkey, 6/68, 6/73
        Uzbekistan, 6/166
Alfama quarter (Lisbon), 5/91
Alfonsín, Raúl, 1/52, 1/65
Alfonso VI (king of Castile), 5/94
Alfonso XII (king of Spain), *5/210*
Alfred the Great, 6/110
Al Fujayrah (United Arab Emirates), 6/84
Al-Gassaniyeh street (Aleppo), *5/269*
Algeria, 1/24–31, **1/27,** *1/27–31*
    climate, 1/28
    culture, 1/25
    economy, 1/30–31, *1/30–31*
    government, 1/26
    history, 1/24–25
    land, 1/28, *1/28*
    people, 1/28–29, *1/28–29*
    Western Sahara, power in, 6/193
Algerian people, in France, 2/239
Alghero (Sardinia), 3/244
Algiers (Algeria), 1/24, 1/26, 1/30
Algonquin Indians, 2/28, 2/29, 6/140,
    7/77
Al-Hafidh, Abd. *See* Abd al-Hafidh
Al-Hajar range, 5/8
Al-Hamadieh suq (Damascus), 5/275,
    5/275
Alhambra (Granada), **5/213**
Al Hillah (Iraq), 3/187
Ali (son-in-law of Muhammad), *1/15,* 5/151
Alia, Ramiz, 1/21
**Alianza Popular Revolucionaria
    Americana,** 5/46
Aliyev, Heydar, 1/111
Aliyev, Ilham, 1/111
Al Jabal al Akhdar, 4/68
Al-Khalifa, Sheik Hamad bin Isa, 1/116
Al-Khalifah Arabs, 1/116
Alkmaar (Netherlands), *4/204*

Al-Krim, Abd. *See* Krim, Abd al-
All Blacks rugby team, *4/220*
All Souls' Day, *4/142–143*
Allah, 5/144, 5/150
Allegheny Mountains, 6/120
Allende Gossens, Salvador, 2/50, 2/51
Allgäu region (Bavaria), 1/100
Allies. *See* World War I; World War II
Alligators, *2/125, 6/149*
Alluvial plains, 1/118, 3/125, 3/130,
    5/18
Almagro, Diego de, 2/49
Almaty (Kazakhstan), 4/8
Almaty Declaration, 4/9
Alphabet
    Chinese, 4/31
    Cyrillic, 1/181, 4/155, 5/164
    han'gul, 4/31
    Khmer, 2/8
    Mongolian, 4/154–155
    Roman, 3/204
    Thai, 6/36
    See also Writing
Alpine Slope (Italy), 3/224
Alps
    Austria, 1/94, 1/96, 1/100, 1/102,
        1/103, 3/112
    Bavaria, 3/24, 3/25
    Croatia, 2/130
    France, 2/236–237, *2/237,* 2/249
    Italy, 3/224, 3/240, 3/241
    Switzerland, 5/255, 5/256–257,
        5/263, *5/263*
    Transylvania, 5/104, *5/104*
    See also Dinaric Alps; Julian Alps
Al Qadhafi, Mu'ammar. *See* Qadhafi,
    Mu'ammar Muhammad al-
Al-Qadir, Abd. *See* Qadir, Abd al-
Al Qahirah (Egypt), 2/188
**Al-Qa`ida.** *See* Qai`da, Al-
Al Qayrawan Mosque (Tunisia), 6/54
Al-Sabah, Sheik al-Ahmad al-Jabir, 4/35
Al-Sabah family, 4/34
Alsace (France), 2/248
Alsace-Lorraine (France), 2/238
Al Shorja market (Baghdad), *3/193*
Altai Mountains, *2/71,* 4/154, *4/154,*
    7/17
Al-Thani, Emir Ahmad bin Ali, 5/99
Al-Thani, Khalifa bin Hamad, 5/99
Al-Thani family, 5/99
Althing, 3/118
Altiplano (Bolivia), 1/152
Altitude sickness, 4/197, 5/51
Alton (Illinois), *6/122*
Aluminum, 2/92, 4/252
Alvarado, Pedro de, 2/194, 3/80
Alvarez, Gregorio, 6/163
Alzette River, 4/81
Amager Island (Denmark), 2/160
Amalienborg Palace (Denmark), *2/157*
Amazon Region
    Brazil, 1/165, 1/166, 1/168, 1/172
    Colombia, 2/104
    early civilization, *7/84,* 7/84–85, **7/85**

Amazon River, 1/161, 1/168, 2/102, 2/106
Amber (India), 3/149
Ambon (Moluccas), 3/165
American Colonization Society, 4/64
American Indians. See Native Americans
American Museum of Natural History (New York City), 7/56–57
American Popular Revolutionary Alliance, 5/46
American Revolution, 1/86, 5/132, 6/112, 6/134, 7/104
American Samoa, 1/32–33, 1/32–33, 1/33, 5/10, 5/138
Americo Liberians, 4/64–65
Amerindians. See Native Americans
Amhara people, 2/209
Amharic language, 2/209.
Ami people, 6/14
Amin Dada, Idi, 6/78
Amman (Jordan), 3/286
Ammonia, 6/47
Ammonite people, 3/282
Amnesty International, 5/102
Amoreiras complex (Lisbon), 5/87
Ampanalana Canal, 4/85
Amstel River, 4/206, 4/208
Amsterdam (Netherlands), 4/200, 4/203, 4/203, 4/204, 4/206, 4/206–207, 4/208
Amsterdam Stock Exchange, 4/206
Amu Darya River, 6/16, 6/74, 6/75, 6/166
Amundsen, Roald, 1/41, 4/254
An Nafud Desert, 5/148
Anacondas, 6/177
Anamur (Turkey), 6/73
Anamur Kalesi, 6/73
Anatolia, 6/57
    history, 6/68, 6/69
    land, 6/60, 6/61, 6/61, 6/64, 6/64–65
    prehistoric settlement, 7/36, 7/36–37, 7/36–37
Anawrahta, 4/178
Ancestor worship, 6/207
Anchorage (Alaska), 6/131
Anchovies, 5/69, 5/69
Ancon, S.S., 5/32
Andalusia (Spain), 5/202, 5/202–203, 5/206
Andaman and Nicobar Islands, 3/127, 3/156–157, 3/156–157, 3/157
    human migration, 7/16, 7/16, 7/16–17, 7/59
Andersen, Hans Christian, 2/160
Andersen Air Force Base, 3/79
Andes Mountains
    Antarctica, 1/43
    Argentina, 1/51, 1/55, 1/56–57, 1/64
    Bolivia, 1/152, 1/152–153
    Chile, 2/48, 2/52, 2/53
    Colombia, 2/98, 2/102, 2/106
    early civilizations, 7/82, 7/82–83, 7/82–83
    Ecuador, 2/172, 2/176

Peru, 5/44, 5/48, 5/49
    Venezuela, 6/176, 6/176
Andine region (Argentina), 1/55
Andorra, 1/34–35, 1/34–35, 1/35
Andorra la Vella (Andorra), 1/34
Andropov, Yuri V., 5/115
Anegada Island, 6/188, 6/189
Angel, Jimmy, 6/177
Angel Falls (Venezuela), 6/177
Angelico, Fra, 6/171
Angkor (Cambodia), 2/8, 2/14–15, 2/14–15, 6/36
Angkor Thom (Cambodia), 2/14
Angkor Wat temple (Cambodia), 2/14, 2/14
Angles, 6/91, 6/104, 6/110
Anglo-Boer Wars, 5/192, 5/193
Angola, 1/36–39, 1/36–39, 1/37, 2/111
    economy, 1/38
    history, 1/36–37, 7/99
    land, 1/38, 1/38–39
    people, 1/39, 1/39
    Portugal, relations with, 5/87
Angoni people. See Ngoni people
Anguilla (St. Kitts and Nevis), 4/58, 5/133, 5/133
Animals
    Alaska, 6/130
    Antarctica, 1/44–45, 1/44/45
    Australia, 1/74, 1/74–75, 1/79, 1/92, 1/92–93
    Cameroon, 2/20
    China, 2/74–75, 2/74–75
    Colosseum in Rome, 3/230
    Congo River, 2/115, 2/115
    coral reef, 4/106–107, 4/106–107
    Costa Rica, 2/125, 2/125
    Ecuador, 2/177–179, 2/179
    Fiji, 2/219
    Himalaya, 4/191
    Kenya, 4/10, 4/10–11
    Madagascar, 4/88–89, 4/88–89
    Mississippi River, 6/149
    Nepal, 4/184, 4/185
    New Zealand, 4/224, 4/224
    Sahara, 4/72, 4/73
    Sri Lanka, 5/222–223, 5/223
    Tanzania, 6/18, 6/24–25, 6/24–25, 6/218–19
    Thailand, 6/31
    Venezuela, 6/176, 6/177
    See also Endangered species; Extinction; and specific places
Animism, 1/146, 2/41, 4/46, 4/173, 6/39, 6/204, 6/207
Anjouan (Comoros), 2/108
Ankara (Turkey), 6/57, 6/59, 6/67
Ankobra River, 3/40
Ankrah, Joseph, 3/44
Anna Amalia, 3/39
Annaba (Algeria), 1/30
Annam, 6/180, 6/181
Annamite Range, 4/40, 4/44, 6/186
Annapurna, 4/190, 4/193
Anne (queen of England), 6/112

Annobón (Equatorial Guinea), 2/200, 2/201
Annunciation, The (Leonardo da Vinci), 3/237
Anseba River, 2/203
Antakya (Turkey), 6/72
Antall, Jozsef, 3/109
Antalya (Turkey), 6/72, 6/73
Antananarivo (Madagascar), 4/84, 4/85
Antarctic Ocean, 1/44–45
Antarctic Peninsula, 1/43, 1/44
Antarctic Treaty, 1/41, 1/46
Antarctica, 1/2, 1/4, 1/40–47, 1/40–47, 1/42
    climate, 1/41
    environment, 1/42–43, 1/42–43
    exploration, 1/41
    problems and challenges, 1/41, 1/46–47, 1/46–47
    wildlife, 1/41, 1/42–43, 1/44–45
Anteaters, 6/177
    pygmy, 2/125
Anti Atlas Mountains, 4/162
Anti-Fascist People's Freedom League (Myanmar), 4/179
Antigua (Guatemala), 3/85
Antigua and Barbuda, 1/48–49, 1/49, 1/49, 3/75, 4/58, 4/59, 4/60, 4/62
Antigua Island. See Antigua and Barbuda
Anti-Lebanon Mountains, 5/270
Antilles. See Greater Antilles; Lesser Antilles; Netherlands Antilles, The
Antioch (Turkey). See Antakya (Turkey)
Antiochus I, 6/70
Antipasti, 3/234
Antony, Mark, 2/192
Antwerp (Belgium), 1/137, 1/138, 1/142, 1/142–143, 1/143
Aoi Matsuri festival (Japan), 3/278
Aoraki/Mount Cook, 4/215, 4/225
Apache Indians, 6/127, 6/140
Apartheid, 5/189–191, 5/193, 5/194, 5/197
Apennine Mountains, 3/224, 5/140
Aphrodite, 2/140, 2/143
    Sanctuary of (Cyprus), 2/143
Apollo, 3/66, 3/66
    Temple of (Cyprus), 2/143
Apollo Belvedere (statue), 6/171
Apollonia (Albania), 1/21
Aponte, José Antonio, 2/132
Appalachian Mountains, 2/27, 6/120, 6/124–125, 6/148
Appalachian National Scenic Trail (United States), 6/121
Appenzell Innerrhoden (Switzerland), 5/256, 5/259
Apple Bay beach (British Virgin Islands), 6/189
APRA, 5/46
Apulia Plain (Italy), 3/225
Apulian Aqueduct (Italy), 3/225
Apulian dialect, 3/233
Aqaba, Gulf of, 3/285

Aqueducts
Italy, 3/225
Madeiras, 1/113
Spain, 5/212
Turkey, 6/63
See also Irrigation
Aquino, Corazon, *5/70,* 5/71
Aquitainian Lowlands, 2/236, 2/237
Arab Empire. *See* Islamic Empire
Arab-Israeli wars, 4/36, 4/52, 5/145
Israel and, 3/209, 3/212–213,
3/216
Jordan and, 3/283, 3/286
Syria and, 5/268–269
Arab Socialist Union, 4/66
Arab Spring, 4/67
Arabesques, 3/185
Arabia, 2/162, 2/206, 3/168, 5/144,
5/150, 5/222
See also Saudi Arabia
Arabia Felix, 6/198
Arabian American Oil Company, 5/149
Arabian Desert, 2/181
Arabian Peninsula, 2/162, 5/8, 5/144,
6/194
human migration route, **7/14,**
7/14–15, *7/15*
Arabian Sea
India on, 3/130, 3/131, 3/132, 3/154,
3/156
Pakistan on, 5/14, 5/15
Arabic language
Bahrain, 1/116
Chad, 2/46
Comoros, 2/109
Djibouti, 2/163
Egypt, 2/186
Ethiopia, 2/209
Farsi and, 3/178
Iran, 3/182
Iraq, 3/187, 3/192
Israel, 3/215
Jordan, 3/282, 3/287
Lebanon, 4/54
Libya, 4/70
Mauritania, 4/122
Morocco, 4/161, 4/164
Nigeria, 4/243
Pakistan, 5/20
Qatar, 5/98
Sudan, 5/230, 5/234
Syria, 5/267
Tunisia, 6/48, 6/50, 6/54
United Arab Emirates, 6/85
Yemen, 6/196
Arabs
Afghanistan, 1/11
Algeria, 1/24, 1/28–29
Armenia, 1/66
Bahrain, 1/116
Chad, 2/46
Comoros, 2/109
Egypt, 2/186, 2/190, 2/192
genetic studies, 7/15
Iran, 3/175, 3/178, 3/182

Iraq, 3/186, 3/187, 3/192–193,
3/192–193
Israel, 3/209, 3/212, *3/214,* 3/215
Italy, 3/232
Jordan, 3/282, 3/283, 3/287
Kenya, 4/10
Kuwait, 4/34
Lebanon, 4/50, 4/54
Libya, 4/66, 4/67, 4/70
Maldives, 4/105
Mali, 4/111
Malta, 4/114, 4/115
Marsh, 3/191, *3/197*
Mauritania, 4/120
Morocco, 4/158, 4/164, 4/166, 4/167
Mozambique, 4/170, 4/173
Oman, 5/8
Pakistan, 5/15
Saudi Arabia, 5/144
Sri Lanka, 5/226
Sudan, 5/228, 5/230, 5/231, 5/234
Syria, 5/267, 5/272, 5/274
Tajikistan, 6/16
Tanzania, 6/18
Tunisia, 6/48, 6/50, 6/54
United Arab Emirates, 6/84–85
Uzbekistan, 6/166
Western Sahara, 6/192
Yemen, 6/200, 6/201
See also Islam; Moors (people)
Aracay River, 5/43
Arafat, Yasir, 3/213
Aragon, Kingdom of, 3/245, 5/214,
5/217
Arakanese people, 4/174
Arakon Yoma mountains, 4/175
Aral Sea, 6/166
Aramco, 5/149
Ararat, Mount, *6/61*
Araucanians. *See* Mapuche Indians
Arawak Indians
Dominica, 2/164
Guyana, 3/92
Haiti, 3/96
Honduras, 3/103
Jamaica, 3/246, 3/248
Lesser Antilles, 4/59
Puerto Rico, 5/96
St. Lucia, 5/134
St. Vincent and the Grenadines, 5/136
Trinidad and Tobago, 6/46
Venezuela, 6/172
Arbat (Moscow), *5/120–121*
Arbenz Guzmán, Jacobo, 3/80
Arc de Triomphe (Paris), 2/246
Arcadiane (Ephesus), *6/70*
Archipelago of Chile, 2/52, 2/53, 2/54
Architecture
by place
Austria, 1/108–109, *1/108–109*
Azerbaijan, 1/110
Belgium, 1/139
Bosnia-Herzegovina, *1/157*
Cambodia, 2/14–15, *2/14–15*
China, *2/67,* 2/68–69, *2/68–69*

Cyprus, 2/143
Czech Republic, 2/150–151,
2/150–151
Denmark, *2/160–161,* 2/161
Finland, 2/228–229, *2/228–229*
France, 2/246–247, *2/246–247,*
2/252, *2/252–253*
Germany, 3/28, *3/28–31,* 3/29
Great Zimbabwe, *6/210–211,*
6/211, 7/101
Hungary, 3/114–115, *3/114–115*
Iran, 3/184–185, *3/184–185*
Italy, *3/236–237,* 3/237, 3/242
London, 6/102–103, *6/102–103*
Mexico, 4/144
Netherlands, 4/204–205
Peru, 5/45, 5/52, *5/52–53*
Russia, 5/122
Tokyo, 3/270, *3/270–271*
Turkey, 6/62–63, *6/62–63*
United States, 6/158
by type
art deco, 2/228
Art Nouveau, *3/114,* 3/115
baroque, 1/109, *1/164,* 1/172,
5/122
Byzantine, 2/229, 3/242, 6/62–63,
6/62–63
Gothic, *2/240,* 2/247, 2/252, 3/31,
3/242
Muslim, 2/190, *2/191*
neoclassical, 5/122
prehistoric, 7/37
pyramid, 2/181, 5/45
Renaissance, *3/236–237,* 3/237
Roman, *3/230–231,* 6/110
See also specific buildings
Arctic
Canadian, 2/26, 2/28, 2/33
human settlement, 7/64–65, *7/64–65,*
7/65, 7/74–75, *7/74–75,* **7/75**
Arctic Circle
Alaska, 6/131
Faroe Islands, 2/212
Finland, 2/222, 2/226
Greenland, 3/72
Iceland and, 3/118
Norway, 4/246, 4/250
Russia, 5/124, 7/64, 7/65
Sweden, 5/248
Arctic Islands (Canada), *2/26, 2/35*
Arctic Ocean, 4/256, 5/124, 6/131
Arctic Small Tool Tradition, 7/75
Ardennes, 1/135, 2/236, 4/78
ARENA party (El Salvador), 2/195
Arévalo, Juan José, 3/80
Argentina, 1/50–65, *1/50–65,* **1/53**
economy, **1/62,** 1/62–63, *1/62–63*
history, 1/52–53, 1/64–65, *1/64–65*
land and climate, **1/54,** 1/54–55,
1/54–55
people, 1/51, 1/56–59, *1/56–59, 1/62*
population, 1/56
relations with
Falkland Islands, 1/52, 2/210, 2/211

Paraguay, 5/39
United Kingdom, 1/52
Uruguay, 6/161
Argentino, Lake, *1/55*
Ari Thorgilsson, 3/122
Ariadne (Greek mythology), 3/68
Arias Sánchez, Oscar, 2/123
Arif, Abdul Salam, 3/188
Aristide, Jean-Bertrand, 3/97
Aristotle, 3/64, 3/65, *3/65*
Arizona (United States), 6/126, 6/127
Arjuna, *3/140*
Arkansas (United States), 6/124
Arles (France), 2/240
Armada, 6/111
Armed Forces Movement (Portugal), 5/86
Armenia, 1/66, 1/66–67, **1/67,** *1/67,*
1/111, 3/14
Armenian Church of the Holy Cross
(Turkey), *6/60–61*
Armenian language, in Lebanon, 4/54
Armenian people, 1/66–67
Azerbaijan, 1/110
Cyprus, 2/142
Iraq, 3/192
Russia, 5/116
Syria, 1/67, 5/272
Turkmenistan, 6/74
Armenian Plateau, 1/66
Armorite people, 3/282
Armory Show, 6/158
Armstrong, Louis, 6/159
Armstrong, Neill, *6/136*
Arno River, 3/226
Árpád, 3/110
Art
by place
Africa, 4/112–113, *4/112–113*
Algeria, 1/25
Argentina, 1/58, *1/60–61*
Austria, 1/108
Belgium, 1/133, 1/141
China, 2/66–67, *2/66–67*
Colombia, 2/98, 2/106–107,
2/106–107
Crete, 3/68
Czech Republic, 2/152–153
Ecuador, 2/172
France, 2/252–253, *2/252–253*
Germany, 3/39
Indonesia, 3/168–169, *3/168–169*
Indus River Valley, 5/25, *5/25*
Iran, 3/184–185, *3/184–185*
Italy, 3/236, *3/236–237,* 3/242
Korea, 4/31
Malaysia, *4/99*
Mali, 4/112–113, *4/112–113*
Mexico, 4/144
Mount Athos, 3/70
Netherlands, 4/202
Nigeria, 4/244–245, *4/244–245*
Peru, *5/53*
Poland, 5/73
Spain, 5/210, *5/210–211*
Taiwan, 6/15

United States, *6/141, 6/142,*
6/158–159, 6/158–159
Vatican City, 6/171
by type
baroque, *1/164, 2/252, 2/252–253*
goldwork, 2/106–107, *2/106–107*
Islamic, 3/184–185, *3/184–185*
lace, *1/140*
Renaissance, 3/236–237, 3/242
prehistoric
cave, 2/240, 7/19, *7/20–21, 7/28,*
7/49, 7/49
Malta carvings, 7/63, *7/63*
See also Crafts; Pottery; Sculpture
Art Deco, 2/229
Art Nouveau, *3/114,* 3/115
Artemis, Temple of (Turkey), 6/70
Artibonite River, 3/96
Artibonite Valley (Haiti), *3/98,* 3/99
Artigas, José Gervasio, 6/161
Aruba, 4/58, 4/60, 4/200, 4/210, **4/211**
Aruwimi River, 2/114
Aryan people
Afghanistan, 1/10
India, 3/128, 3/138, 7/20
Iran, 3/178, **3/181**
Nepal, 4/188
Arzew (Algeria), 1/30
As Sulamaniyah (Iraq), 3/198
Asados, 1/59
Ascot Races, 6/108, *6/109*
ASG, 5/65
Ash Shariqah (United Arab Emirates),
6/84, 6/87
Ashanti Empire, 3/44
Ashanti people, 3/42, 3/44, *3/44–45*
Ashkenazim, 3/214
Ashmore and Cartier Islands (Australia),
1/89
Ashoka, 3/128
Ashura, *4/55*
Asia, 1/2, 1/4
See also Human migrations; *and specific*
countries
Asia Minor, 3/65, 3/66, 6/57
Asian Americans, 6/138
Asian Development Bank, 5/66
Asian people
Canada, 2/23
Jamaica, 3/246
South Africa, 5/189, 5/191, 5/195
United States, 6/129, 7/105
Zimbabwe, 6/207
See also specific Asian countries
Asir (Saudi Arabia), 5/148
Asmara (Eritrea), 2/202
Aso, Mount, *3/256–257*
Aspen (Colorado), 6/127
Aspendos (Turkey), 6/73
Assad, Bashar al-, 5/268
Assad, Hafez al-, 5/268
Assal, Lake, *2/163*
Assam (India), 1/151, 3/126, 3/137
Assamese language, 3/138
Assiniboine Indians, 2/28–29

Association of Southeast Asian Nations,
4/43, 4/101, 6/37, 6/183
Assyria, 3/198, 3/212, 4/50, 7/40
Assyrian people, 3/192, 3/193, 4/54,
5/272, 5/274
Astana (Kazakhstan), 4/8, *4/9*
Asunción (Paraguay), 5/38, 5/40, *5/41,*
5/43
Aswan High Dam, 2/184, 2/187
Atacama Desert, 2/48, 2/52, *2/53,* 2/56
Atacama Indians, 2/48, 2/54
Atahualpa, 5/53
Atambaev, Almazbek, 4/38
Atatürk, Kemal, 3/199, 6/58, *6/59,* 6/65,
6/69, *6/69*
Atatürk Dam (Turkey), *6/66*
Athabaskan Indians, 2/28
Athabaskan language, 2/28
Athena, 3/58
Athena Nike, Temple of, *3/64*
Athens (Greece), 3/51, 3/54, 3/58–59,
3/58–59, **3/59,** 3/61
ancient, 3/58–59, 3/62, 3/64, 3/68
Athos, Mount. *See* Mount Athos
Atlanta (Georgia), 6/125
Atlantic Ocean, 1/4
Angola on, 1/38, *1/38*
Argentina on, 1/51
Azores in, 1/112
Bermuda in, 1/148
Brazil on, 1/161, 1/169, *1/169, 1/171*
Canary Islands in, 5/218
Côte d'Ivoire on, 2/126
Cuba on, *2/138–139*
Equatorial Guinea in, 2/200
Faroe Islands in, 2/212
France on, 2/236
Greenland in, 3/72, 5/85
Guyana on, 3/94
Madeiras in, 1/112, *1/113*
Martinique on, 4/118
Norway on, 4/246, 4/250
Panama on, 5/28, 5/32–33
Portugal on, 5/90, 5/93
prehistoric migration, 7/71
Suriname on, 5/238
Sweden on, 5/242
United States on, 6/124, 6/151
Uruguay on, 6/164, *6/165*
Virgin Islands, British, on, 6/188
Western Sahara on, 6/192, 6/193
Atlantic Provinces (Canada), 2/27, 2/35,
2/38
Atlas Mountains, 1/26, 1/28, 4/162,
4/162–163, 4/168–169, *4/168–*
169, 6/52
Atlatls, 7/46, 7/49, 7/68
Atolls
Cook Islands, 2/118
French Polynesia, 2/259
Indonesia, 3/165
Kiribati, 4/16
Maldives, 4/104, *4/105,* 4/106
Marshall Islands, 4/116, *4/117*
Seychelles, 5/170

Tuvalu, 6/76
Atomic bomb, 3/265, 6/136
See also Nuclear weapons and testing
Atomium (sculpture, Brussels), 1/133,
1/140–141
Attica peninsula, 3/58
Attu Island (Alaska), 6/118
AUC, 2/101
Auckland (New Zealand), *4/223*
Audience Hall (Iran), *3/184–185*
August coup (Soviet Union), 1/66, 1/110,
1/131, 4/146, 6/74–75, 6/167
Augustus, 2/192, 3/226, *3/227*
Aung San, 4/177, 4/179
Aung San Suu Kyi, 4/176–177, *4/179*
Aunuu Island (American Samoa), 1/32
Aurangzeb, 3/128–129
Aurelian, 5/266
Aurignacian industry, 7/39, 7/44, 7/46,
7/49, 7/49–51, 7/64
Aurora borealis, 2/222
***Aurora* cruiser,** *5/123*
Austin (Texas), *6/127*
Austral Islands, 2/258
Australasia, 5/10
Australia, 1/4, 1/68–93, *1/68–93,* **1/71**
economy, 1/71
agriculture, 1/80–81, *1/80–81,*
1/81
manufacturing, 1/83, 1/85
mining, **1/82,** 1/82–83, *1/82–83,*
1/88
Fiji aid, 2/216–217
government, 1/70
history, 1/86–87, *1/86–87,* **1/87**
human migration
aboriginal, 7/23, **7/26,** 7/26–27,
7/27
European, 7/106, *7/106–107,*
7/107
land and climate, **1/72,** 1/72–73,
1/72–73
outback, **1/78,** 1/78–79, *1/78–79*
people, 1/90–91, *1/90–91*
population, 1/70, 1/72, 1/77, **1/78**
power in
Nauru, 4/182, 4/183
Pacific Islands, 5/12
Papua New Guinea, 5/34–35
Solomon Islands, 5/185
states and territories, 1/70, 1/88–89,
1/88–89
wildlife, 1/69, 1/74–75, *1/74–75,*
1/92, *1/92–93*
Australian Alps, 1/72
Australian Antarctic Territory, 1/89
Australian Capital Territory, 1/70, 1/88,
1/91
Australopithecus boisei, 7/9
Austria, 1/94–109, *1/94–109,* **1/97**
economy, 1/104–105, *1/104–105*
government, 1/96
history, 1/98–99, *1/98–99,* **1/99**
land, 1/96
Danube River, 5/180

Liechtenstein, relations with, 4/75
music, 1/101, 1/106–107, *1/106–107*
people, 1/97
power in
Bohemia, 2/144, 2/145, 2/152
Italy, 3/229
Luxembourg, 4/79
Poland, 5/72, 5/76, 5/80
Sardinia, 3/245
Slovakia, 5/178
Slovenia, 5/182
provinces
north and east, **1/102,** 1/102–103,
1/102–103
south and west, **1/100,** 1/100–
101, *1/100–101*
Austria-Hungary, 1/98, **1/99,** 2/130,
3/111, **3/111,** 5/169, 5/178
Dual Monarchy of, 1/98, 3/111, 3/114
Austronesians, 7/91–93, *7/92–93,* **7/93**
New Zealand, 7/96–97, *7/96–97,*
7/97
Philippines, 7/59
Polynesia, 7/94–95, *7/94–95,* **7/95**
Taiwan, *7/58,* 7/58–59
Automobiles
Albania, 1/22
Belarus, 1/130
Bermuda, 1/148
Brazil, 1/174, 1/176
Canada, 2/34
China, *2/93*
Cuba, *2/133*
Germany, 3/36
Italy, *3/240*
Japan, *3/262*
Korea, South, 4/28–29
Mexico, *4/134*
Spain, 5/208, 5/209
Thailand, *6/35*
United States, *6/123,* *6/153–154*
Avar people, 5/116
Avars, 4/146
Avarua (Rarotonga), 2/118
Avenida 9 de Julio (Buenos Aires), 1/60
Aventine Hill (Rome), 3/230
Averroës, *5/212*
Avignon (France), *2/252*
Ávila Camacho, Manuel, 4/141
Awami League, 1/120
Axim (Ghana), 3/40
Aya, *3/265*
Ayers Rock. *See* Uluru
Ayion Oros. *See* Mount Athos
Aylwin Azócar, Patricio, 2/50
Aymara Indians, 1/152, 1/155, 2/54,
5/54–55, 5/55
Aymara language, 5/44
Ayolas, Juan de, 5/38
Ayutthaya kingdom, 6/36, *6/37*
Azande people, 5/229, 5/236
Azawad, 4/109
Azem Palace (Damascus), 5/275
Azerbaijan, 1/67, 1/110–111, *1/111,*
1/111, 3/14

Azerbaijani people, 1/67, 1/111, 3/178
Aznar, José María, 5/201
Azores, 1/112, *1/112,* **1/112,** 5/87,
5/95
Azrou (Morocco), 4/168
Aztec, 4/130, 4/136, 4/138, 4/142,
4/144, 7/78, *7/80*
history, 4/140, 7/102, *7/102–103*

B

Baalbek (Lebanon), 4/50
Baath Party (Iraq), 3/188, 3/192, 3/193
Baath Party (Syria), 5/268, 5/273
Bab el-Mandeb Strait, 7/14
Babangidi, Ibrahim Badamosi, 4/239
Babenberg family, 1/98
Bábi faith, 3/179
Babisuri Cave, 7/78
Babur, 1/18, 3/128, 5/15
Babylonia, 3/186–187, 3/212, 4/50,
7/40
Bacar, Mohamed, 2/108
Bach, Johann Sebastian, 3/17
Bachelet, Michelle, 2/51
Bactria (Afghanistan), 1/10
Baden (Austria), 1/102
Badjao people, 5/64–65, *5/64–65*
Baedeker's Guide to Russia, 5/129
Baga (India), *3/154–155*
Bagaza, Jean-Baptiste, 1/189
Baghdad (Iraq), 3/187
history, 3/188, 3/189, 3/197
people, 3/192, *3/192–193*
Bagpipes, 6/96
Baguettes, *1/27*
Baguio market (Manila), 5/68
Baguirmi kingdom, 2/45
Bahadur, Jung, 4/186
Bahá'í faith, 3/179
Bahamas, 1/114–115, **1/115,** *1/115,*
4/58, 4/59
Bahasa Indonesia language, 3/166
Bahasa Malaysia language, 1/179, 4/98,
5/176
Bahr al Jabal river, 5/232
See also White Nile River
Bahrain (country), 1/116–117, **1/117,**
1/117
Bahrain (island), 1/116
Bahutu people. *See* Hutu people
Baiji (Iraq), 3/195
Baikal, Lake, 5/125, *5/125–126,*
7/62–63
Baikonur (Kazakhstan), 4/9
Bailiwicks, 6/100
Bainimarama, Frank, 2/217, 2/221
Bairiki (Kiribati), 4/17
Baixa (Puerto Rico), 5/84
Baixa district (Lisbon), 5/90
Baja California (Mexico), 4/130
Bakdash ice cream shop, *5/271*
Bakhtiari people, 3/178, *3/179–181*
Bakiev, Kurmanbek, 4/38
Bakr, Ahmad Hassan al-. *See* Al-Bakr,
Ahmad Hassan

Baku (Azerbaijan), 1/110, *1/111*
Balance of trade, 3/263
Balanta people, 3/90
Balaton, Lake, 3/112, *3/112*
Balboa, Vasco Núñez de, 5/32, 5/214
Balearic Islands, 5/199, 5/202, 5/205, 5/216–217, *5/216–217*, **5/217**
Balewa, Sir Abubakar Tafawa, 4/237, 4/238
Balfour Declaration, 3/212
Bali (Indonesia), 3/164, *3/164*, 3/169, *3/170–171*
Balkan Mountains, 1/182, 1/183
Balkan Peninsula, 1/20, 1/156, 1/180, 2/130, **3/53**, 5/166
Balkan Wars
    First, 1/20, 3/63, 4/82, 4/157, 5/157, 5/158
    Second, 3/63, 4/82
Ballet, *5/116*, 5/120
Ballet Folklórico, *4/143, 4/145*
Balochi language, 5/20
Balochi people, 1/15, 3/178
Balochistan (Pakistan), 5/15, 5/16, 5/20
Balochistan Plateau (Pakistan), 5/18, 5/19
Balsa wood, 2/176
Balsas River, 4/130
Baltic Lakes region (Poland), 5/78
Baltic Sea
    Czech Republic and, 2/154
    Denmark and, 2/156
    Estonia on, 2/204
    Finland on, 2/222, 2/226
    Germany on, 3/24, *3/25*
    Latvia on, 4/48
    Lithuania on, 4/76
    Poland on, 5/72, 5/76, *5/78*
    Russia and, 5/112, 5/118
    Sweden on, 5/248, *5/249*, 5/250
    Ukraine on, 6/82
Baltimore (Maryland), 6/121, 6/124, *6/153*
Balzac, Honoré de, 2/252
Bamako (Mali), *4/110–111*, 4/111
Bambara people, 4/110, 4/113, 4/122
Bamboo, 2/74, 3/268, 5/68
Bamian (Afghanistan), 1/11, *1/16–17*
Bamiléké people, 2/16
Banaba Island, 4/16, 4/17
Banana republics, 3/100, 3/104
Bananas
    Brazil, 1/172, 1/174
    Canary Islands, *5/218*
    Honduras, 3/100, 3/104, *3/105*
    India, 3/134
    Jamaica, *3/246*
    Panama, *5/31*
    Réunion, *5/100*
    Uganda, *6/81*
Banaras Hindu University, 3/147
Banat (Romania), 5/104, 5/105, 5/107
Banda, Hastings Kamuzu, 4/91
Banda, Joyce, 4/91
Banda, Rupiah, 6/202

Bandar Seri Begawan (Brunei), 1/178, *1/179*
Banff National Park (Canada), *2/26*, 2/38, 2/39
Bangka (Indonesia), 3/170
Bangkok (Thailand), 4/93, 6/27, *6/31*, 6/34–35, *6/34–35*
    history, 6/28, 6/31, 6/36
    people, 6/31
Bangladesh, 1/118–127, *1/118–127*, **1/121**, 5/15, 5/16
    economy, 1/118, 1/121, *1/122*
    history, 1/118–120, 1/126
    land, 1/120, 1/122
    natural disasters, 1/120, 1/126–127, *1/126–127*, **1/127**
    people, 1/124–125, *1/124–125*
        river life, 1/122–123, *1/122–123*
    population, 1/125
    relations with
        India, 1/118
        Pakistan, 1/118, 1/120, 1/124, 1/126
Bangladeshis, in United Kingdom, 6/104, *6/104*
Bangui (Central African Republic), 2/42
Banihal Pass, *3/159*
Banjul (Gambia), 3/12, *3/13*
Bank of Sweden, 5/247
Banking and finance
    Argentina, 1/53, *1/63*
    Frankfurt, 3/28
    Hong Kong, 2/84
    Luxembourg, 4/79, 4/80, *4/81*
    Switzerland, 5/255, *5/260*, 5/263, *5/264–265*
    United States, 6/152
Bannockburn, Battle of, 6/110
Bantenese people, 3/166
Banto faros, 3/11
Bantu languages, 1/36, 1/39, 2/16, 2/112, 6/81, 6/205, 7/98–99
Bantu people, 2/16, 4/90, 4/170, 4/173, 5/192
    Bantu Migration, 7/98–99, **7/99**
    Great Zimbabwe, 7/100–101, *7/100–101*
Baqubah (Iraq), *3/189*
Bar (Montenegro), 4/156
Barada River, 5/274
Barbados, 1/128–129, *1/128–129*, **1/129**, 3/75, 4/60
Barbaresco wine, 3/239
Barbarossa (king of Germany). *See* Frederick I (king of Germany)
Barbarossa (sea captain), 1/24
Barbary apes, *3/46, 3/47*
Barbecues, 1/56, 156–57
Barbuda Island. *See* Antigua and Barbuda
Barcelona (Spain), *5/204–205*, 5/209
Barcelona (Venezuela), *6/176*
Barents, Willem, 4/256
Barents Island (Norway), 4/256
Barentsburg (Norway), 4/257
Bariba people, 1/147, *1/147*

Barnaul (Siberia), 5/127
Barolo wine, 3/239
Barong tagalogs, 5/63
Baroque architecture, 1/101, 1/109, *1/164*, 1/172
Baroque art, 2/252, *2/252–253*, 3/236
Baroque music, 2/252, 3/237
Barranquilla (Colombia), 2/102
Barrier reefs, 1/69, 1/92–93, *1/92–93*
    See also Coral and coral reefs
Barrios, *1/60–61*, 1/61, 6/179
Bartók, Béla, *3/110*
Bartolomé Island (Ecuador), *2/178–179*
Basalt, 2/212, 3/121
Baseball, 3/276, *3/276–277*
Basel (Switzerland), 5/261
Basescu, Traian, 5/103
Bashkir people, 5/116
Basian Cave, 7/58
Basilica of Our Lady of Guadalupe (Mexico), 4/142
Basilica of St. Mark (Venice), 3/242
Basilique du Sacré Coeur (Paris), 2/233, *2/233*
Basotho nation, 4/57
Basotho people, 4/56, 4/57
Basques, 2/238, 5/200, 5/201, 5/205
Basra (Iraq), 3/186, 3/195
Bass Saxophone, The (Skvorecky), 2/152
Bassa people, 2/16, 4/64
Basse-Terre (Guadeloupe), 3/76
Basse-Terre Island (Guadeloupe), 3/76, 3/77
Basseterre (St. Kitts and Nevis), 5/133
Basters, 4/181
Bastille, 2/244
Basutoland, 4/57
Basutoland National Party, 4/57
Bata (Equatorial Guinea), 2/200
Batak people, 3/166
Batéké people, 2/117
Batéké Plateau, 2/116
Batik cloth, 4/102
Batinawaks, 5/63
Batista, Fulgencio, 2/132
Batlle Ibáñez, Jorge, 6/163
Batlle y Ordóñez, José, 6/162
Batman, John, 1/85
Baton Rouge (Louisiana), 6/149
Battle of ... *See under key word, as in* Pyramids, Battle of the
Batu Caves (Malaysia), *4/97*
Batutsi people. *See* Tutsi people
Bauhaus, 3/39
Bauxite, 3/89, *3/89*, 3/113, 3/248, *3/249*, 4/252
Bavadra, Timoci, 2/220
Bavaria, 1/100, 3/26, *3/35–36*
Bavarian Alps, 3/24, 3/25
Bay Islands (Honduras), 3/100, *3/104*
Bay of Bengal
    Bangladesh and, 1/118, 1/122, 1/126
    India and, 3/127, 3/130, 3/142, 3/156
    Myanmar and, 4/174, 4/175
Bay of Biscay, 2/236, 5/93, 5/205

Bay of Islands (New Zealand), *4/221*
Bay of Naples, 3/224
Bay of Pigs invasion, 2/133
Bayon temple (Cambodia), 2/14, *2/15*
Bazaars, 2/190
Bear Island (Norway), 4/250, 4/257
Bears, *2/38*
   brown, *1/35*
   polar, *4/256*
   trained, *3/117*
Beavers, *2/38*
Bechuanaland Protectorate, 1/158
Beckett, Samuel, 3/201
Bečkovic, Matija, 5/165
Bédié, Henri Konan, 2/127
Bedouin people, 2/187, 3/191, *3/286*,
   3/287, 5/144, 5/147, 5/148, *5/149*
Beech trees, 4/225
Beef Island, *6/188*
"Beehive" building (Wellington), *4/217*
Beer, 3/27
Beer Cart restaurant, *5/107*
Beer Hall Putsch, 3/26
Beethoven, Ludwig van, 1/106, 1/108,
   3/17
Begin, Menachem, 3/213
Beijing (China), 2/69, 2/72, 2/80,
   2/82–83, *2/82–83*, **2/83**
   capture of, 2/64
   demonstrations, 2/60–61, *2/65*, 2/93
   Forbidden City, 2/60, *2/63*, 2/82
   Olympic Games, 2/61, 2/83
Beijing Airport, 2/83
Beijing National Stadium, 2/83
Beijing opera, 2/67, 6/15
Beijing Zoo, *2/82*
Beira (Portugal), 5/89
Beirut (Lebanon), 4/50, *4/53*
Beja people, 5/234
Bekaa Valley, 4/50
Belarus, 1/130–131, *1/131*, **1/131**
Belarusian people, 5/116
Belau. *See* Palau
Belaúnde Terry, Fernando, 5/46
Bélep Islands, 4/212
Belfast (Northern Ireland), 6/99, *6/99*
Belfast Agreement, 6/98
Belgium, 1/132–143, *1/132–143*, **1/135**
   economy, 1/142–143, *1/142–143*,
      1/143
   government, 1/134
   history, 1/134
   land, 1/133, 1/135
   medieval Flanders cities, 1/138–139,
      1/138–139
   people, 1/136–137, *1/136–137*
   population, 1/133, 1/136, **1/137**
   power in
      Burundi, 1/188
      Congo, 2/110
      Rwanda, 5/130
   See also Flanders (Belgium); Wallonia
      (Belgium)
Belgrade (Serbia), 4/156, 5/154, *5/164–*
   165, 5/168–169, *5/168–169*

economy, *5/162–163*, 5/163
history, 5/158–159, *5/159*
Belgrade, University of, 5/168
Belgrade Fortress, *5/168–169*
Belgravia (London), 6/93
Belitung (Indonesia), 3/170
Belize, 1/144–145, **1/145**, *1/145*
Belize City (Belize), 1/144
Bellini, Giovanni, 3/242
Belmint Bay (British Virgin Islands), 6/189
Belo, Carlos Felipe Ximenes, 2/168,
   2/169
Belovezha Forest (Belarus), *1/131*
Belváros (Budapest), 3/114
Belvedere Palace (Vienna), 1/109
Ben Ali, Zine el-Abidine, 6/49, 6/50
Ben Bella, Ahmed, 1/26
Ben-Gurion, David, 3/212, *3/213*
Bendjedid, Chadli, 1/26
Benedict XVI (pope), *6/170*
Benelux, 1/133
Benelux economic union, 4/79
Beneš, Eduard, 2/148
Benga people, 2/201
Bengal, 1/119, 1/124
Bengal, Bay of. *See* Bay of Bengal
Bengali language, 3/138
Bengali people, 1/124, 5/15
Bengaluru (India), 3/144
Benghazi (Libya), 4/68, 4/69
Benguela Current, 3/9
Benin (country), *1/146*, 1/146–147,
   1/147, *1/147*
Benin, Kingdom of, 4/236, 4/245, *4/245*
Benue River, 4/240, 4/244
Beograd, 5/168
Beqa Island, *2/220–221*
Berber language, 4/164
Berber people
   Algeria, 1/24, 1/25, 1/27–29
   Libya, 4/66, *4/68*, 4/70
   Mauritania, 4/120, 4/122
   Morocco, 4/158, 4/164, *4/164–165*,
      4/166, 4/169
   Niger, 4/232
   Tunisia, 6/48, 6/54–55, *6/55*
   Western Sahara, 6/192, *6/192*
Berbice River, 3/94
Berdimuhammedov, Gurbanguly, 6/75
Berengaria of Navarre, 2/143
Berg, Alban, 1/106
Bergen (Norway), 4/254, *4/254*, 4/255
Bergenhus fortress (Norway), 4/255
Bering Land Bridge
   North American migration and, 7/67–
      70, **7/68**, *7/68–69*, 7/70
   South American migration and, 7/82,
      7/86, 7/88
Bering Strait, 7/67, 7/86
Beringia. *See* Bering Land Bridge
Berlin (Germany), 3/17, 3/18, 3/32–33,
   3/32–33, **3/33**
   history, 3/22, 3/33
Berlin Wall (Germany), 3/17, 3/23, *3/23*,
   3/32

Berlioz, Hector, 2/253
Berlusconi, Silvio, 3/223
Bermuda, *1/148*, **1/148**, 1/148–149,
   1/149
Bermuda Island, 1/148
Bermúdez, Juan de, 1/149
Bern (Switzerland), 5/255, *5/257*, 5/260,
   5/261
Bernadotte, Jean Baptiste, 5/243
Bernini, Gian Lorenzo, *3/237*
Berry, Chuck, 6/159
Beskid mountains, 5/79
Beslan (Russia), 5/111
Bessarabia (Moldova), 4/146
Béthencourt, Jean de, 5/220
Betio (Kiribati), 4/17
Bette Peak, 4/68
Beyin, Amanuel, 7/14
Beys, 6/48
Bhaktapur (Nepal), 4/185
Bhilai (India), 3/136
Bhils, 3/138
Bhimbetka Rock Shelters, 7/19, *7/20–21*
Bhumibol Adulyadej, 6/29
Bhutan, *1/150*, 1/150–151, **1/151**,
   1/151
Bhutia people, 1/150, *1/150*, 1/151,
   3/153
Bhutto, Benazir, 5/17
Bhutto, Zulfikar Ali, 5/16–17
Bia, Mount, 4/44
Biafra, 4/239
Białowieska Forest (Poland), 5/78
Bicycling, *3/148–149*, 4/27, 4/199,
   4/206, *4/206*
Bidan people, 4/122
Bidonvilles, 4/166
Bielecki, Jan Krzysztof, 5/75
Big Ben (London), 6/103
Big Bend National Park (United States),
   6/126
Big Four, 3/22
Big Hole mine, *5/197*
Big Wild Goose Pagoda (China), 2/80
Bihac (Bosnia-Herzegovina), 1/157
Bihar (India), 3/133, 3/136
Bihari people, 1/124
Bihor Mountains, 5/104
Bijagós Island, 3/91
Bikini Island, 4/117
Bikkhus, 5/226
Bilbao (Spain), 5/205
Bill of Rights (United Kingdom), 6/112
Bin Laden, Osama, 1/13, 5/17
Bingham, Hiram, 5/44
Binnenhof (The Hague), 4/205
Bió-Bió River, 2/53
Bioko (Equatorial Guinea), 2/200–201
Birch trees, *5/249*
Birdman Cult, 7/95
Birds
   Amazon Region, 1/168
   Antarctica, 1/45, *1/45*
   Australia, 1/74, *1/75*

Canada, 2/38, *2/38–39*, 2/39
China, *2/74–75*
Galapagos Islands, 2/178, *2/178*, 2/179
Nauru, *4/183*
New Zealand, 4/224, 7/96–97
Birendra, 4/186–187
Birmingham (England), 6/93
Biscay, Bay of, 2/236, 5/93, 5/205
Bishkek (Kyrgyzstan), 4/39
Bishop, Maurice, 3/75
Bislama language, 6/168
Bismarck, Otto von, 3/21, *3/21*
Bismarck Archipelago (Papua New Guinea), 5/36, 7/93
Bison, *1/131*
Bissau (Guinea-Bissau), *3/91*
Biya, Paul, 2/19
Bizerte (Tunisia), 6/53
Bizimungu, Pasteur, 5/131
Bjul (India), 3/127
Black Africans
    Antigua and Barbuda, 1/48
    back to Africa movement, 3/251
    Bahamas, 1/114–115
    Barbados, 1/129
    Belize, 1/145
    Bermuda, 1/148, 1/149
    Brazil, 1/166–167
    Colombia, 2/99, 2/104
    Comoros, 2/109
    Costa Rica, 2/120, *2/121*
    Cuba, 2/132, 2/136
    Dominica, 2/165
    Ecuador, 2/174
    Equatorial Guinea, 2/200
    French Guiana, 2/257
    Grenada, 3/74
    Guyana, 3/92, 3/93, 3/95
    Haiti, 3/98
    Jamaica, 3/246
    Lesser Antilles, 4/59, 4/60
    Martinique, 4/118
    Nicaragua, 4/229
    Panama, 5/30
    Puerto Rico, 5/96
    St. Kitts and Nevis, *5/132*
    St. Vincent and the Grenadines, 5/137
    São Tomé and Príncipe, 5/143
    South Africa, 5/189, 5/190, 5/191, 5/194, 5/197
    Trinidad and Tobago, 6/47
    Venezuela, 6/178
    See also African Americans; Coloureds; *and specific African countries and peoples*
Black Caribs, 3/103
Black Communities Act (South Africa), 5/190
Black Death, 3/119
Black Forest, 3/24, *3/24*, 3/25, 5/180
Black Moors, 4/122
Black Nazarene image, 5/67
Black Sea, 5/180–181, 5/242
    Bulgaria on, 1/182, *1/182*, 1/184

Georgia on, 3/14
    Romania on, 5/104, 5/105
    Russia on, 5/118
    Turkey on, 6/57, 6/61, 6/62, 6/70
    Ukraine on, 6/82, 6/83
Black Volta River, 3/40
Blackfeet Indians, 2/28–29
Blair, Tony, 6/107, 6/112
Blaize, Herbert A., 3/75
Blanco Party (Uruguay), 6/162
Blankets, Ceremonial, *2/28–29*
Blarney Stone, 3/206
Blenheim Palace (England), *6/93*
Bligh, William, 5/137
Blitz, The, *6/113*
Bloomington (Minnesota), *6/156*
Blue Bird, The (Maeterlinck), 1/133
Blue Mosque (Afghanistan), *1/10–11*, 1/11, *1/14–15*
Blue Mosque (Turkey). *See* Sultan Ahmet Mosque (Turkey)
Blue Mountains, 3/248
Blue Nile River, 2/208, 5/232, 5/234
Blue Ridge Mountains, 6/124, 6/132
Blue-weeds, 5/219
Blues, 6/159
Bluford, Guion Stewart, Jr., 6/145
Boat people, 6/182
Bob Marley Museum, 3/251
Bobo people, 1/186, 1/187, *1/187*
Boca, La (Buenos Aires), *1/60–61*, 1/61
Bocauniors, 1/61
Bode Museum, *3/32*
Bodrum (Turkey), *6/72–73*, 6/73
Boers, 4/57, 5/192–193, 5/240
Bogle, Paul, 3/247
Bogotá (Colombia), 2/98, 2/99, *2/105*, *2/107*
Bohemia (Czech Republic), 1/98, 2/144, 2/148, 2/149
Bohemian Basin, 2/154, 2/155
Bohemian Forest, 2/154–155
Bohemian-Moravian Highlands, 2/154, 2/155
Bohemian Mountains, 2/154
Bohíos, 2/138
Bohol Island (Philippines), 5/61
Bohuslän (Sweden), *5/248*
Boii people, 2/144
Boiling Lake, 2/164
Bokassa, Jean-Bedel, 2/43
Bokmål Norwegian language, 4/248
Bolan Pass, *5/18–19*
Bolas, 1/58, *1/159*
Bolesław I, 5/76
Bolger, Jim, 4/217
Bolívar, Simón, 1/153, 2/99, 2/104, *2/105*, 5/45, 6/173
Bolivia, *1/152–153*, 1/152–155, **1/155,** *1/155*, 5/42
Bolkiah, Sir Muda Hassanal, 1/178, *1/178*
Bolovens Plateau, 4/44
Bolsheviks, 5/114, 5/121, 5/123, 6/82
Bolshoi Ballet, *5/116*, 5/120

Bom Jesus Basilica, 3/154
Bombay (India). *See* Mumbai (India)
Bon Pays region (Luxembourg), 4/78, 4/80, 4/81
Bonaire (Netherlands Antilles), 4/60, 4/200, 4/210, 4/211, *4/211*
Bonaparte, Napoleon. *See* Napoleon I
Bonaventure Island (Canada), 2/38
Bondi Beach (Australia), *1/84*
Bongo, Ali Ben, 3/9
Bongo, Bernard-Albert, 3/9
Boni, Yayi, 1/146
Bonin Islands (Japan), 3/258, *3/258–259*
Bonn (Germany), 3/32
Bonriki (Kiribati), 4/17
Book of Kells, 3/202
Bora Bora (French Polynesia), *2/258–259*, 2/259
Bordaberry, Juan Maria, 6/163
Bordeaux (France), *2/238*, 2/248
Borges, Jorge Luis, 1/64
Borneo, 1/178, 3/161, 3/164, 3/170, 4/93, 4/96, 7/23
Bornholm Island (Denmark), 2/158, *2/159*
Borno (Nigeria), 4/236
Borrowed scenery, 3/280
Boruca people, 2/120
Bosnia-Herzegovina, 1/156–157, *1/157*, **1/157,** 5/155, 5/159
Bosniaks, 1/157, 4/32, 5/160
Bosporus, Sea of, 6/66–67
Bosporus Bridge, 6/62, 6/66
Bosporus Strait, 6/57, 6/62, 6/63, *6/63*, 6/69
Boston (Massachusetts), 6/120, 6/121, 6/146
Boston Massacre, 6/134
Boston Tea Party, 6/134
Botanical Gardens (St. Vincent and the Grenadines), 5/136–137
Botany Bay (Australia), 1/86, 7/106
Bothnia, Gulf of, 2/226, 2/227, 5/248, 5/249
Botswana, 1/158–159, *1/159*, **1/159,** 7/99
Botticelli, Sandro, 2/236, 3/221
Bottom, The (Saba), 4/62
Bougainville (Papua New Guinea), 5/35, 5/36, 5/37
Boulder Valley (Montana), *6/127*
Boulez, Pierre, 2/253
Boumedienne, Houari, 1/26
Bounty **mutineers,** 2/259
Bourbon colony, 5/100
Bourguiba, Habib, 6/49, 6/50
Boussance people, 1/186, 1/187
Bouteflika, Abdelaziz, 1/26
Bowler, Jim, 7/26–27
Boxer Rebellion, 2/63
Bozizé, François, 2/43
Bradley, Thomas, 6/145
Bradman, Sir Donald, *1/87*
Brahma, 4/193
Brahmans, 3/140, 3/146

Brahmaputra River, 1/118, 1/122, 2/94, 3/142, 4/190, 4/191, 4/194
Brahms, Johannes, 1/106
Brahui people, 1/15, 5/20
Brain, in early humans, 7/6, 7/34
Bramante, Donato, 3/237
Brandenburg Gate (Berlin), *3/23*, 3/32
Brandt, Willy, *3/22*, *5/247*
Braque, Georges, 2/253
Brasília (Brazil), 1/161, 1/166, 1/172, *1/172–173*
Bratislava (Slovakia), *5/179*, *5/180–181*, 5/181
Bratislava Castle, *5/181*
Brazil, 1/160–179, *1/160–179*, **1/163**, 5/95
   cities, 1/166, 1/172–173, *1/172–173*
   early settlement, 7/84
   economy, 1/162
      agriculture, 1/174–175, *1/174–175*, **1/175**
      industry, 1/166–167, *1/166–167*
   government, 1/163
   history, 1/163–165, *1/164–165*, 7/102–103
   land and climate, 1/168–169, *1/168–169*, **1/169**
   Paraguay, relations with, 5/39
   people, 1/166–167, *1/166–167*
   population, 1/166, **1/172**
   Uruguay, power in, 6/160, 6/161
Brazilian people, in Japan, 3/274
Brazzaville (Congo), 2/116
Breadfruit, 5/137
Brecon Beacons National Park, 6/94
Breg River, 5/180
Bregenz (Austria), 1/100, 1/107
Bremen (Germany), **3/30**, *3/30–31*, 3/31
Bremerhaven (Germany), 3/31
Brenner Pass (Germany), 1/100, 3/26
Brest (Belarus), *1/131*
Breton language, 2/238
Breton people, 1/112
Brezhnev, Leonid I., 5/115
Brian Boru, 3/204
Bride price, 6/205
Bridgetown (Barbados), 1/128, *1/128*
Brigach River, 5/180
Brimstone Hill fortress (St. Kitts and Nevis), 5/132
Brisbane (Australia), 1/72, 1/88
Bristol Channel, 6/94
Britain. *See* United Kingdom
Britannia, 6/110
British Antarctic Territory, 2/210
British Columbia (Canada), 2/22, 2/26, 2/33, 2/35, 2/37, 2/39
British East Africa, 4/10
British Guiana, 3/92–93
British Honduras, 1/144
British India, 3/129, 5/15
British Museum, 3/59
British North America Act (Canada), 2/24–25, 2/31

British people
   Australia, 1/69, 1/84, 1/85, 1/90
   Bahamas, 1/114
   Canada, 2/23, 2/38
   South Africa, 5/194
   United States, 6/138
   See also United Kingdom
British Somaliland, 5/186
British Togoland, 6/41
Brittany (France), 2/238, 2/248, *7/48–49*
Brittany-Normandy Hills (France), 2/236
Brive (France), *2/239*
Brno (Czech Republic), 2/155
Broadway (New York City), 6/158, 6/159
Bromo, Mount, *3/165*
Bronze Age, 5/216
Brooks Range, 6/130
Brown, Gordon, 6/107
Brown, Linda, 6/144
Brown v. Board of Education of Topeka, 6/144
Bruckner, Anton, 1/107
Bruegel, Pieter, the Elder, 1/133
Bruges (Belgium), 1/138, *1/138–139*, 1/139
Brunei, 1/178–179, *1/178–179*, **1/179**
Brunei Darussalam, University of, 1/179
Brunelleschi, Filippo, 3/237
Brussels (Belgium), 1/133, *1/135*, 1/136, 1/140–141, *1/140–141*
   economic region, 1/134, 1/137
Bryggen (Norway), 4/255
BSPP (Myanmar), 4/176
Bu Craa (Western Sahara), 6/192
Bubi people, 2/200, 2/201
Bubonic plague, 5/32
Bucaram Ortiz, Abdala, 2/173
Bucharest (Romania), 5/103, *5/103*, 5/105, *5/107*
Buckingham Palace (London), 6/103
Buckland, William, 7/50
Buda (Hungary), 3/114, 3/115
Budapest (Hungary), 3/107, *3/109*, *3/111*, 3/114–115, *3/114–115*, 5/180
Buddha, 2/66, *2/67*, 3/128, 3/147, 4/30, 4/47, *4/188*, 5/227, 6/38
Buddhism
   Afghanistan, 1/11
   Bangladesh, 1/124
   Brunei, 1/179
   Cambodia, 2/8, 2/13
   China, 2/62, 2/66, *2/67*, 2/80, 2/81
      Macau, 2/86
      Tibet, 2/94
   Himalaya people, 4/192, 4/193
   India, 3/128, *3/140*, 3/141, 3/147, 3/149, 3/153, 3/158
   Indonesia, 3/168–169
   Japan, 3/266, 3/268–269, *3/268–269*, 3/275, 3/278–279, *3/278–279*
   Korea, 4/22, 4/30, 4/31, *4/31*
   Laos, 4/41, 4/46–47, *4/46–47*
   Malaysia, 4/103
   Mongolia, 4/151
   Myanmar, 4/174, 4/178, *4/178–179*

   Nepal, *4/188*, 4/189
   Pakistan, *5/24*
   Russia, 5/117
   Singapore, 5/176
   Sri Lanka, 5/226–227
   Taiwan, *6/15*
   Thailand, 6/31, *6/34–35*, *6/37*, 6/38–39, *6/38–39*
   United States, 6/156
Buenos Aires (Argentina), 1/51, *1/53*, 1/56, 1/60–61, *1/60–61*, 1/63
   Colón Theater, 1/51, *1/56*
   history, *1/64*, 1/64–65
   Pampas and, 1/55
   River Plate Stadium, 1/57
Buffalo, 2/29, *2/29*, 4/185, 6/141, 7/76, 7/77
Buganda kingdom, 6/78, 6/81
Bugi people, 3/166
Buhari, Mohammed, 4/239
Building stone, 5/92
Buka (Papua New Guinea), 5/36
Bukhara (Uzbekistan), *6/167*
Bukit bintang (Kuala Lumpur), 4/99
Bukovina (Romania), 5/104
Bulgar people, 1/181
Bulgaria, 1/180–185, *1/180–185*, **1/181**
   economy, 1/184–185, *1/184–185*
   land and climate, 1/182–183, *1/182–183*
   Macedonia, power in, 4/82
Bulgarian Empire, 4/82
Bulgarian language, 1/181
Bulgarian Socialist Party, 1/180
Bulge, Battle of the, 4/79
Bullfighting, 4/143, 5/207
Bulls, running of the, *5/206–207*, 5/207
Bumiputras, 4/98
Bundesrat (Austria), 1/96
Buñuel, Luis, 5/211
Burakumin people, *3/274*, 3/275
Burbank (California), *6/159*
Bureau of Indian Affairs, 6/141
Burek, 5/165
Burgas resort (Bulgaria), 1/183
Burgenland (Austria), 1/103, *1/105*
Burggarten (Vienna), 1/109
Burgher people, 5/226
Burgundians, in France, 2/240
Burgundy, 4/79
Burials, Prehistoric, 7/32, *7/33*, 7/34, 7/48, 7/50, *7/52*
Burj Khalifa (Dubai), *6/85*
Burke, Robert O'Hara, 1/86
Burkina Faso, 1/186–187, *1/187*, **1/187**, 2/128
Burma. *See* Myanmar
Burma Socialist Programme Party, 4/176
Burman people, 4/174, 4/178
Burmese language, 4/174
Burnham, Forbes, 3/93
Burnooses, 4/165
Burqas, 5/20–21
Bursa (Turkey), 6/65, 6/66, 6/68, 6/70

Burundi, 1/188–189, *1/189*, **1/189**,
2/111
Buryat people, 5/126, *5/126*
Bush, Australian. *See* Outback, Australian
Bush, in Zambia, 6/204
Bushidan warriors, 3/266
Bushido code, 3/266
Butterflies, Monarch, *6/149*
Buxoro (Uzbekistan), *6/167*
By Fiord (Norway), 4/255
Byelarus. *See* Belarus
Byelorussia. *See* Belarus
Byelorussian Republic, 1/130
Bygdøy (Norway), 4/254
Byzantine Empire, 3/242, 5/151
    power in
        Albania, 1/20
        Algeria, 1/24
        Balearic Islands, 5/217
        Crete, 3/68
        Cyprus, 2/140
        Greece, 3/63
        Jordan, 3/282
        Libya, 4/70
        Macedonia, 4/82
        Morocco, 4/158
        Serbia, 5/158
        Syria, 5/266, 5/267, 5/274
        Tunisia, 6/48
        Turkey, 6/62, 6/63, 6/68, 6/70

C

Cabinda (Angola), 1/38
Cabinet, of United Kingdom, 6/90
Cabo de Honduras, 3/100
Caboclos, 1/167
Caboet, Lord of, 1/34
Cabot, John, 2/30, 6/134
Cabot, Sebastian, 1/64
Cabot Trail highway, 2/38
Cabral, Amilcar, 3/90, 3/91
Cabral, Gonzalo, 1/112
Cabral, Luis, 3/91
Cabral, Pedro Álvares, 1/164, *1/165*,
5/85, 7/102, **7/103**
Cabrera Island, 5/216, 5/217, **5/217**
Cacao
    Brazil, 1/174, 1/175, *1/175*
    Colombia, 2/99
    Côte d'Ivoire, *2/128*, 2/129
    Equatorial Guinea, 2/200
    Ghana, 3/40, *3/40*, 3/41, 3/43, 3/44,
        3/45
    Haiti, 3/99
    Nigeria, 4/241
Cactus Hill site, 7/70, *7/71*
Cadia mine, *1/83*
Cádiz (Spain), 5/206
Cádiz, Gulf of, 5/202
Caelian Hill (Rome), 3/230
Caernarfon Castle (Wales), *6/95*
Caesar, Julius, 2/233, 3/226, *3/227*
CAFTA-DR, 2/121, 2/195
Caftans, 4/165
Cagliari (Italy), 3/245

Cahokia site, 7/76–77, 7/77
Cahuita National Park (Costa Rica), 2/125
Caibarién (Cuba), 2/137
Caicos Islands. *See* Turks and Caicos
    Islands
Caimans, Black, 2/177
Cairenes, 2/190–191
Cairo (Egypt), *2/183, 2/186, 2/187,*
    2/190–191, 2/190–191, **2/191**
    dialect, 2/186
    origins, 2/188
Cairo (Illinois), 6/148, 6/149
Cairo Road (Lusaka), *6/205*
Cajamarca (Peru), 5/53
Cakobau, 2/220
Calabria (Italy), 3/232, *3/233*
Calabrian dialect, 3/233
Calahorra, Castle of (Spain), *5/202*
Calama (Chile), 2/52
Calcutta (India). *See* Kolkata (India)
Caldera de Taburiente, 5/219
Caldera Rodríguez, Rafael, 6/174
Calderas, *3/164*
Calderón, Felipe, 4/129
Calderón de la Barca, Pedro, 5/215
Calendar, Egyptian, 2/181
Calgary (Alberta), 2/37
Calgary Exhibition and Stampede
    rodeo, 2/37
Cali (Colombia), 2/100
California (United States), 6/118, 6/128–
    129, 6/153, *7/105*
    agriculture, 6/154, *6/154*
    Hispanic Americans, 6/142–143
    Native Americans, 6/140–141
Caliphs, 2/188, 2/192, 5/144, 5/151,
    6/58
Callampas, 2/54
Callao (Peru), 5/50
Calligraphy, 2/66–67, *2/66–67,* 3/185,
    4/31, *5/176,* 6/15
Calvary, 3/217
Calving, by glaciers, *1/47*
Calypso, 3/250, 4/61, *4/61*
Cambodia, 2/8–15, *2/8–15,* **2/11,** 4/41
    history
        Angkor, 2/14–15, *2/14–15*
        early, 2/8–9
        modern, 2/9–11
    land and climate, 2/12
    people and economy, 2/8, 2/12–13,
        2/12–13
    relations with
        Thailand, 6/36
        United States, 2/9
        Vietnam, 2/9, 2/10, 6/182
Cambodian People's Party, 2/11–12
Cambrian Mountains, 6/94
Cambridge, University of, 6/93, 7/14,
    7/26
Camels, 1/18, *2/45, 2/71, 4/151,* 5/186,
    5/220–221
    Algeria, *1/30–31,* 1/31
    Libya, *4/68–69*
    Niger, *4/233*

    Sahara, 4/73, *4/73*
Cameron, David, 7/107, *7/107*
Cameroon, 2/16–21, *2/16–21,* **2/19,**
    2/47, 7/98
    economy, 2/20–21
    government, 2/18
    history, 2/18–19
    land and climate, 2/20, *2/20–21,* 2/21,
        2/21
    people, 2/16–17, *2/16–17*
Cameroon, Mount, 2/20
Camogie, 3/207
Camorta Island (India), 3/156
Camp David Accords, 3/213
Campesinos, 1/152
Campidano plain, 3/245
Campine (Belgium), 1/135
Camus, Albert, 2/253
Canaan, 3/209
Canada, 2/22–39, *2/22–39,* **2/25**
    cities, 2/36–37, *2/36–37*
    climate, 2/27
    conflicts, 2/32–33, *2/32–33*
    economy and resources, 2/34–35,
        2/34–35
    government, 2/24–25
    history
        first Canadians, 2/28–29, *2/28–29*
        since European settlement, **2/30,**
        2/30–31, *2/30–31*
    land, 2/26–27, *2/26–27,* **2/27**
    people, 2/23
    population, 2/22
    relations with
        Mexico, 2/34
        United States, 2/23, 2/32, 2/34
    wilderness, **2/38,** 2/38–39, *2/38–39*
Canadian Guard Band, *2/30*
Canadian Pacific Railway, 2/31
Canadian Shield, 2/26, 2/35, 2/38
Canals
    Bangkok, 6/34, *6/34*
    Bangladesh, *1/126–127*
    Cambodia, *2/14–15*
    Denmark, *2/161*
    Greece, 3/54, *3/61*
    Guyana, *3/95*
    Madagascar, 4/85
    Netherlands, 4/199, *4/203,* 4/206–
        207, 4/208
    Russia, 5/122
    Sweden, 5/249
    Turkmenistan, 6/74, *6/75*
    Venice, 3/242–243, *3/242–243*
    Volga River, 5/118
    See also Panama Canal; Suez Canal
Canary Islands, 5/199, 5/202, 5/203,
    5/218–221, *5/218–221,* **5/219**
    history and economy, 5/220–221,
        5/220–221
    land, 5/218–219, *5/218–219*
Canberra (Australia), 1/70, 1/88
Cancún (Mexico), 4/133
Cannibalism, 7/38
Cannons, *2/138–139*
Canoes, of Austronesians, 7/91, *7/93*

Cantabrian Mountains, 5/202, *5/202–203, 5/209*

Canton (China). *See* Guangzhou (China)

Cantonese dialect, 2/76

Cantons, Swiss, 5/256

Cão, Diogo, 2/114

Cap Vert (Senegal), 5/152

Cape Breton Highlands National Park (Nova Scotia), 2/38

Cape Colony (South Africa), 5/192

Cape Horn, 2/53, 2/210

Cape of Good Hope, 5/95

Cape Sounion (Greece), *3/64*

Cape Town (South Africa), 5/189, 5/192, 5/196

Cape Verde, 1/164, 2/40–41, *2/40–41,* **2/41,** 3/90–91, 5/87, 5/143

Cape Wrangell (Alaska), 6/118

Capet, Hugh, 2/241, *2/241*

Capetian dynasty (France), **2/240,** 2/241

Capitol, United States, 6/120–121

Capitoline hill (Rome), 3/230

Cappucino, 3/235

Caprivian people, 4/181

Car Nicobar Island (India), 3/156

Carabao, 5/62–63

Carabobo (Venezuela), 6/173

Caracas (Venezuela), 6/172, 6/174, 6/176, *6/178,* 6/179

Caracol (Belize), *1/145*

Caral-Supe civilization, 7/82–83, *7/83*

Caramanlis, Constantine, 3/52

Caravans, 1/18, 1/31, 2/71, **4/110,** *5/270–271*

Carbon-14, 7/21

Carbon dioxide, 2/21, *2/21*

Cardamomes Range, 2/12

Cárdenas, Lázaro, 4/141, *4/141*

Cardiff (Wales), 6/94–95, *6/95*

Cardinals (birds), *2/38*

Cardoso, Fernando, 1/163

Cargados Carajos Shoals (Mauritius), 4/124

Carías Andino, Tiburcio, 3/100

Carib Indians
    Antigua and Barbuda, 1/48
    Costa Rica, 2/120
    Dominica, 2/164
    Greater Antilles, 3/48
    Grenada, 3/74
    Guadeloupe, 3/76
    Guyana, 3/92
    Lesser Antilles, 4/59
    Puerto Rico, 5/97
    St. Kitts and Nevis, 5/132
    St. Lucia, 5/134
    St. Vincent and the Grenadines, 5/136
    Trinidad and Tobago, 6/46
    Venezuela, 6/172
    Virgin Islands, U.S., 6/190

Caribbean Sea
    Antigua and Barbuda on, 1/48
    Belize on, 1/144
    Colombia on, 2/98, 2/102, *2/102–103*
    Costa Rica on, 2/122

    Cuba on, 2/134
    Dominica on, 2/165
    Greater Antilles on, 3/48
    Grenada on, 3/74
    Haiti on, 3/96
    Jamaica on, 3/246
    Lesser Antilles on, 4/58
    Martinique on, 4/118, *4/119*
    Netherlands Antilles on, 4/211
    Nicaragua on, 4/228–229
    Panama on, 5/28
    St. Kitts and Nevis on, 5/132, *5/133*
    St. Vincent and the Grenadines on, 5/136
    Trinidad and Tobago on, 6/46
    Venezuela on, 6/172, 6/176

Caribou, *2/39*

Carinthia (Austria), 1/101, *1/101*

Cariocas, 1/170, *1/170,* 1/171

Carleton, Sir Guy, 2/30

Carlos III (king of Spain), *5/211*

Carlsbad Caverns, 6/126, *6/133*

Carniola, 5/182

Carnivals
    Argentina, 1/57
    Brazil, 1/171, *1/171*
    Italy, 3/233
    Lesser Antilles, 4/61, *4/61*
    Spain, 5/206

Caroline Islands, 2/214, 5/26, 5/27

Carolingian dynasty, 2/240–241

Carpaccio, Vittore, 3/242

Carpathian Mountains, 5/78, *5/78–79,* 5/79, 5/104–105, 5/178, 6/82

Carpets
    Afghanistan, *1/15, 1/17*
    Belgium, *1/142*
    China, 2/96
    Iran, *3/180, 3/185*
    Morocco, *4/162*
    Pakistan, 5/23, *5/23*
    Turkmenistan, 6/75

Carrara (Italy), *3/241*

Carrera, José Miguel, 2/49

Carrera, Rafael, 3/80

Carriacou (Grenada), 3/74

Carrillo, Braulio, 2/123

Cars. *See* Automobiles

Cartagena (Colombia), 2/102

Cartels, Drug, 2/100–101

Carter, Jimmy, 3/176, 3/213

Carthage, 6/48

Carthaginians
    Balearic Islands, 5/217
    Corsica, 2/255
    Italy, 3/226
    Libya, 4/67, 4/70
    Portugal, 5/94
    Spain, 5/212
    Tunisia, 6/48

Cartier, George, 2/31

Cartier, Jacques, 2/30, **2/30**

Casa de las Américas, 2/139

Casa Rosada (Buenos Aires), *1/60,* 1/61

Casablanca (Morocco), 4/160, 4/161, 4/167

Casablanca Valley (Chile), *2/57*

Casamance (Senegal), 5/152

Casbahs, 4/166, 4/168, 4/169, *4/169*

Cascades d'Ouzoud waterfall, 4/168

Cashews, 4/172

Cashmere, 3/158

Casimir the Great, 5/76

Casinos, *2/87,* 4/148–149

Caspian Sea, 1/4, 1/66, 4/8, 5/242
    Azerbaijan on, 1/110, 1/111, *1/111*
    Iran on, 3/180
    Volga River and, 5/118, 5/119

Cassava, 5/37

Caste system of India, 3/140

Castellana (Madrid), 5/210

Castile (Spain), 5/94, 5/204, 5/213
    kingdom of, 5/214

Castilian Spanish, 2/104, 5/204

Castillo de San Marcos (St. Augustine), *7/103*

Castle Geyser, *6/133*

Castle Hill. *See* Hradčany (Prague)

Castle Hill (Budapest), 3/114

Castle of St. Peter (Turkey), *6/72–73*

Castries (St. Lucia), *5/135*

Castro, Fidel, 2/132–133, *2/133,* 2/135, 2/138

Castro, Raúl, 2/133, 2/135

Catacombs, 3/231

Catalan language, 5/205

Catalonia (Spain), 5/202, *5/203,* 5/205, 5/206, 5/207, *5/207*

Catamarans, *2/221,* 5/12–13, 6/8–9

Cathedral of Florence (Italy), 3/237

Cathedral of St. John (Warsaw), 5/81

Cathedral Square (Christchurch), *4/221*

Catherine II, the Great, 5/112, 5/113

Catholicism. *See* Roman Catholicism

Catskill Mountains, 6/120

Cattle
    Argentina, 1/55, 1/58–59, *1/58–59,* *1/62,* 1/62
    Australia, 1/78, *1/79,* 1/80
    Brazil, *1/168–169,* 1/174
    Burkina Faso, 1/186
    Canada, 2/37
    Chad, 2/44
    Channel Islands, 6/100
    Denmark, *2/158*
    Great Zimbabwe, 7/101
    Guinea, *3/88*
    India, 3/134, *3/140*
    Kenya, *4/14*
    Madagascar, 4/84
    Malaysia, *4/97*
    New Zealand, 4/218
    Nigeria, *4/240*
    Romania, *5/105*
    Swaziland, 5/241
    Switzerland, 5/262
    Tanzania, 6/22
    Tunisia, *6/53*
    United States, 6/122, 6/126, 6/154, *6/155*
    Venezuela, 6/176

Cauca River, 2/102
Caucasus Mountains, 1/110, 1/111
Caudillos, 6/173
Cave of Ghar Dalam (Malta), 4/115
Caverna de Pedra Pintada, 7/84
Caviar, 5/119
Cayenne (French Guiana), *2/257*
Cayman Brac (Cayman Islands), 3/48
Cayman Islands, 3/48, *3/48*, **3/48**
Caymans, *6/177*
CCM. *See* Chama Cha Mapinduzi party
Ceauşescu, Nicolae, 5/102
Cebu (Philippines), 5/61, 5/69
Cebu Island (Philippines), 5/61, 5/69
Celebes. *See* Sulawesi (Indonesia)
Celebes Sea, 5/60
Celtic language, 6/101
Celts
    Britain, 6/90, 6/104, 6/110, 7/50
    Czech Republic, 2/144
    France, 2/238, 2/240
    Ireland, 3/204
    Portugal, 5/94
    Spain, 5/212
Center Party (Finland), 2/224
Central African Empire, 2/43
Central African Republic, 2/42–43,
    2/42–43, **2/43**
Central America, 5/32–33, **5/33**
    See also Human migrations; *and specific
    countries*
Central Intelligence Agency, 2/133
Central Railroad (Peru), 5/50–51, *5/50–
    51*, **5/51**
Central Siberian Plateau, 5/125
Central Valley (Chile), 2/48, 2/49, 2/52–
    54, 2/57
Central Valley (United States), 6/128
CERN, *5/265*
Cerro Tololo Inter-American
    Observatory, 2/48
Cerros de Celaque mountains, 3/102
Cervantes, Miguel de, 5/210, 5/215
Cetshwayo, *5/192*, 5/193
Cevapcici, 5/165
Cewa people. *See* Chewa people
Ceylon. *See* Sri Lanka
Chaco. *See* Gran Chaco
Chaco War, 1/154, 5/42
Chad, 2/44–47, *2/44–47*, **2/45**, 4/67,
    4/112
    history, 2/44–45
    land and people, 2/44, 2/46–47,
    2/46–47
Chad, Lake, 2/20, 2/44, 2/46, *2/47*,
    4/234
Chad Basin, 4/240
Chadors, 3/176
Chagga people, 6/22
Chagos Archipelago, 4/124
Chagres River, 5/32
Chahar Aimak people, 1/15
Chain Bridge (Budapest), *3/115*
Chakmas tribe, 1/124
Chalong Bay, *6/33*

Chama Cha Mapinduzi party, 6/20
Chamber of the Sacred Relics (Istanbul),
    6/63
Chamorro, Violeta Barrios de, 4/227,
    4/231
Chamorro people, 3/77, 3/78
Champa kingdom, 6/180
Champagne (France), 2/248
Champasak (Laos), 4/40
Champlain, Samuel de, 2/30, *2/31*, 2/36
Champs Élysées (Paris), 2/246, 2/247
Chan Chan, 5/45
Chandni Chowk (India), *3/144*
Changamire Empire, 6/210, 7/100
Chang'an (China), 2/80
Changgo drums, 4/30, 4/31
Channel Islands, **6/100**, 6/100–101,
    6/100–101
Channel Tunnel, 2/245, 6/89, 6/113,
    6/113
Chao Phraya River, 6/27, 6/30, 6/34,
    6/35
Charlemagne, 3/227, 4/75
    France and, 2/233, *2/240*, 2/241,
    2/255
    Germany and, 3/20, 3/34, 3/38, *3/38–
    39*
Charles I (king of England), 6/111
Charles I (king of Spain), 3/228
Charles II (king of England), 6/111
Charles III (king of Spain), 5/38
Charles IV (Capetian king), 2/241
Charles IV (Holy Roman emperor), 2/148,
    2/150, 4/78
Charles V (Holy Roman emperor), 4/115
Charles VII (king of France), 2/241
Charles VIII (king of France), 3/228
Charles XIII (king of Sweden), 4/247
Charles XIV (king of Sweden), 5/243
Charles Bridge (Prague), *2/150–151*
Charles Francis University, 1/103
Charles Martel, 2/240, 5/151
Charles the Bald, 2/241
Charles the Great. *See* Charlemagne
Charles University, 2/150
Charlotte Amalie (U.S. Virgin Islands),
    6/191, *6/191*
Charrúa Indians, 6/160
Chars (islands), 1/122
Charter 77, 2/153
Charter of Rights and Freedoms
    (Canada), *2/32*
Chartered cities (Philippines), 5/58
Château d'Azay-le-Rideau (France), *2/242*
Château de Blois, *2/243*
Château de Chambord (France), *2/242–
    243*, 2/243
Château de Chaumont (France), *2/242*
Château de Chenonceau (France), *2/242*
Château de Villandry (France), *2/242*,
    2/243
Château d'Ussé (France), *2/242*
Châteaux, 2/233
    Loire Valley, 2/242–243, *2/242–243*,
    2/243, 2/252

Chaudhry, Mahendra, 2/220
Chaves, Federico, 5/40
Chávez, Carlos, 4/145
Chavez Frías, Hugo, 6/174–175
Chavín Indians, 5/44, 7/83
Chechen people, 5/116
Chechnya, 5/111
Chedis, *6/38*
Cheese, 2/249, *2/249*, 3/60, 4/204,
    5/262, 5/263
Cheju Island (South Korea), 4/26
Chelsea Flower Show, 6/109
Chen Shui-bian, 6/13
Chenab River, 5/18, 5/22
Chengdu (China), *2/80*
Chenla kingdom, 2/8
Chennai (India), 3/144
Chernenko, Konstantin U., 5/115
Chernobyl nuclear power plant, 1/131
Cherry blossoms, 3/253, *3/280–281*
Cherry trees, Black, 6/149
Cherskiy range, 5/125
Chetniks, 5/158
Chewa people, 4/90
Cheyenne Indians, 6/141
Chi River, 6/30
Chiang Ching-kuo, 6/10, 6/13, 6/15
Chiang Kai-shek, 2/64, *2/64*, 6/10, 6/12,
    6/15, *6/15*
Chiang Kai-shek Memorial Hall, *6/13*
Chiang Mai (Thailand), *6/37*
Chianti wine, 3/239
Chiapas (Mexico), 4/135, 4/137
Chiapas Highlands (Mexico), 4/130,
    4/130
Chiba (Japan), 3/270, 3/272
Chibcha Indians, 2/98–99, 2/106–107
Chicago (Illinois), 6/123, *6/123*, 6/142,
    6/158, *6/158–159*
Chichewa language, 4/90–91
Chichicastenango (Guatemala), 3/85
Chihuahua (Mexico), *4/130–131*
Chile, 2/48–57, *2/48–57*, **2/51**
    Bolivia, war with, 1/153
    early settlement, 7/70, **7/86**, 7/86–87,
    7/87
    Easter Island, power in, 2/171
    economy, 2/56–57, *2/56–57*
    history, 2/48–51
    land and climate, 2/48, 2/52–53,
    2/52–53
    people, 2/54–55, *2/54–55*
Chili peppers, 6/154
Chiluba, Frederick, 6/202
Chilwa, Lake, 4/91
Chimanimani Mountains, *6/206–207*
Chimu Indians, 5/44, 5/45
Chin people, 4/174, 4/178
China, 1/2, 2/58–97, *2/58–97*, **2/61**
    cities, 2/80–81, *2/80–81*
    economy
        agriculture, 2/70–71, 2/73, 2/90–
        91, *2/90–91*, **2/91**
        industry, 2/92–93, *2/92–93*,
        2/93

120

government, 2/59, 2/60
history
    Australia, emigration to, 7/107
    cultural, 2/66–67, *2/66–67*
    Great Wall, **2/68,** 2/68–69,
        2/68–69
    imperial China, **2/62,** 2/62–63,
        2/62–63
    Khyber Pass, 1/18
    modern, **2/64,** 2/64–65, *2/64–65,*
        6/15, **6/15**
    Mongol Empire, 4/150
    Myanmar settlement, 4/178
    United States, migration to, 7/105,
        7/105
Ladakh, claim on, 4/193
land regions
    eastern, 2/72–73, *2/72–73,* **2/73**
    western, 2/70–71, *2/70–71,* **2/71**
people
    Han, 2/76–77, *2/76–77,* **2/77**
    minority groups, 2/78–79, *2/78–79,*
        2/79
political regions
    Hong Kong, 2/63, **2/84,** 2/84–85,
        2/84–85
    Macau, 2/86–87, *2/86–87,* **2/87**
population, 2/59, 2/76, **2/77**
power in
    Kyrgyzstan, 4/38
    Macau, 5/87
    Vietnam, 6/180
prehistoric human settlement, 7/54,
    7/56–59, *7/56–59,* **7/57, 7/59**
relations with
    Brunei, 1/178
    Japan, 2/63, 2/64, 3/253, 3/258,
        3/264
    Korea, North, 4/18–19
    Korea, pre-partition, 4/23, 4/30
    Malaysia, 4/102
    Portugal, 2/87
    Soviet Union, 4/151
    Taiwan, 6/10, 6/11, 6/12–13
    United Kingdom, 2/62–63
    United States, 2/63
    Vietnam, 6/182, 6/185
Trans-Siberian Railroad and, 5/128–129
wildlife, 2/74–75, *2/74–75,* **2/75**
China Central Television, *2/83*
China World Trade Center Tower 3, *2/83*
Chinatown (New York), *6/139*
Chinchilla, Laura, 2/123
Chinese-Japanese War, 3/264
Chinese exclusion acts (U.S.), 7/105
Chinese language, 2/76
    Brunei, 1/179
    Malaysia, 4/98
    Singapore, 5/176
    Taiwan, 6/14
Chinese people
    Australia, 7/107
    Brunei, 1/179
    Cambodia, 2/8
    French Polynesia, 2/259

Hong Kong, 2/84
Indonesia, 3/166
Jamaica, 3/246
Japan, 3/272, 3/274
Korea, South, 4/27
Lesser Antilles, 4/60
Macau, 2/86
Malaysia, 4/93, 4/98–99, 4/102
Mauritius, 4/125
Mongolia, 4/154
Papua New Guinea, 5/34
Philippines, 5/62
Singapore, 5/174, 5/176, 5/177
Taiwan, 6/14
Thailand, 6/31
Trinidad and Tobago, 6/47
United Kingdom, 6/105
United States, 6/138, *6/139,* 7/105,
    7/105
Vietnam, 6/180, 6/182
See also China
Chinese Teahouse, *3/39*
ChiShona language, 6/207, 6/210
Chissano, Joaquim, 4/171
Chitlada Palace (Bangkok), 6/35
Chitral (Pakistan), 5/18
Chittagong Hills (Bangladesh), 1/124
Chitwan National Park (Nepal), 4/185
Chobe National Park, *1/159*
Chocó Indians, 5/30
Chocolate, 1/143, 3/235, 5/262,
    5/262
Chogyal, 3/152
Choluteca River, 3/102
Chongmyo Shrine (South Korea), *4/30*
Chongqing (China), 2/89
Chopin, Frédéric, 5/73, *5/76,* 5/216
Chorotega Indians, 2/120
Chorus Line, A (musical), 6/159
Chosen Soren, 3/274
Chosica (Peru), 5/50
Choson, 4/22, 4/23
Chotts, 6/53
Christchurch (New Zealand), 4/217,
    4/221, 4/222
Christian IV (king of Denmark), 2/161,
    4/254
Christian Democratic Union (Germany),
    3/18, 3/23
Christian Democrats (Italy), 3/222
Christianity
    Angola, 1/39
    Australia, 1/91
    Azerbaijan, 1/111
    Bosnia-Herzegovina, 1/156
    Brunei, 1/179
    Chad, 2/44, 2/47
    Cook Islands, 2/118
    Côte d'Ivoire, 2/128
    Egypt, 2/186–187, 2/190, 2/192
    Georgia, 3/14
    Germany, 3/35
    Greece, 3/70–71, *3/70–71*
    Hungary, 3/106
    India, 3/141

Indonesia, 3/166
Ireland, 3/201, 3/204
Islam, struggle with, 5/151, **5/151**
Israel, role of, 3/209, 3/216, 3/217
Jamaica, 3/246
Japan, 3/264
Lebanon, 4/50–52
Madagascar, 4/86
Mauritius, 4/125
Morocco, 4/158
Mozambique, 4/173
Nauru, 4/182
Nigeria, 4/243, *4/243*
Pacific Islands, 5/12
Philippines, 5/57, 5/63
Portugal, 5/94
Rome, Ancient, 3/231
Russia, 5/109, 5/117
Samoa, 1/32
Serbia, 5/164
Singapore, 5/176
Spain, 5/206, 5/210, 5/213
Sri Lanka, 5/227
Sudan, 5/228, 5/230
Sweden, 5/243
Syria, 5/266, 5/267, 5/272, 5/273
Togo, 6/40
Tonga, 6/44
Uganda, 6/81
Ukraine, 6/82
United States, 6/156
Zambia, 6/205
See also Eastern Orthodox Churches;
    Missionaries; Protestantism;
    Roman Catholicism
Christiansborg Castle (Denmark), 2/161
Christiansted (St. Croix), 6/191
Christmas Island, 1/89
Chromite, 1/23
Chubut River, *1/54–55*
Chukchi people, 7/75
Chulalongkorn. *See* Rama V
Chun Doo Hwan, 4/24–25
Chunnel. *See* Channel Tunnel
Chuquicamata mine (Chile), 2/56
Church of England, 6/111
Churchill, Sir Winston, 4/167, *5/115,*
    6/113
Churún River, *6/177*
Chuuk Islands, 2/214–215
Chuvash people, 5/116
Cibao (Dominican Republic), 2/166
Ciboney Indians, 4/58
Cid, El, *5/212,* 5/213
Cienfuegos (Cuba), 2/134, 2/137
Cimbri tribe, 3/19, 3/34
Cinema. *See* Motion pictures
Circus of Diocletian, *3/230–231*
CIS. *See* Commonwealth of Independent
    States
Cisterns, 6/63
Citadel (Azerbaijan), 1/110
Citadel (Quebec), 2/30, 2/36
Cité Berlaymont (Brussels), *1/143*
Cité Souterrain (Montreal), 2/36

CITES program, 3/150
City of the Dead (Cairo), 2/191
City Palace (India), *3/136–137*
City-states
  Greece, 3/62, 3/65
  Italy, 3/221, 3/228, 3/236
  Lebanon, 4/50
  Nigeria, 4/236, 4/245
  Sumer, 3/196–197
Ciudad Guayana (Venezuela), 6/177
Ciudadela (Minorca), *5/216*
Civic Democratic Party (Czech Republic),
  2/147
Civic Forum (Czech Republic), 2/146
Civic Platform (Poland), 5/75
Civil rights, in United States, 6/136,
  6/144, 6/156
Civil War, American, 1/114, 6/121,
  6/124, 6/134, *6/135,* 6/136, 7/104
Clans in Scotland, 6/96
Clapton, Eric, 3/251
Clark, Helen, 4/217
Clark Air Base (Philippines), 5/58
Class structure, in Hawaii, 7/95
Classic Period (Mexico), 4/140, *4/140*
Claudel, Paul, 2/253
Claus, Hugo, 1/133
Clement V (pope), *2/252*
Cleopatra VII, 2/192
Cleveland (Ohio), 6/123
Cliff Palace (Colorado), *6/126–127*
Climate, and human migration, 7/8
  See also Ice Age
Clontarf (Ireland), 3/204
Closely Watched Trains (film), *2/152,*
  2/153
Clothing industry, 3/136, 3/240, *3/240,*
  3/241
Cloves, 4/84
Clovis, 2/240
Clovis culture, 7/68, 7/68–69, *7/68–69*
  Amazon region, 7/84–85
  animal extinctions and, 7/72–73
  migrations pre-dating, 7/70–71, **7/71,**
    7/71, 7/85–86
Clovis points, 7/68–69, *7/68–69,* 7/72,
  7/84–85
Clown fish, 4/106, *4/107*
Cluny (Paris), 2/240
CN Tower (Toronto), *2/37*
Coal
  Austria, 1/105
  Belgium, 1/142
  China, 2/92
  India, 3/136–137
  Norway, *4/257*
  Poland, *5/83*
  United Kingdom, 6/93, 6/95, 6/106
  United States, 6/150
  Zimbabwe, 6/207
Coalition government, 2/224
Coast Mountains (Canada), 2/26, 2/37
Coastal Range, 2/196
Coca, 2/103, *2/103*
Cocaine, 2/100–101, 2/103

Cochin (India), *3/156*
Cochin China (Vietnam), 6/181, 6/184
Cockatoos, 1/74
Cockpit Country (Jamaica), 3/247, 3/248
Cocoa, *5/135*
Cocody, University of, 2/129
Coconuts, *3/104, 3/170,* 4/172, *5/171,*
  6/77
  Maldives, 4/104, 4/105
  Puerto Rico, 5/96
Cocos de mer, *5/171*
Cocos Islands, 1/89, 2/124–125, *2/125*
Cocos tectonic plate, *4/132*
Cocullo (Italy), 3/233
Coffee
  Brazil, 1/162, 1/165, 1/169, 1/174,
    1/175
  Colombia, 2/103, *2/103*
  Costa Rica, 2/120, *2/121*
  Côte d'Ivoire, *2/128–129,* 2/129
  Cuba, 2/136
  El Salvador, 2/196, 2/198
  Equatorial Guinea, 2/200
  Guatemala, 3/82
  Haiti, 3/99
  Honduras, 3/104
  Italy, 3/235
  Kenya, 4/15
  Laos, 4/44
  Mexico, *4/135*
  Nicaragua, *4/230–231*
  Rwanda, 5/130
  Tanzania, *6/22–23*
  Yemen, 6/197
Cognac, 1/66
Coimbra (Portugal), *5/88,* 5/91
Coir, 3/156
Colbert, Jean Baptiste, 2/244
Cold War, 3/22, 5/246, 6/136
Cole (warship), 6/201
Collective farms, 1/23, 5/93
Collectives, in China, 2/91
Collectivization, 4/42
Collor de Mello, Fernando, 1/163, 1/176
Cologne (Germany), 3/35
Colombia, 2/98–107, *2/98–107,* **2/101**
  economy, 2/103, *2/103*
  government, 2/100
  history, 2/98–101
    El Dorado myth, 2/106–107,
      2/106–107
  land and climate, 2/102, *2/102–103*
  Panama, power in, 5/29, 5/32
  people, 2/104–105, *2/104–105*
  United States, relations with, 2/101
  U.S. immigrants, 6/143
Colombian people, in Venezuela, 6/178
Colombo (Sri Lanka), *5/225,* 5/227
Colón (Panama), 5/31
Colón Theater (Argentina), 1/51, *1/56*
Colonia (Uruguay), 6/160
Colonias, 4/138
Colony of the Straits Settlements, 4/102
Colorado (United States), 6/126, *6/126–*
  127, 6/127

Colorado Party (Paraguay), 5/41
Colorado Party (Uruguay), 6/162, 6/163
Colorado River, 6/127, *6/150*
Coloreds. *See* Coloureds
Colosseum (Rome), 3/221, *3/226,* 3/230,
  3/230
Colossus of Rhodes, *3/69*
Coloureds
  Namibia, 4/181
  South Africa, 5/189, 5/193, 5/195
  Zimbabwe, 6/207
Coltan, 2/113
Columbia (South Carolina), 6/124
Columbus, Christopher, 1/164, 3/161,
  3/242, 5/199, 5/214, 6/134, 7/102,
  7/103
  voyages
    Antigua and Barbuda, 1/48
    Bahamas, 1/114
    Costa Rica, 2/123
    Cuba, 2/132
    Dominica, 2/164, 2/165
    Dominican Republic, 2/167
    Greater Antilles, 3/48
    Grenada, 3/74
    Guadeloupe, 3/76
    Guyana, 3/96
    Honduras, 3/100, 3/102
    Jamaica, 3/246
    Lesser Antilles, 4/59, 4/60
    Martinique, 4/118
    Nicaragua, 4/226
    Puerto Rico, 5/97
    St. Kitts and Nevis, 5/132
    St. Vincent and the Grenadines,
      5/136
    St. Lucia, 5/134
    Suriname, 5/238
    Trinidad and Tobago, 6/46
    Venezuela, 6/172
    Virgin Islands, U.S., 6/190
Column of King Sigismund (Warsaw),
  5/80
Combites, 3/99
Comercio Square (Lisbon), *5/90*
Comercio Square (Puerto Rico), 5/84
Comet, as cause of extinctions, 7/73
Comino Island, 4/114
Cominotto Island, 4/114
Comité de Organización Política
  Electoral Independiente, 6/173
Commission for Gender Equality, 5/191
Commission for the Promotion and
  Protection of Cultural, Religious
  and Linguistic Communities, 5/191
Common law, 6/90
Commonwealth Games, *4/223*
Commonwealth of Independent States
  Armenia in, 1/66
  Azerbaijan in, 1/110
  Belarus in, 1/130, 1/131
  Kazakhstan in, 4/8–9
  Russia in, 5/109, 5/110
  Turkmenistan in, 6/75
  Uzbekistan in, 6/166, 6/167

Commonwealth of Nations, 6/89, 6/105, 6/112
  Australia in, 1/70
  Belize in, 1/144
  Canada in, 2/24
  Fiji in, 2/221
  Gambia in, 3/12
  Ghana in, 3/42
  Greater Antilles in, 3/49
  Jamaica in, 3/249
  Nauru in, 4/183
  New Zealand in, 4/215
  Papua New Guinea in, 5/35
  Tonga in, 6/42
  Tuvalu in, 6/77
Commonwealth of the United States, 3/78, 5/96
Communes, *2/90–91,* 2/91
Communications industry, 6/152
Communism
  Afghanistan, 1/12
  Albania, 1/20–21, 1/23
  Angola, 1/36–37
  Azerbaijan, 1/111
  Bulgaria, 1/180, 1/184
  Cambodia, 2/9, 2/12
  China, 2/59, 2/60, 2/64–65, 2/68, 2/76
    agricultural policies, 2/90, 2/91
    Beijing, 2/82, 2/83
    Hong Kong, 2/85
    industrial policies, 2/92, 2/93
    Taiwan, 6/10, 6/12, 6/13, 6/15
    Tibet, 2/97
  Colombia, 2/100
  Congo, Republic of the, 2/117
  Croatia, 2/130–131
  Cuba, 2/133, 2/136, 2/138
  Czech Republic, 2/145, 2/146, 2/149, 2/152, 2/153
  Germany, 3/22, 3/36
  Grenada, 3/75
  Hungary, 3/106, 3/108, 3/109, 3/111
  Italy, 3/222
  Korea, North, 4/18, 4/21
  Kosovo, 4/33
  Laos, 4/41, 4/42, 4/47
  Latvia, 4/48–49
  Lithuania, 4/77
  Macedonia, 4/83
  Moldavia, 4/146, 4/147
  Mongolia, 4/151, 4/152
  Mozambique, 4/170, 4/173
  Myanmar, 4/176
  Poland, 5/72–75, 5/77, 5/82
  Roma and, 3/116
  Romania, 5/102
  Slovakia, 5/178
  Slovenia, 5/182–183
  Soviet Union, 5/109
    history, 5/110, 5/114–115, 6/136
    way of life, 5/116–117
  Ukraine, 6/82
  Vietnam, 6/181, 6/182, 6/183, 6/184–185, 6/187

  Yemen, 6/199
  Yugoslavia, 5/155, 5/156, 5/158–159, 5/162
  See also Socialism
Communist Party
  Albania, 1/21
  Bulgaria, 1/180
  China, 2/59, 2/64, 2/78, 2/93
  Croatia, 2/130–131
  Cuba, 2/135, 2/138, 2/139
  Czechoslovakia, 2/146, 2/149
  Hungary, 3/108, 3/109, 3/111
  Japan, 3/254
  Korea, North, 4/21
  Mongolia, 4/152
  Poland, 5/74, 5/75
  Romania, 5/102
  Russia/Soviet Union, 5/114, 5/117
  Yugoslavia, 5/158–159, 5/162, 5/182
Comodoro Rivadavia (Argentina), *1/63*
Comorian language, 2/109
Comoros, 2/108–109, *2/108–109,* **2/109**
Compaoré, Blaise, 1/187
Compounds (Nigeria), 4/242
Computers, *6/152*
Conakry (Guinea), 3/88, 3/89, *3/89*
Concepcion. *See* Grenada
Concepción (Chile), 2/49, 2/52, 2/56
Concepcion Island, 3/74
Concertgebouw Orchestra, Royal, 4/206
Condé, Alpha, 3/87
Confederacy (United States), 6/124, 6/134
Confucianism
  China, 2/59, 2/62, 2/66, 2/67
  Korea, 4/21, 4/22, 4/30, 4/31, *4/31*
  Malaysia, 4/98
Confucius, 2/62, *2/63,* 4/30
Congo, Democratic Republic of the, 1/38, 2/110–115, **2/111,** *2/112–115*
  Congo River in, 2/114, **2/115**
  economy, 2/112–113, *2/112–113*
  government, 2/111
  people, 2/112, *2/112–113*
  Red Cross, *5/264*
  relations with
    Rwanda, 5/131
    Uganda, 6/79
Congo, Republic of the, 2/110, 2/116–117, *2/116–117,* **2/117**
Congo Free State, 2/110
Congo River, 1/38, 2/110, 2/114–115, *2/114–115,* **2/115,** 2/116
Congo River Basin, 2/116
Congress (United States), 6/116, 6/132
Congress Building (Buenos Aires), 1/61, *1/61*
Congress of Berlin, 4/157
Congress of Tucumán, 1/65
Congress of Vienna, 3/21, 5/258
Connecticut (United States), 6/120
Connemara (Ireland), *3/207*

Conquistadors, 2/49, 2/106, 2/172, 5/53, 5/214, 6/172
  See also Spain
Conservation
  China, 2/75, **2/75**
  Ecuador, 2/177
  Himalaya, 4/195
  India, 3/150–151, *3/150–151*
  Indonesia, 3/172–173
  Netherlands, 4/200
  Slovakia dam, criticism of, 5/181
  Sri Lanka, 5/223
  Tanzania, 6/24
  See also Endangered species
Conservative Party (Colombia), 2/99, 2/100
Conservative Party (France), 2/250
Conservative Party (United Kingdom), 6/107
Constance, Lake, 1/100, 1/107, 5/257
Constantine (Algeria), 1/30
Constantine I, the Great, 6/68, 6/170
Constantine II (king of Greece), 3/52
Constantine VII (Byzantine emperor), 5/164
Constantinople, 3/63, 3/242, 6/62, 6/68
  See also Istanbul (Turkey)
Constitution
  British, 6/90
  Canadian, 2/25
Constitution, United States, 6/116–117, 6/121, 6/135
  14th Amendment, 6/144
  15th Amendment, 6/144
Constitution Act of 1982 (Canada), 2/24–25, 2/31, *2/32*
Constitution Plaza (Mexico City), 4/138
Constitutional Act of 1791 (Canada), 2/30
Constitutional Convention (United States), *6/134*
Construction industry, 6/152–153, *6/153*
Consumer goods, 6/153
Conté, Lansana, 3/87
Continental drift, 1/46
Continents, 1/4
Contorni, 3/235
Contras, 4/230–231
Convention on International Trade in Endangered Species of Wild Fauna and Flora, 3/150–151
Convicts. *See* Prisoner settlements
Cook, James
  Australia voyages, 1/86, **1/87,** 7/100
  Pacific Islands voyages, 2/119, 2/220, 4/212, 5/12, 6/42, 6/169, 7/95
Cook, Mount. *See* Aoraki/Mount Cook
Cook Island Christian Church, 2/118, *2/119*
Cook Islands, 2/118–119, *2/118–119,* **2/119,** 5/11
Cooper, James Fenimore, 6/158
Copacabana Beach (Rio de Janeiro), 1/161, 1/170, *1/171*

Copán (Honduras), 3/100, *3/100–101*
Copenhagen (Denmark), *2/157,* 2/158, *2/160–161, 2/160–161,* 2/213
Copley, John Singleton, 6/158
Copper
    Chile, 2/50, 2/56
    Congo, 2/113
    Papua New Guinea, 5/37
    Philippines, 5/69
    production, *2/56–57*
    Uganda, 6/80
    Zambia, 6/203, *6/203*
Copper Mine Point (British Virgin Islands), 6/189
Copra, 5/27, 6/77
Coptic Christianity, 2/186–187, 2/190
Coral and coral reefs, 4/106, *4/106–107*
    American Samoa, 1/32
    Bermuda, 1/148
    Cuba, 2/134
    Fiji, 2/219
    Indonesia, 3/165
    Kiribati, 4/16
    Maldives, 4/104
    Marshall Islands, 4/116, *4/117*
    Netherlands Antilles, 4/211
    Papua New Guinea, 5/36–37
    Seychelles, 5/170, *5/170*
    See also Atolls; Great Barrier Reef
Coral Sea Islands, 1/89
Corbett National Park, 3/150, *3/151*
Corcyra Island, *3/55*
Cordillera Central (Hispaniola), 2/166, *2/167*
Cordillera Central (Puerto Rico), *5/96–97*
Cordillera Septentrional (Hispaniola), 2/166
Cordilleras, 2/122, 2/125
Córdoba (Argentina), 1/63
Córdoba (Spain), *5/213*
Córdoba Mountains, 1/54
Core-and-flake tools, 7/60
Corinth Canal (Greece), 3/54, *3/61*
Corisco (Equatorial Guinea), 2/200
Cork, 5/92, *5/93*
Cork (Ireland), 3/201
Cormorants, 2/179
Corn
    Bulgaria, *1/183,* 1/185
    Cambodia, 2/8
    early Indians, 7/77, 7/78, 7/80, 7/81
    Honduras, 3/104
    Kenya, 4/15
    Mexico, 4/134
    United States, *6/122,* 6/154, *6/155*
    Zambia, 6/205
    Zimbabwe, *6/206, 6/207,* 6/209
Cornaro Chapel, *3/237*
Corneille, Pierre, 2/252
Coro (Venezuela), 6/172
Corobici people, 2/120
Correa, Rafael, 2/173
Corridos, 4/145
Corroborees, *1/76–77*
Corsairs, 1/24, 4/158

Corsica (France), 2/237, 2/238, **2/254,** 2/254–255, *2/254–255*
Corsican dialect, 3/233, 3/237
Corte (Corsica), *2/254–255*
Cortes (Spain), 5/200
Cortés, Hernán, 4/138, 4/140, *4/141,* 5/214, 6/172, 7/102
Cossacks, 5/126
Costa (Ecuador), 2/176
Costa Blanca (Spain), 5/203
Costa Brava (Spain), 5/203, *5/203*
Costa del Sol (Spain), 5/203
Costa Rica, 2/120–123, *2/120–123,* **2/123**
    economy, 2/120–121, *2/121*
        tourism, **2/124,** 2/124–125, *2/124–125*
    government, 2/122–123
    history, 2/120, 2/123
    land and climate, 2/122, **2/123**
Costa Smeralda (Italy), 3/245
Côte d'Ivoire, 1/187, 2/126–129, **2/127,** *2/127–129*
    economy, *2/128–129,* 2/129
    history, 2/126–127
    land, 2/126
    people, 2/128–129, *2/128–129*
Cotton
    Benin, 1/147
    Egypt, 2/184
    Mali, 4/111
    Pakistan, *5/22–23*
    Sudan, *5/231,* 5/232, *5/232*
    Syria, 5/270
    Turkey, 6/64–65
    United States, 6/124, 6/126, 6/154
    Uzbekistan, 6/167
Council of Europe, 5/264, 6/57
Council of Guardians (Iran), 3/176
Council of States (Switzerland), 5/256
Country music, 6/159
Couperin, François, 2/252
Courantyne River, 3/94
Couscous, 4/165
Cozumel (Mexico), 4/133
Crafts
    Haiti, 3/99
    Inca, *5/53*
    Indus Valley, 7/20
    Mexico, 4/134, *4/145*
    Navajo, *6/141*
    Turkey, 6/65
    United States, 6/158
    Yemen, 6/200
Cranes (birds), 2/74
    black-throated, *2/75*
Crassus, Marcus, 3/226
Crater Lake National Park (United States), 6/133
Credit Suisse Bank (Switzerland), *5/264–265*
Creole language
    Guyana, 3/95
    Martinique, *4/119*
    Mauritius, 4/125

Seychelles, 5/170
Creoles
    Belize, 1/145
    French Guiana, 2/257
    Mauritius, 4/124
    Mexico, 4/140
    São Tomé and Príncipe, 5/143
    Sierra Leone, 5/172
    Suriname, 5/239
Crete (Greece), *3/56–57, 3/62, 3/63, 3/68–69, 3/68–69,* 3/242
Cricket, *1/90–91,* 3/246
Crime, Organized, 3/223, 3/275
Crinoids, *4/106–107*
Crnjanski, Miloš, 5/165
Cro-Magnon rock shelter, 7/48
Cro-Magnons, 7/42, 7/48–49, *7/48–49*
Croatia, 2/130–131, *2/131,* **2/131**
    relations with
        Bosnia-Herzegovina, 1/156, 1/157
        Hungary, 2/130
        Serbia, 5/155, 5/159, 5/160, 5/164
        Slovenia, 5/183
Croatian language, 1/156, 2/130
Croats, 1/156, 2/130, 5/155
Crocodiles, *1/74–75*
Crocuses, *1/35*
Cromwell, Oliver, 6/111, *6/111*
Crusades
    Egypt, 2/192
    Jordan, 3/282
    Lebanon, 4/51
    Spain, 5/213
    Syria, 5/267
    Turkey, 6/68
Cuango River, 1/38
Cuanza River, 1/38
Cuauhtémoc, 4/140
Cuba, 2/132–139, *2/132–139,* **2/135,** 3/48, 4/58
    economy, 2/137, *2/137*
    government, 2/135
    history, 2/132–133, 7/103
    land and climate, 2/134, *2/135*
    music, 3/250
    people, *2/136, 2/136*
        daily life, 2/138–139, *2/138–139*
    relations with
        Angola, 1/36–37
        Grenada, 3/75
        Nicaragua, 4/231
        United States, 2/132–133
Cuban Americans, 6/142, 6/143
Cuban Institute of Cinematographic Art, 2/139
Cuban missile crisis, 2/133
Cubas Grau, Raúl, 5/41
Cucumbers, *3/195*
Cuddle Springs (Australia), 7/28
Cuenca (Ecuador), *2/175*
Cukorova (Turkey), 6/65, 6/67
Cultural pluralism, 6/139
Cultural Revolution, 2/65, 2/87
Culture and Science, Palace of (Warsaw), *5/80*

Culture Plaza (Costa Rica), 2/124
Cumbia, La (dance), 2/104
Cuna Indians, 5/30, *5/31*
Cuneiform writing, 3/186, 3/196
Curaçao (Netherlands Antilles), *4/58,* 4/60, *4/60, 4/210,* 4/210–211
Curragh boats, *3/204*
Cusco (Peru), 5/49, 5/52
Cyclades Islands (Greece), 3/66
Cyclones
Bangladesh, 1/120, *1/126,* 1/126–127
Eline, in Mozambique, 4/171
Nargis, 4/177
Tonga, 6/42
Vanuatu, 6/169
Cypress trees, *3/224*
Cyprus, 2/140–143, **2/141,** *2/141–143,* 3/242
history, 2/140–141, 3/52
land, 2/142, *2/142*
tourism, 2/142–143, *2/142–143*
Cyrenaica, 4/67
Cyrillic alphabet, 1/181, 4/155, 5/164
Cyrus the Great, 3/182, 3/184
Czars, 2/222–223, 5/109, 5/112–113, 5/126
Czech people, 2/144, 2/145, 5/178
Czech Republic, 2/144–155, *2/144–155,* **2/147**
history, 2/144–145, 2/148–149, *2/148–149*
art and politics, 2/152–153, *2/152–153*
modern state, 2/144, 2/146–147
Prague, 2/150–151, *2/150–151*
land and economy, 2/154–155, *2/154–155*
people, 2/145
prehistoric human migration, 7/44
See also Czechoslovakia
Czechoslovak People's Party, 2/146
Czechoslovak Socialist Party, 2/146
Czechoslovakia, 2/144, 2/145
art and politics, 2/152–153, *2/152–153*
dissolution, 2/146–147, 5/178
history, 2/148–149, *2/149*
origin, 5/178
See also Czech Republic; Slovakia
Czestochowa (Poland), *5/75*

D

Dacia, 5/106
Dacian people, 5/106
Dacko, David, 2/43
Dadiwan site, 7/56–57
Da Gama, Vasco, 3/242, 5/85, 5/95
Dagestan, 5/111
Dahomey, 1/146
Dai Co Viet, 6/180
Dai Viet, 6/180–181
Dáil (Ireland), 3/202
Daimyos, 3/266, 3/267
Daisenin Temple (Kyoto), 3/281
Dakar (Senegal), *4/113,* 5/152, *5/153*
Dal bhatu, 4/197

Dalai Lama, 2/97
Dalarna (Sweden), 5/249
Dalí, Salvador, 5/211
Dalian (China), 2/75, *2/92*
Dalmatia, *2/131,* 3/242
Dama dance, *4/112–113,* 4/113
Daman (India), 3/155
Damara people, 4/181
Damascus (Syria), 5/267, 5/271, *5/273,* 5/274–275, *5/274–275*
Damnoen Saduak market (Thailand), *6/34*
Damrak (Amsterdam), 4/206
Dams, 4/199, 4/208, 4/209, 5/119
See also Hydroelectric power
Dan Fodio, Usuman, 4/236
Dance
by group
Ami people, *6/14*
Australian Aborigines, *1/77*
Roma, 3/117
by place
Africa, 4/112–113, *4/112–113*
Colombia, 2/104
France, *2/239*
Indonesia, 3/168, *3/169*
Korea, 4/31
Mexico, *4/143, 4/144*
Peru, 5/52
Russia, *5/116*
Serbia, 5/165
Spain, 5/206, *5/206–207*
Thailand, *6/35, 6/36*
United States, 6/158
Venezuela, 6/179
Zimbabwe, *6/207*
by type
ballet, *4/143,* 5/116, 6/158
calypso, 3/250
crane dance, *3/268–269*
dama, *4/112–113,* 4/113
flamenco, 3/117, *5/206–207*
lakhon, *6/35*
modern, 6/158
morris, *6/108*
pencak silat, 3/167
reggae, 3/250–251
salsa, 3/250
samba, *1/171*
tango, *1/61*
Dance of the Condors, 5/52
Danes, in Britain, 6/91, 6/104
Danish language, 2/156, 2/213
Danish West Indies, 6/190
Danube Basin, 3/110
Danube Delta, 5/181
Danube River, 3/24, 5/180–181, *5/180–181,* **5/180–181**
Austria, 1/97, 1/102, 1/103, *1/104*
Bulgaria, 1/182
Hungary, 3/106, *3/109,* 3/110, 3/112, 3/114
Romania, 5/104, 5/105
Serbia, 5/154, 5/163, 5/166, *5/166,* 5/167, 5/168, *5/168*

Danubian Lowlands, 5/178
Danubian Plateau, 1/182, 1/183, *1/183*
Daoud, Muhammad, 1/12
Dar es Salaam (Tanzania), 6/20
Darangan Epic, 5/64
Dardanelles Strait, 6/69
Darfur (Sudan), 5/229, 5/231, *5/234–235*
Dari language, 1/15
Darién Mountains, 5/28
Darius I, 3/182
Darjeeling tea, *3/134–135*
Dark Age (Greece), 3/62
Darnah (Libya), 4/68
Darwin (Australia), 1/88
Darwin, Charles, 2/178, 7/88, *7/89*
Dasht-e Kavir Desert, 3/180
Dasht-e Lut Desert, 3/180
Date palms, *1/31,* 3/190, *3/190–191,* 4/67
Daugava River, *4/49*
Davao (Philippines), 5/69
David (king of Israel), 3/216, 3/282
Da Vinci, Leonardo. *See* Leonardo da Vinci
Davis, John, 2/210
Dayak people, 3/166, 4/99
Dead Chest (British Virgin Islands), 6/189
Dead Sea, 3/209, 3/211, 3/286
Deadman Bay (British Virgin Islands), *6/188–189*
Deák, Francis, 3/111
Death of a Salesman (Miller), 6/159
Deccan (India), 3/130, 3/131
Decorative Arts, Museum of (Copenhagen), 2/161
Deep Skull (fossil), 7/23
Deer, *4/185*
chital, *4/185*
musk, 2/75
Père David's, 2/75
sambar, *4/185*
swamp, *4/185*
white-tailed, *6/149*
Defenestration of Prague, The (etching), *2/148*
Defoe, Daniel, 2/53
Deforestation
Australia, 7/28
Brazil, *1/168–169*
Greece, 3/55
Haiti, *2/97*
Himalaya, 4/194–195, *4/195*
India, 3/150
Indonesia, 3/172–173, *3/172–173*
Madagascar, 4/84, 4/89
Malaysia, 4/100
New Zealand, 7/97
Paraguay, 5/43
Sri Lanka, 5/223
Zambia, 6/204
See also Forests and forest products
Deganya settlement, 3/218
Degas, Edgar, *2/253, 2/253*
De Gaulle, Charles, 2/245, *2/245*
De Klerk, Frederik Willem, 5/190, 5/193, *5/193*

De la Rua, Fernando, 1/53
Delaware (United States), 6/124
De Leon Carpio, Ramiro, 3/81, 3/83
Delft (Netherlands), 4/204, 4/205
Delgado, José Matías, 2/194
Delhi (India), 3/141, 3/144, 3/148
Delhi Sultanate, 5/15
Delian League, 3/65
Delphi (Greece), 3/63
Delta Project (Netherlands), 4/199, 4/208
Demerara River, 3/94, 3/95
Demilitarized zone (Korea), 4/19, 4/23
Demirel, Süleyman, 6/58
Democracy, Greece and, 3/51, 3/64
Democratic Constitutional Rally Party, 6/49, 6/50
Democratic Forum (Hungary), 3/109
Democratic Labor Party (Barbados), 1/129
Democratic Left (Ireland), 3/202
Democratic Left Alliance (Poland), 5/75
Democratic National Salvation Front (Romania), 5/103
Democratic Party (Albania), 1/21
Democratic Party (Serbia), 5/156
Democratic Party of Japan, 3/254, 3/255
Democratic Party of Serbia, 5/156, 5/161
Democratic Party of Socialists (Montenegro), 4/157
Democratic People's Republic of Korea. See Korea, North
Democratic Progressive Party (Taiwan), 6/13
Democratic Republic of the Congo. See Congo, Democratic Republic of the
Democratic Republican Party (South Korea), 4/24
Democratic Union (Poland), 5/75
Denbigh people, 7/75
Deng Xiaoping, 2/60, 2/65
Denisovan people, 7/17, 7/17
Denmark, 2/156–161, 2/156–161, 2/157
   climate, 2/159
   history, 2/157, 2/160
   land and economy, 2/158–159, 2/158–159, 2/159
   people, 2/156
   power in
      Faroe Islands, 2/213
      Ghana, 3/44
      Greenland, 3/72
      Iceland, 3/119
      Lesser Antilles, 4/59
      Norway, 4/247
      Virgin Islands, U.S., 6/190–191
   Sweden, relations with, 5/242–243
Denpasar (Indonesia), 3/164
D'Entrecasteaux Islands (Papua New Guinea), 5/36
Deodar trees, 5/18
Deodoro da Fonseca, Manoel, 1/165
Dependencies, 1/2

Depression, Great. See Great Depression
Dervishes, 6/69
Desalination, 6/86
Desemparados Station (Peru), 5/50
Desertas (Madeira Islands), 1/113
Desertification, 4/120
Desmans, Pyrenean, 1/35
Dessalines, Jean Jacques, 3/96
Desuaguadero River, 5/54
Detroit (Michigan), 6/123
Deuba, Sher Bahadur, 4/187
Devils Island (French Guiana), 2/256
Dhilos (Greece), 3/66, 3/66–67
Dhlakama, Afonso, 4/173
Dhofar (Oman), 5/8
Dhotis, 3/148
Dhows, 6/197
Diaguita Indians, 2/48
Diamonds
   Angola, 1/38, 1/38–39
   Belgium, 1/143
   Central African Republic, 2/42
   Congo, 2/113
   Israel, 3/211
   Lesotho, 4/56
   Namibia, 4/181
   Sierra Leone, 5/172
   South Africa, 5/196, 5/197
Dias, Bartolomeu, 5/85
Díaz, Porfirio, 4/141
Díaz de Solís, Juan, 1/64, 6/160
Dien Bien Phu (Vietnam), 6/184
Dien Bien Phu, Battle of, 6/184
Diet (Japan), 3/254
Dijon (France), 2/251
Dikes, 3/94, 4/208, 4/209
Dilmun civilization, 1/116
Dimitrovgrad (Bulgaria), 1/185
Dinaric Alps, 2/130, 3/54
Dinka people, 5/229, 5/236, 5/237
Dinkins, David, 6/145
Dinosaur National Park (France), 7/48–49
Diocletian, 3/226
Dionysus, Theater of, 3/64
Diori, Hamani, 4/233
Diouf, Abdou, 5/153
Dipendra, 4/187
Dir (Pakistan), 5/18
Dirty war (Argentina), 1/52
Discrimination, 6/139
   African Americans, 6/144–145
   Hispanic Americans, 6/143
Disney, Walt, 6/159
District of Columbia, 6/118
Diu (India), 3/155
Divorce, 3/223, 6/157
Diyarbakir (Turkey), 3/198, 3/199
Djenné Mosque, 4/112–113
Djerma-Songhai language, 4/235
Djerma-Songhai people, 4/234
Djibouti, 2/162–163, 2/163, 2/163
Djibouti (city, Djibouti), 2/162, 2/163
Djurgården Island, 5/251
DNA. See Genetic studies

Dnieper River, 1/130, 6/82
Dniester River, 4/146, 6/82
Doams, 3/146
Dobruja (Romania), 5/105
Dodecanese Islands (Greece), 3/66
Dodoma (Tanzania), 6/20
Doe, Samuel K., 4/65
Doges' Palace (Venice), 3/243
Doggerland, 7/50–51
Dogon people, 4/110, 4/112–113, 4/113
Doha (Qatar), 5/98, 5/99
Dolci (food), 3/235
Dolgoruki, Yuri, 5/120
Dollfuss, Engelbert, 1/98
Dolmabahçe Palace (Turkey), 6/70
Dolmeh, 4/55
Dolmus, 6/63
Dolphins, Chinese, 2/74
Dome of the Rock (Israel), 3/216–217, 3/217
Dominica, 2/164, 2/164–165, 2/165, 3/75, 4/58
Dominican Republic, 2/121, 2/166–167, 2/167, 2/167, 3/48, 3/49, 6/142, 6/143
Dominican Republic-Central American-United States Free Trade Agreement, 2/121, 2/195
Dominion of Canada, 2/31, 2/36
Domtoren (Utrecht), 4/205
Don Quixote (Cervantes), 5/210
Don River prehistoric settlement, 7/44–47, 7/44–47, 7/62
Dongting Lake, 2/74
Donkeys, 4/68–69
Dorian culture, 3/62
Dorset culture, 7/75
Dos Santos, José Eduardo, 1/37
Dostam, Abdul Rashid, 1/13
Douala (Cameroon), 2/21
Douala people, 2/16, 2/18
Douro River, 5/88, 5/89, 5/89, 5/91, 5/92, 5/92–93
Dover (England), 6/92
Dovre Mountains, 4/250
Downs, 6/92
Dracula, 5/105, 5/105
Dragon Fountain (Copenhagen), 2/160–161
Dragon trees, 5/219
Drake, Sir Francis, 6/111
Drakensberg range, 4/56
Drammen (Norway), 4/254
Drašković, Vuk, 5/165
Drava River, 2/130
Dravidian language, 3/138, 3/156
Dravidians, 3/128, 3/138, 3/149, 7/20
Dreadlocks, 3/251
Dreamtime, 1/76, 7/29
Dromedary camels, 2/45
Drought
   Antigua and Barbuda, 1/48–49
   Argentina, 1/51
   Australia, 1/72, 1/79

Eritrea, 2/203
Ethiopia, 2/207–209
India, 3/133
Mali, 4/110
Mauritania, 4/120, 4/122, 4/123
Niger, 4/233, 4/235
Nigeria, 4/240
Senegal, 5/153
Somalia, 5/187
Sudan, 5/230
Zimbabwe, 6/209
Drug trafficking, 1/155, 2/98,
    2/100–101, 3/275, 4/129, 5/28
Drums
Africa, *2/128–129,* 4/112, *4/113*
Cuba, *2/139*
Korea, *4/30,* 4/31
Lesser Antilles, *4/60,* 4/61
Morocco, *4/165*
Druses, 4/54, 4/55, 5/272
Dry gardens, 3/280–281
Dry valleys, *1/43*
Dual Monarchy of Austria-Hungary,
    1/98, 3/111, 3/114
Duarte, José Napoleón, 2/195
Duarte, Juan Pablo, 2/167
Dubai (United Arab Emirates), 6/84, 6/85,
    6/85, 6/86–87, 6/87
Dubček, Alexander, 2/149, 2/153, *2/153*
Dublin (Ireland), 3/201, 3/205
Dublin Bay (Ireland), 3/201
Duchy of Warsaw, 5/80
Ducks, 2/90
mallard, *2/39*
Duhalde, Eduardo, 1/53, 1/65
Duma, 5/113, 5/114
Dunant, Jean Henri, 5/264
Dunav River. *See* Danube River
Dunay River. *See* Danube River
Dungan people, 4/39
Dunkerque (France), 2/251
Durban (South Africa), 5/196, *5/196*
Dürer, Albrecht, 3/39
Durgapur (India), 3/136
Durham Miners' Gala, 6/108
Durrani, Ahmad Shah, 1/11
Durrani people, 1/14
Dürrenmatt, Friedrich, 5/255
Dušan, Stefan, 5/158
Dussehra festival, *3/141*
Dutch language
Belgium, 1/133, 1/136, 1/137, 1/140,
    1/141
Lesser Antilles, 4/60
Netherlands, 4/202
Suriname, 5/239
Dutch people
Brazil, 1/167, 1/172
South Africa, 5/194
United States, 6/120
See also Netherlands, The
Duvalier, François, 3/97, *3/97*
Duvalier, Jean-Claude, 3/97
Dyak people. *See* Dayak people
Dynamite, 5/247

Dyukhtai Culture, 7/63
Dzongkha language, 1/151

E

EAC. *See* East African Community
Eagles
imperial, *1/35*
Verreaux's, *6/25*
Earth
physical map, 1/4, **1/4–5**
political map, 1/2, **1/2–3**
population, 1/2
surface area, 1/4
Earthquakes
Armenia, 1/66
Azores, 1/112
Chile, 2/52
Cyprus, 2/143
El Salvador, 2/195
Guadeloupe, 3/77
Haiti, 3/97, *3/98–99*
India, 3/127
Indonesia, 3/163
Italy, 3/224
Japan, 3/255, 3/256, *3/257,* 3/270
Mexico, 4/132, *4/132,* **4/133**
Mongolia, 4/150
New Zealand, 4/217
Nicaragua, 4/230
Philippines, 5/57, 5/60
Portugal, 5/88, 5/90
Turkey, 6/59, 6/61
East African Community, 5/131, 6/79
East Cameroon, 2/19
East Chicago (Indiana), 6/150–151
East China Sea, 2/81, 2/88, 2/89
East Falkland Island, 2/210, 2/211
East Germany, 3/17, 3/22–23, 3/32,
    3/36
East India Company
British, 1/119, 3/129, **3/129,** 4/178,
    4/186, 5/15, 5/174
Dutch, 4/178, 5/192
French, 5/100
East Kalimantan (Indonesia), 3/170
East Pakistan, 1/119, 1/120, 1/124,
    1/126, 3/157, 5/15, 5/16
East Roman Empire, 3/226
    See also Byzantine Empire
East Sea, 4/22
East Siberian Highlands, 5/125
East Slavic people, 1/130
East Timor, 2/168–169, *2/169,* **2/169,**
    3/163, 7/24
Easter Island, 2/52, 2/53, 2/170–171,
    2/170–171, **2/171,** 7/95, *7/95*
Easter Rebellion (Ireland), *3/204,* 3/205
Eastern Ghats range, 3/131
Eastern Orthodox Churches
Albania, 1/22
Cyprus, 2/143
Greece, 3/69
Kosovo, 4/32, 5/157
Lebanon, 4/54
Moldova, 4/147

Romania, *5/107*
Russia, 5/109, 5/117, *5/117*
Serbia, 5/164, *5/164*
Ukraine, 6/83
United States, 6/156
Eastern Slavs, 5/109
Eastern woodlands agriculture, 7/76–77
Ebla, 5/266
Ebro River, 5/202
Ecevit, Bülent, 6/59
Echidnas, 1/74
Economic crisis (2007-)
France, 2/245
Greece, 3/53, 3/60
Iceland, 6/119
Ireland, 3/201
Italy, 3/241
Portugal, 5/87, 5/92
Spain, 5/208
United Kingdom, 6/107
Uruguay, 6/163
Ecosystems, 4/106
Ecuador, 2/172–179, *2/172–179,* **2/175**
history, 2/99, 2/172–173
land and economy, 2/174, 2/176–177,
    2/176–177
people, 2/174–175, *2/177–178*
U.S. immigration, 6/143
Ecumenical Centre (Nigeria), *4/243*
Edam (Netherlands), 4/204
Edge Island (Norway), 4/256
Edict of Nantes, 2/244
Edinburgh (Scotland), 6/96–97, 6/97
Edinburgh Castle, 6/96–97
Edinburgh International Festival, 6/109
Edmonton (Alberta), 2/37
Edo (Japan), 3/270
Edo people, 4/243, 4/245
Edomite people, 3/282
Education
Cambodia, 2/13
Cameroon, 2/17
China, 2/76
Cuba, *2/137,* 2/138
El Salvador, 2/199
Finland, 2/224–225
France, 2/235
India, 3/139
Indonesia, 3/167
Israel, 3/211, 3/215
Kenya, 4/12
Laos, 4/47
Morocco, 4/161
New Zealand, 4/220
Nicaragua, 4/230
Niger, 4/232
Philippines, 5/63
Russia, 5/117
Singapore, 5/177
Sweden, 5/245
Tunisia, 6/51
Zambia, 6/205
See also specific places
Eduskunta, 2/224
Edward I (king of England), 6/110

Edward II (king of England), 6/110
Edward, Lake, 6/80
Efate (Vanuatu), 6/168
Efik people, 4/243
Egbert (king of Wessex), 6/110
Egede, Hans, 3/72
Egmond aan Zee resort (Netherlands), 4/199
Egrets, *4/185*
Egypt, 2/180–193, *2/180–193,* **2/183**
  economy, 2/182
    agriculture, **2/184,** 2/184–185, *2/184–185*
  government, 2/182–183
  history, 2/182–183, 2/188–189, 2/192–193, *2/192–193*
  land, 2/181
  people, 2/181, 2/186–187, *2/186–187*
  population, 2/181, 2/191
  power in
    Lebanon, 4/51
    Sudan, 5/229
    Yemen, 6/198
  relations with
    Israel, 3/212, 3/213
    Sudan, 5/229
    Syria, 5/268
Egypt, Ancient, 2/181, 2/192, *2/193*
  human migration, *7/32,* **7/32,** 7/32–33, 7/40–41, *7/40–41*
  power in
    Lebanon, 4/50
    Sudan, 5/228–229
Ehrenström, Johan Albrekt, 2/228
Eid al-Fitr, *3/178*
Eiffel, Alexandre Gustave, 2/246
Eiffel Tower (Paris), 2/246, *2/246*
Eisenhower, Dwight D., *6/137*
Eisteddfod festival, 6/94, *6/95,* 6/108
Ejido system of agriculture, 4/134
Elahi, Chaudhri Fazal, 5/16
Elamite people, 3/175, 3/178, 3/182
Elbe River, 2/154, 2/155, 3/24, 3/30, 3/106
Elburz Mountains, 3/178, 3/180
El Cedral site, 7/78
El Chichón, 4/133
Elcibey, Abulfaz, 1/111
El-Dahab, Abdul Rahman Suwar, 5/230
El Dorado (legend), 2/99, 2/106–107, *2/106–107*
Electoral Commission (South Africa), 5/191
Electrical Town (Japan), *3/270*
Electronics, 1/184, *3/270,* 4/100, 6/10, *6/152*
Elephant birds, *4/88,* 4/89
Elephants, *1/159,* 4/185, 6/24, *6/25,* 6/31, *6/32,* 7/20–21
El Escorial (Spain), 5/199
Elfstedentocht ice skating, 4/199
El Gezira (Sudan), *5/231,* 5/232, 5/233
Elgin, Lord, 3/59
Elgin Bridge, *5/175*
Elgin Marbles, 3/59

Elgon, Mount, 6/80
El Greco. *See* Greco, El
Elite (Panama), 5/30
Elizabeth I (queen of England), 3/204, 6/111, *6/111,* 6/188
Elizabeth II (queen of United Kingdom), *2/32*
Elk (Poland), *5/78–79*
Ellice Islands. *See* Tuvalu
Ellis Island (New York), *6/138*
El Moro Fortress (Puerto Rico), 5/97
ELN (Colombia), 2/100, 2/101
El Obeid (Sudan), 5/234
Elobey Chico (Equatorial Guinea), 2/200
Elobey Grande (Equatorial Guinea), 2/200
El Rinconcillo restaurant, *5/204*
El Rocío (Spain), 5/207
El Salvador, 2/194–199, *2/194–199,* **2/197**
  history, 2/194–195, *2/194–195*
    civil war, *2/194,* 2/195, 2/197, *2/198*
  land and economy, 2/196–197, *2/196–197, 2/199*
  people, *2/194,* 2/198–199, *2/198–199*
El Salvador, Church of (Canary Islands), 5/220
El Teniente mine (Chile), 2/56
Embraer aircraft plant, *1/176–177*
Emerald Buddha, Temple of the (Bangkok), 6/35
Emeralds, 2/99, 2/103
Emigration, from Ireland, 3/201
Emilian dialect, 3/233
Emirates
  Bahrain, 1/116
  Kuwait, 4/34
  Qatar, 5/99
  United Arab Emirates, 6/84
Emmentaler cheese, 5/262, *5/263*
Empty Quarter. *See* Rub al Khali
Ems River, 3/24
Emus, 1/74, *1/75,* 1/79
Enchanted Isles, 2/178
Endangered species
  Australia, 1/75
  China, 2/74–75
  Ecuador, 2/177
  Madagascar, 4/89
  Nepal, 4/185
  Sri Lanka, 5/223
  tigers, 3/150–151, *3/150–151*
  whales, 4/253
Enewetak (Marshall Islands), 4/117
Engel, Carl Ludwig, 2/228–229
England, **6/92,** 6/92–93, *6/92–93*
  Danish invasion, 2/157
  early history, 6/110–111, *6/110–111*
  economy, 6/93
  land, 6/92–6/93, *6/92–6/93*
  Norman invasion, 2/238
  See also United Kingdom
English Channel, 2/236, 2/238, 6/89, 6/92, 6/100
  See also Channel Tunnel

English Harbour (Antigua), *1/49, 4/59*
English language
  Australia, 1/69, 1/90–91
  Belize, 1/145
  Brunei, 1/179
  Cameroon, 2/16
  Canada, 2/23, 2/30–31
  Channel Islands, 6/100
  Costa Rica, 2/120
  Ethiopia, 2/209
  Fiji, 2/216
  Ghana, 3/42
  Grenada, 3/74
  Guam, 3/78
  Guyana, 3/95
  Hispanic Americans, 6/143
  Honduras, 3/103
  India, 3/139
  Iraq, 3/192
  Ireland, 3/202–203
  Jamaica, 3/246
  Jordan, 3/287
  Kiribati, 4/16
  Lebanon, 4/54
  Lesser Antilles, 4/60
  Libya, 4/70
  Madagascar, 4/86
  Malawi, 4/90–91
  Mauritius, 4/125
  Mozambique, 4/173
  Nauru, 4/182
  New Zealand, 4/220, 4/223
  Nigeria, 4/243
  Pakistan, 5/20
  Panama, 5/30
  Philippines, 5/62, 5/63, 5/71
  St. Vincent and the Grenadines, 5/137
  Samoa, 5/138
  Sierra Leone, 5/172
  Singapore, 5/176, 5/177
  Solomon Islands, 5/184–185
  Sudan, 5/234
  Sweden, 5/244
  Tanzania, 6/22
  Tonga, 6/45
  Trinidad and Tobago, 6/47
  Uganda, 6/81
  United Kingdom, 6/90
  United States, 6/139
  Wales, 6/94
  Zambia, 6/205
  Zimbabwe, 6/207
Enlightenment, 2/252
Ennahda party, 6/50
Ennedi Plateau, *2/45*
Enns River (Austria), 1/102
Enosis, 2/140
Enriquillo, Lake, 2/166
Ensenada (Mexico), 4/143
Entre Ríos. *See* Mesopotamia region (Argentina)
EOKA, 2/140
Ephemeral plants, 4/72
Ephesus (Turkey), *6/67,* 6/68, 6/70, *6/70*
Epirus, 3/63

Equatorial Guinea, 2/200–201, *2/201,* **2/201**
Erbakan, Necmettin, 6/58–59
Erdogan, Recep Tayyip, 6/59
Erebus, Mount, *1/42,* 1/43
Ergs, 1/28, 4/72
Erik the Red, 3/72
Eritrea, 2/202–203, *2/203,* **2/203,** 2/206
Eritrean People's Liberation Front, 2/202
Erosion, 1/118, 4/73, 4/89, *4/181, 4/194, 4/194*
Erzberg (Austria), *1/102,* 1/103
Erzurum (Turkey), 6/70
Esbjerg (Denmark), *2/159*
Esdraelon plain (Israel), 3/209
Eshkaft-e Gavi site, 7/38
Eskimos. *See* Inuit; Paleo-Eskimos
Espresso, 3/235
Esquiline Hill (Rome), 3/230
Essaouira (Morocco), *4/162–163*
Essential oils, 5/101
Essequibo River, 3/94
Estancias, 1/58, 1/59, 1/62
Estates-General (France), 2/244
Estonia, 2/204–205, *2/204–205,* **2/205**
Estonian people, 2/204
Estrada, Joseph, 5/71
ETA, 5/201
Ethanol, *1/177*
Ethiopia, 2/206–209, *2/206–209,* **2/207**
  economy, 2/207
  history, 2/206–207
    famine, 2/209, *2/209*
  human fossils, 7/6
  land and people, 2/208–209, *2/208–209*
  Rastafarian movement in Jamaica, 3/246, 3/251
  relations with
    Djibouti, 2/162
    Eritrea, 2/202–203
    Somalia, 5/187
Ethiopian Orthodox Church, *2/208,* 2/209
Ethiopian People's Revolutionary Democratic Front, 2/206
Ethiopian Plateau, 2/208
Etna, Mount, 3/224, 3/225, 3/244
Etruscans, 2/234, 2/254, 3/226, 3/238
EU. *See* European Union
Eucalyptus trees, 5/241
Eucalyptuses, 1/74–75, 2/177
Euphrates River, 3/190–191, 3/196, 5/270, 6/61, 7/38
Eurafricans, 5/241
Eurasia, 1/4, 7/42, 7/44
  See also Asia; Europe
Eurasian Plate, 3/120
Euro
  Estonia and, 2/205
  Finland and, 2/224, 2/230
  France and, 2/234, 2/245, 2/251
  Germany and, 3/23

Italy and, 3/323
Portugal and, 5/92
Euronext, *4/202–203*
Europe, 1/2, 1/4
  See also Human migrations; *and specific countries*
European Court of Auditors, 4/81
European Court of Justice, 4/79, *4/79,* 4/81
European Economic Community. *See* European Union
European Free Trade Association, 5/264
European Investment Bank, 4/81
European Organization for Nuclear Research, *5/265*
European Parliament, 4/81
European people
  Antigua and Barbuda, 1/48
  Argentina, 1/56
  Australia, 1/90
  Bahamas, 1/114
  Barbados, 1/129
  Bermuda, 1/148
  Brazil, 1/166–167
  Canada, 2/29, 2/30
  Chile, 2/54
  Colombia, 2/104
  Ecuador, 2/174
  Jamaica, 3/246
  Mali, 4/111
  Mozambique, 4/173
  Namibia, 4/181
  Native American relations with, 6/140
  New Caledonia, 4/212
  New Zealand, 4/223
  Panama, 5/30
  Papua New Guinea, 5/34
  Swaziland, 5/241
  Trinidad and Tobago, 6/47
  Tunisia, 6/54
  Venezuela, 6/178
  See also specific European countries
European Union
  Austria in, 1/95
  Belgium in, 1/133, 1/140, 1/143
  Bulgaria in, 1/181
  Croatia and, 2/131
  Cyprus in, 2/141
  Czech Republic in, 2/147
  Denmark in, 2/157
  Estonia in, 2/205
  Finland in, 2/224, 2/230
  France in, 2/234, 2/245, 2/251
  Germany in, 3/23
  Greece in, 3/53, 3/60
  Hungary in, 3/113
  Ireland in, 3/202
  Italy in, 3/223, 3/240
  Latvia in, 4/49
  Lithuania in, 4/77
  Luxembourg in, 4/79
  Malta in, 4/115
  Norway and, 4/249
  Poland in, 5/75
  Portugal in, 5/87, 5/92

Romania in, 5/103
Serbia and, 5/161
Slovakia in, 5/178
Slovenia in, 5/183
Spain in, 5/199, 5/208
Sweden in, 5/244
Switzerland and, 4/264
United Kingdom in, 4/216, 6/107, 6/112
Eurozone. *See* Euro
Euskara language, 2/238, 5/205
Evangelical Lutheran Church, 2/156
Evans, Sir Arthur, 3/68
Evenki people, 5/126
Everest, Mount. *See* Mount Everest
Evolution, *2/178,* 4/88
Evvoia (Greece), 3/66
Ewe people, 3/42, 6/40
Ewondo people, 2/16
Executive Yuan, 6/12
Exeter (England), *6/108*
Exploration
  by Portugal, 5/85, 5/90, 5/95
  by Spain, 5/214, 7/102
  See also Human migrations
Export-processing zones, 5/69
Extended families, 2/129
Extinction, 1/75, 4/89
  prehistoric humans as cause of, 7/28–29, *7/29,* 7/72–73, *7/73*
  See also specific animals
Exxon Mobil Corporation, 2/47
Exxon Valdez, 6/131
Eyadéma, Gnassingbé, 6/41
Eyre, Lake, *1/73*
Eysturoy (Faroe Islands), 2/212

F
Fado, 5/91
Fagg, Bernard, 4/244
Fair American (schooner), 7/94
Fairbanks (Alaska), 6/131
Fairmont Château Frontenac (Quebec), 2/36
Faisal (king of Saudi Arabia), 5/145
Faisal I (king of Iraq), 3/187
Falconry, 5/99
Falkland, Viscount, 2/210
Falkland Islands, 1/52, 1/65, **2/210,** *2/210–211, 2/210–211*
Fall Line (United States), 6/124
Falmouth Harbor (Antigua and Barbuda), *4/59*
Famagusta (Cyprus), 2/143
Family life, in United States, 6/157, *6/157*
Famine, 1/126, 2/44, 5/187
  Ethiopia, 2/207, 2/209, *2/209*
Fan Si Pan Mountain, 6/186
Fang language, 2/200
Fang people, 2/16, 2/200, 2/201, 3/8
Fantasias (Morocco), *4/164–165*
Fante people, 3/42
Faqih, 3/176

Farabundo Martí National Liberation Front, 2/197
Farafangana (Madagascar), 4/85
FARC (Colombia), 2/100, 2/101
Farm cooperatives, 2/136
Faroe Islands, 2/212–211, *2/212–211,* **2/213**
Faroese language, 2/213
Farsi language, 1/15, 1/116, 3/178, 6/17
Faruk, 2/193
Fascism, 3/229, 5/200
Fatih Sultan Mehmet Bridge, 6/62
Fatima (daughter of Muhammad), 2/188, 5/151
Fatimids, 2/188
Fault lines, 3/120
Favelas, 1/162, 1/170–171, *1/172*
Faya-Largeau (Chad), *2/46–47*
Fazendas, 1/169
Fdérik (Mauritania), 4/123
Fearn, John, 4/182
Feast of the Holy Savior of the World, 2/199
February Revolution, 5/109, 5/114, 5/122
Federal District of Mexico, 4/138
Federal Republic of Germany. *See* West Germany
Federal Territory of Malaysia, 4/96
Federated States of Micronesia, 2/214–215, *2/215,* **2/215**
Federation of Malaya, 4/94
Federation of Mali, 4/109
Feldberg peak, 3/25
Fellahin, 2/185, *2/185,* 2/187
Fencing, 3/106
Fennecs, 4/73
Fens, 6/92
Ferdinand III (king of Spain), *5/212,* 5/214
Ferdinand V (king of Spain), 3/116
Fergana Valley (Uzbekistan), 6/167
Feria, 2/172, 5/206–207, *5/206–207*
Ferlach (Austria), 1/101
Fernández de Kirchner, Cristina, 1/53, 1/65
Fernández de Lizardi, José Joaquin, 4/145
Fernandino people, 2/200
Ferns, *1/44*
Ferris wheel, in Prater, 1/109
Fertile Crescent, 5/266, 5/270, 7/38, 7/40
Fertile Triangle (China), 2/89
Festival of Lights, 6/105
Festival of the Ages (Japan), *3/266–267*
Fethiye (Turkey), 6/73, *6/73–74*
Fethiye, Gulf of, 6/72
Feudalism, 2/240–241, 3/267, 3/275
Fez (Morocco), 4/165, 4/166, *4/166–67,* 4/168
Fez, Treaty of, 4/159
Fezes, 4/165

Fezzan, 4/67
Fianna Fáil, 3/202
Fiefs, 2/240
Fierro, José Hernández, 1/58–59
Fiestas, 4/143, 5/206–207, *5/206–207*
Fifth Republic (France), 2/234, 2/245
Fiji, 2/216–221, **2/217,** *2/217–221,* 5/10, 5/11
  economy, 2/216–217
  government, 2/217
  history, 2/217, 2/220–221
  land, 2/218–219, *2/218–219*
  people, 2/216, *2/217, 2/220,* 2/220–221
Fijian language, 2/216
Filfla Island, 4/114
Filipino language, 5/62
Filipinos, 5/57
  See also Philippines
Films. *See* Motion pictures
Finance. *See* Banking and finance
Financial crisis. *See* Economic crisis (2007-)
Fincas, 2/196
Finches, 2/178, *2/178*
  snow, *1/35*
Fine Arts, Museum of (Brussels), 1/141
Fine Arts, Palace of (Mexico City), 4/138, 4/142, *4/143*
Fine Gael, 3/202
Finland, 2/222–231, *2/222–231,* **2/225,** 5/243
  economy, 2/230–231, *2/230–231*
  government, 2/224
  history, 2/222–223
  land and climate, 2/226–227, *2/226–227*
  people, 2/222, 2/225
Finland, Gulf of, 2/204, 2/222, 2/226, 2/228, 5/122
Finlandia Hall, 2/229, *2/229*
Finnish language, 2/225
Fiords, 2/26, 3/72, 4/246, *4/250–251,* 4/251, 4/257, 6/130
Fire, Human use of, 7/9, 7/28–29, 7/96, 7/97
  See also Deforestation
Fire walking, *2/220–221*
Firemen's Ball, The (film), 2/152
Firestone Company, 4/65
First Balkan War, 1/20, 4/82, 4/157, 5/157, 5/158
First Fleet, **7/107**
First Nations. *See* Native Americans
First-past-the-post electoral system, 4/217
First Triumvirate, 3/228
FIS, 1/26, 1/27
Fischer, Jan, 2/147
Fish
  coral reef, 4/106–107
  Great Barrier Reef, 1/92, *1/92*
Fishing
  Alaska, 6/130
  American Samoa, 1/32

  Angola, 1/38, *1/39*
  Bangladesh, *1/122*
  Bermuda, *1/148*
  Canada, 2/34, 2/35, *2/35*
  Canary Islands, 5/221
  Cape Verde, 2/41
  Chad, *2/47*
  Chile, *2/54*
  China, 2/91
  Cuba, 2/137
  Cyprus, *2/142*
  Denmark, *2/159*
  Easter Island, *2/171*
  Egypt, ancient, 7/40
  El Salvador, *2/199*
  Falkland Islands, 2/211
  Faroe Islands, 2/213
  Finland, 2/229
  Germany, *3/31*
  Gibraltar, *3/47*
  Greece, 3/60
  Greenland, 3/72
  Haiti, *3/98–99*
  Iceland, 3/118, 3/119, *3/121*
  Indonesia, 3/170
  Iran, *3/181*
  Ireland, 3/204
  Italy, 3/235, 3/239
  Japan, 3/258, 3/260–261, *3/261*
  Korea, South, 4/26, 4/29
  Madagascar, 4/84
  Madeira Islands, *1/113*
  Malawi, 4/91
  Maldives, 4/105
  Mali, *4/110–111,* 4/111
  Martinique, *4/119*
  Mauritania, *4/122–123,* 4/123
  Morocco, *4/162–163,* 4/163
  Namibia, 4/181
  New Caledonia, *4/213*
  New Zealand, 4/219
  Norway, 4/246, *4/253*
  Pacific Islands, 5/11
  Peru, 5/48
  Philippines, 5/69, *5/69*
  Portugal, 5/90, *5/90,* 5/93, *5/93*
  prehistoric, 7/46, 7/54, 7/59, **7/77**
  Puerto Rico, 5/84
  Romania, *5/105,* 5/107
  Saudi Arabia, 5/149
  Scotland, 6/97
  Sweden, 5/253
  Thailand, 6/33
  Turks Islands, 3/49
  United Arab Emirates, 6/87
  United States, 6/151
  Yemen, 6/197
Five-Year plans (Soviet Union), 5/114, 5/115
Flamenco, 3/117, 5/206, *5/206–207*
Flanders (Belgium), 1/133, 1/134, 1/137, *1/137*
  economy, 1/142, 1/143
  medieval, 1/138–139, *1/138–139*
Flaubert, Gustave, 2/253

Fleming people, 1/112, 1/133, 1/134, 1/136–137, *1/136–137*
Flinders, Matthew, 1/86
Flinders Ranges, *1/73*
FLN (Algeria), 1/25, 1/26
Floods
  Australia, 1/78, 1/79, 1/87
  Bangladesh, 1/118, 1/120, 1/122, *1/123, 1/126,* 1/126–127, **1/127**
  Cambodia, 2/12
  Chile, 2/52
  China, 2/89, 2/90
  Dominican Republic, 2/167
  Egypt, 2/184
  Gambia, 3/11
  Himalaya, *2/194–195, 4/195*
  India, 3/133
  Iraq, 3/190, 3/196
  Mesopotamia, 3/196
  Mozambique, 4/171
  Netherlands, 4/199, 4/209
  Russia, 5/122
  Sudan, 5/232
  Venezuela, 6/178
  Venice, *3/242,* 3/243
  Vietnam, 6/183
Florence (Italy), 3/221, *3/228*
  as city-state, 3/228, 3/236, 3/237
  tourism, 3/241
Flores (Indonesia), 7/25
Florida (United States), 2/138, 6/124, 6/125
Flowers, 1/75, 1/143, *6/101,* 6/109
Flowers, Battle of the, *6/101*
Flute, Ancient, *7/46*
Flycatchers, Asian, *2/75*
Flying buttresses, 2/247
FMLN (El Salvador), 2/197
FNLA (Angola), 1/36–37
Foehn, 5/257
Foix, Count of, 1/34
Folklore District (Sweden), 5/249
Folsom points, 7/68, 7/72
Fon people, 1/147
Fono, 5/139
Fonseca, Gulf of, 3/102
Fontvieille (Monaco), 4/149
Food chain, *1/45*
Food processing, 6/66
Football, *1/32–33,* 6/157
Forbidden City (Beijing), 2/60, *2/63,* 2/82
Ford, Richard, 5/206
Forests and forest products
  Belarus, 1/130
  Canada, 2/26, *2/33,* 2/35, 2/39
  Chile, *2/57*
  China, 2/91
  China bamboo, 2/74
  Czech Republic, 2/154–155
  Finland, 2/230
  France, 2/236
  Gabon, 3/9, *3/10–11, 3/11*
  Germany, 3/24, 3/25
  Greece, 3/55

  Himalaya, 4/191
  Indonesia, 3/170, 3/173
  Korea, South, 4/26
  Malaysia, 4/93, 4/100, *4/101*
  Mexico, 4/134
  Morocco, 4/168
  Myanmar, 4/175
  New Zealand, 4/219
  Norway, 4/253, *4/253*
  Papua New Guinea, 5/37
  Philippines, 5/68
  Poland, 5/78, 5/79
  Portugal, 5/92
  Russia, 5/118
  St. Lucia, 5/134
  Serbia, *5/167*
  Siberia, 5/124
  Spain, *5/202–203*
  Sweden, *5/248,* 5/249, *5/252, 5/253*
  Thailand, 6/31, *6/32,* 6/32–33
  Turkey, 6/64
  United States, 6/151, *6/151*
  See also Deforestation; Rain forests; *and specific places*
Forman, Milos, 2/152
Formentera Island, 5/216, 5/217, *5/217,* **5/217**
Former Yugoslav Republic of Macedonia, 4/83
Formosa. *See* Taiwan
Fort-de-France (Martinique), 4/119
Fort San Lorenzo (Panama), *5/30*
Fort Sumter (South Carolina), *6/135*
Forum of Trajan, 3/230–231
Fos (France), 2/251
Fossils, in Antarctic, 1/44, *1/44,* 1/46
  See also Human migrations; Prehistoric humans
Fountain of Neptune (Rome), *3/230–231*
Fountain of Trevi (Rome), 3/231
Four Apostles (Dürer), 3/39
Fouta Djallon (Guinea), 3/86, 3/88
Fox Quesada, Vicente, 4/129
France, *2/232–255, 2/232–259,* **2/236**
  arts, 2/252–253, *2/252–253*
  economy
    agriculture, 2/248–249, *2/248–249,* **2/249**
    industry, **2/250,** 2/250–251, *2/250–251,* **2/251**
    land, **2/236,** 2/236–237, *2/236–237*
    Paris, 2/247
  government, 2/234, 2/250
  history
    early, 2/240–241, *2/240–241,* **2/241**
    Panama Canal construction, 5/32
    settlement of North America, 7/104, *7/104*
    since 1500, *2/244–245*
  people, 2/238–239, *2/238–239*
  power in
    Algeria, 1/24–25, *1/27*
    Andorra, 1/34

  Benin, 1/146
  Burkina Faso, 1/187
  Cambodia, 2/9
  Cameroon, 2/16, 2/18–19
  Canada, 2/29–32
  Central African Republic, 2/42
  Chad, 2/44, 2/45
  Comoros, 2/108
  Congo, Republic of the, 2/117
  Côte d'Ivoire, 2/126
  Djibouti, 2/162
  Dominica, 2/164
  Egypt, 2/192–193
  French Guiana, 2/256
  Gabon, 3/8–3/10
  Germany, 3/22
  Greater Antilles, 3/48
  Grenada, 3/74
  Guadeloupe, 3/76–77
  Guinea, 3/86
  Guyana, 3/92
  Haiti, 2/166, 2/167, 3/96
  Italy, 3/228–229
  Laos, 4/40, 4/41
  Lebanon, 4/51
  Lesser Antilles, 4/59, 4/60, 4/62, 4/63
  Liechtenstein, 4/75
  Luxembourg, 4/79
  Madagascar, 4/87
  Mali, 4/108–109
  Martinique, 4/118–119
  Mauritania, 4/120
  Mauritius, 4/125
  Mexico, 4/141
  Morocco, 4/158–159
  Netherlands, 4/200
  New Caledonia, 4/212, 4/213
  Niger, 4/232–233
  Pacific Islands, 5/12
  Portugal, 1/164, 5/95
  Réunion, 5/100, 5/101
  Russia, 5/114, 5/121
  St. Kitts and Nevis, 5/132, 5/133
  St. Lucia, 5/134
  St. Vincent and the Grenadines, 5/136
  Senegal, 5/152, 5/153
  Seychelles, 5/170
  Spain, 5/215
  Syria, 5/267, 5/274
  Tahiti, 6/8
  Togo, 6/41
  Tunisia, 6/48, 6/49
  United States, 6/134
  Vanuatu, 6/169
  Vietnam, 6/181, 6/184
  Virgin Islands, U.S., 6/190
  prehistoric humans, 7/48
  relations with
    Britain, 6/112
    Germany, 3/26
    Greece, 3/63
    Monaco, 4/148, 4/149
    Spain, 7/103

131

Switzerland, 5/258
United States, 6/135
Francis I (king of France), 2/242, 2/243
Francis II (Holy Roman emperor), 1/98
Franco, Francisco, 5/199, 5/200, 5/206, 5/211, 5/215, *5/215*
Franco, Itamar, 1/163
Franco-Prussian War, 3/21
Frank, Anne, 4/207
Frankfurt (Germany), 3/28–29, *3/28–29*
Frankfurt Book Fair, 3/28, *3/29*
Frankfurt Cathedral, 3/28
Frankfurt Opera House, *3/28–29*
Frankfurters (sausages), 3/29
Frankincense, 5/8, 6/198, *6/199*
Franklin, Benjamin, 6/135, *6/135*
Franks, 4/202, 5/151, 5/217
France, 2/233, 2/240
Germany, 3/19, 3/20, 3/28, 3/34
Franz Ferdinand, 1/98, 5/158
Franz Joseph, 1/98, *1/99*
Frauenkirche cathedral (Munich), *3/26, 3/27*
Frederick I (king of Germany), 3/20, 3/38
Frederick I (king of Prussia), 3/33
Frederick II, the Great (king of Germany), 3/21, *3/21*, 3/39
Frederick the Wise, 3/38
Frederick William, 3/20, 3/33
Frederiksted (St. Croix), 6/191
Free association (government), 2/119, 4/117, 5/13, 5/27
Free Church of Scotland, 4/91
Free Democrats (Germany), 3/18, 3/23
Free enterprise, 1/142, 2/131
Free ports, 2/85, 5/175
Free University of Brussels, 1/141
Freedoms, in United States, 6/156
Freetown (Sierra Leone), 5/172
Frei Montalva, Eduardo, 2/50
Frelimo, 4/170, 4/171, 4/173
Fremantle, Charles, 1/86, 7/107
French and Indian War, 2/30, 6/112, 7/104
French Cameroon, 2/18
French Canadians, 2/23, 2/30, 2/32
French Community, 2/126
French Equatorial Africa, 2/42, 2/117, 3/9
French Guiana, 2/256, **2/256,** *2/256–257*
French Guinea, 3/86
French Indochina, 4/40, 6/181
French language
Algeria, 1/25
Belgium, 1/133, 1/136, 1/137, 1/140, 1/141
Cameroon, 2/16
Canada, 2/23, 2/30–31
Central African Republic, 2/42
Channel Islands, 6/100
Comoros, 2/109
Congo, 2/112
Côte d'Ivoire, 2/128
Djibouti, 2/163
Equatorial Guinea, 2/200

France, 2/238, 2/239
French Guiana, 2/256, 2/257
Grenada, 3/74
Guinea, 3/88
Haiti, 3/98
Lebanon, 4/54
Luxembourg, 4/80
Madagascar, 4/86
Martinique, *4/119*
Mauritius, 4/124
Monaco, 4/149
Morocco, 4/161, 4/164
Niger, 4/235
St. Lucia, 5/135
St. Vincent and the Grenadines, 5/137
Switzerland, 5/260
Trinidad and Tobago, 6/47
Tunisia, 6/54
French patois language, *4/119*
French people
Italy, 3/232
Puerto Rico, 5/96
Réunion, 5/101
United Kingdom, 6/104
United States, 6/138
See also France
French Polynesia, 2/258, 2/258–259, *2/258–259,* 5/11
See also Tahiti
French Revolution, 2/233, 2/234, 2/244, 2/247, 2/256, 3/22–21, 3/229, 4/148, 5/258
French Riviera, 4/148
French Somaliland, 2/162
French Sudan, 4/109, 5/153
French Territory of the Afars and Issas, 2/162
French Togoland, 6/41
French West Africa, 4/109, 4/232, 5/153
FRETILIN (East Timor), 2/168
Freud, Sigmund, 1/108
Frisch, Max, 5/255
Frisian language, 4/202
Frisian people, 4/202
Friuli-Venezia Giulia (Italy), 3/232
Frobisher, Martin, 3/72
Frolinat (Chad), 2/45
Front de Libération Nationale (Algeria), 1/25, 1/26
Front de Libération Nationale (Chad), 2/45
Front for the Liberation of Angola, 1/36–37
Front for the Liberation of Mozambique. *See* Frelimo
Front for the Liberation of Occupied South Yemen, 6/199
FSM. *See* Federated States of Micronesia
Fua'amotu (Tonga), 6/45
Fuentes, Carlos, 4/145
Fuerteventura (Canary Islands), 5/218, 5/219, *5/219,* 5/220, 5/221
Fuji, Mount, 3/253, 3/256, 3/280
Fujian (China), 6/14
Fujimori, Alberto, 5/47

Fujiwaras family, 3/264
Fukiage Garden (Tokyo), 3/280
Fula people, 3/12
Fulani people
Burkina Faso, 1/187
Cameroon, 2/16, *2/16*
Guinea, 3/88
Guinea-Bissau, 3/90
Mali, 4/110
Mauritania, 4/122
Nigeria, 4/234, 4/235, 4/236, 4/243
Fumaroles, *3/121*
Funafuti (Tuvalu), 6/76
Funan kingdom, 2/8, 6/180
Funchal (Madeira Islands), 1/113
Funj people, 5/228
Fur people, 5/234
Fur trade, 2/31, 2/34, *7/104*
Futas, 6/200

G

Ga people, 3/42
Gabcikovo (Slovakia), 5/181
Gabon, 3/8–11, *3/8–11,* **3/9**
history, 3/8–3/9
land and economy, 3/10–11, *3/10–11*
people, 3/8
Gabriel (angel), 5/150
Gaelic football, 3/207
Gagauz people, 4/147
Gaillard Cut, 5/32
Gairy, Eric M., 3/74–75
Galapagos Islands, 2/174, 2/176, **2/178,** 2/178–179, *2/178–179*
Galdhøpiggen Mountain, 4/250
Galera Tunnel (Peru), 5/51
Galicia (Spain), 5/205
Galician Mountains, 5/202
Galilee (Israel), 3/209
Gallegos, Rómulo, 6/173
Gallen-Kallela, Akseli, 2/222
Galliard Cut, *5/33*
Galway Bay (Ireland), 3/201
Gama, Vasco da. *See* Da Gama, Vasco
Gambia, *3/12,* 3/12–13, **3/13,** 5/152
Gambia River, 3/12, *3/13*
Gambier (French Polynesia), 2/258
Gamelan orchestras, *2/168–169,* 3/169
Gamla Stan (Sweden), 5/250–251, *5/250–251*
Gamsakhurdia, Zviad K., 3/15
Ganda people, 6/78, 6/81
Gandan monastery (Tibet), *2/94–95*
Gandhi, Indira, 3/126, *3/128,* 3/150
Gandhi, Mohandas K., *3/128,* 3/129
Gandhi, Rajiv, 3/126
Ganesh, *3/141,* 5/227
Ganga, 3/142–143
Ganges Delta, 1/118, *1/121,* 1/122
Ganges River, 1/10, 4/194
Bangladesh, 1/118, 1/122
India, 3/142–143, **3/142–143,** *3/142–143,* 3/146–147
Hindu ritual, 3/141, 3/143, 3/146, *3/146*

Gangtok (India), *3/152–153*
Gansu province (China), 2/69
Ganwa people, 1/188
Gaoua region (Burkina Faso), 1/187
Garay, Juan de, 1/61, 1/64
García Lorca, Federico, 5/211
García Moreno, Gabriel, 2/173
García Perez, Alan, 5/47
Garda, Lake, *3/224–225*
Garden City (Cairo), 2/190
Garden of Love, 2/242
Gardens, 3/280–281, *3/280–281*
Garibaldi, Giuseppe, *3/229*
Garifuna, 1/145, 3/103
Garvey, Marcus, 3/246, 3/251
Gaspé Peninsula (Canada), 2/38
Gastarbeiter, 6/65
Gate of Hadrian, 6/73
Gate of Heavenly Peace (Beijing), 2/82
Gateway Arch (St. Louis), *6/149*
Gatun Lake, *5/33*
Gauchos, 1/51, 1/58–59, *1/58–59, 1/62,
*   *6/165*
Gauguin, Paul, 6/9
Gaul, 2/240, 3/226
Gault site, 7/68
Gavi wine, 3/239
Gaza Strip, 3/209, 3/212, 3/213
Gbagbo, Laurent, 2/127
GCC, 6/85
Gdańsk (Poland), 5/74, *5/77*
Geese, *6/149*
    Canada, *2/39*
    snow, *6/149*
Ge'ez language, 2/209
Gehry, Frank, *6/159*
Geisel, Ernesto, 1/174
Gelato, 3/235
Gelderland (Netherlands), 4/201
Gellért Baths (Budapest), *3/114*
General People's Committee (Libya),
    4/66
Genetic studies, 7/11
    African migration, 7/11–15, 7/31,
        7/42, 7/54
    Andaman islanders, 7/16–17
    Arctic people, 7/75
    Australian Aborigines, 7/26
    Austronesians, 7/91
    China and Taiwan, 7/54, 7/58
    Denisovans, 7/17
    European migration, 7/42, 7/45
    Neandertals, 7/17, 7/34–35
    North American migration, 7/67, 7/69
    Tierra del Fuego settlers, 7/88–89
Geneva (Switzerland), 5/255, 5/261,
    5/263, 5/265
Geneva, Lake, 5/257
Geneva Accords, 5/265, 6/184
Geneva Conventions, 5/265
Genghis Khan, 1/11, 3/182, 4/150,
    5/126, 6/74, 6/166
Genoa (Italy), 2/255, 3/242
    as city-state, 3/228, 4/148
Genocide, 1/157, 5/131

Genographic Project, 7/11
Genome, Neandertal, 7/34–35
Gentile da Fabriano, 3/242
George I (king of England), 6/112
George I (king of Greece), 3/63
George V (king of Tonga), *6/43*
George, Lake, 6/80
George Town (Cayman Islands), 3/48
George Town (Malaysia), 4/96
Georgetown (Guyana), *3/93, 3/94*
Georgia (country), 1/111, 3/14–15, *3/15,*
    3/15
Georgia (United States), 6/124, 6/132
Georgian Menshevik Party, 3/14
Geothermal energy, *3/121,* 4/218,
    4/218–219
Ger (Tibet), 2/94
Gerasa (Jordan), *3/286*
German Confederation, 3/21
German Democratic Republic.
    See East Germany
German language
    Belgium, 1/137
    Bohemia, 2/144
    dialects
        Alamannic, 1/100
        High, 3/19, 3/34
        Low, 3/19, 3/34
        Old High, 3/34
        Old Low, 4/202
        Standard, 3/19, 3/34
        Swiss-German, 2/238
    Germany, 3/34
    Luxembourg, 4/80
    Switzerland, 5/260, *5/261*
German Museum (Munich), 3/26
German people
    Belgium, 1/134
    Canada, 2/23
    Czech Republic, 2/148–150
    Italy, 3/232
    Luxembourg, 4/80
    Romania, 5/107, *5/107*
    Russia, 5/116
    Sweden, 5/245
    United States, 6/123, 6/138, 6/139
    See also Germany
Germani people, 3/34
Germanic tribes
    Britain, 6/110
    France, 2/233, 2/240
    Germany, 3/34
    Netherlands, 4/202
    Norway, 4/246
    Portugal, 5/94
    Roman Empire, 3/227
    Slovenia, 5/182
Germany, 3/16–38, *3/16–38, 3/19*
    Alsace-Lorraine and, 2/238
    economy, 3/36–37, *3/36–37,* **3/37**
    historic sites, 3/38–39, *3/38–39*
    history
        before 1945, 3/17, 3/20–21,
            3/20–21, **3/21,** 3/38–39
        reunification, 3/18

        since 1945, 3/17, 3/18, 3/22–23,
            3/22–23
        land and climate, 3/24–25, *3/24–25,*
            3/25
            Danube River, 5/180
        people, 3/18–19, 3/34–35, *3/34–35*
            roots and traditions, 3/34–35,
                3/34–35
        power in
            Albania, 1/20–21
            Algeria, 1/25
            Austria, 1/98–99
            Belarus, 1/131
            Bohemia, 2/148
            Burundi, 1/188
            Cameroon, 2/18
            Czech Republic, 2/148–149,
                2/152–153
            Denmark, 2/157
            Estonia, 2/204
            Federated States of Micronesia,
                2/215
            France, 2/245
            Latvia, 4/48
            Luxembourg, 4/79
            Mariana Islands, 3/79
            Marshall Islands, 4/116
            Namibia, 4/180
            Nauru, 4/182
            Pacific Islands, 5/12
            Palau, 5/26
            Papua New Guinea, 5/34
            Poland, 5/72, 5/77, 5/80–81
            Rwanda, 5/130
            Samoa, 1/32–33, 5/139
            Serbia, 5/158
            Slovenia, 5/182
            Tanzania, 6/18–19
            Togo, 6/41
            Ukraine, 6/82
        prehistoric human settlement, 7/46,
            7/47
        relations with
            Italy, 3/229
            Soviet Union, 5/77
            Sweden, 5/243
            Turkey, 6/65
            United States, 6/136
Gershwin, George, *6/159*
Gesellius, Herman, 2/229
Getreidegasse (Salzburg), *1/106*
Gettysburg (Pennsylvania), 6/121, 6/133
Geysers, *2/53,* 3/118, 3/120, *5/219,*
    6/133
Géza, 3/110
Gezelligheid, 4/200
Ghana, 3/40–45, *3/40–45,* **3/43**
    economy, 3/40, *3/41, 3/41*
    history, 3/44–45, *3/44–45*
    land and climate, 3/40
    people, 3/42–43, *3/45–47*
Ghana Empire, 4/108, 4/120, 5/152
Ghats, 3/146, *3/146*
Gheg people, 1/22
Ghent (Belgium), 1/138, *1/139, 1/142*

Ghetto (Warsaw), 5/80
Ghilzais people, 1/14
Gia Long, 6/181
Giacometti, Alberto, 5/255
Giant Jaguar, Temple of the (Guatemala), *3/80–81*
Gibbons, *3/173*
Gibraltar, 3/46–47, *3/46–47,* **3/47**
    Neandertals, 7/53, *7/53*
Gibraltar, Strait of, 3/46, 4/160, 5/199
Gide, André, 2/253
Gilani, Yousaf Raza, 5/17
Gilani people, 3/178
Gilbert and Ellice Islands Colony, 6/77
Gilbert Islands, 4/16, 4/17, 5/12, 6/77
Gilbertese language, 4/16
Gimie, Mount, 5/134
Ginastera, Alberto, 1/58
Ginnungagap, 3/122
Gion Festival, *3/268–269*
Giorgione, 3/242
Giraffes, *6/25*
Giza (Egypt), 2/181, 7/41
Glacier Bay National Park (Alaska), *6/130*
Glaciers, 7/52, 7/52–53
    Alaska, 6/130, *6/131*
    Antarctica, 1/41, *1/47*
    Denmark, 2/158
    Faroe Islands, 2/212
    Finland, 2/226
    France, 2/237
    Greenland, 3/72
    Iceland, 3/118, 3/120
    Norway, 4/250
    Perito Moreno, *1/54–55*
    Sweden, 5/249
Glåma River, 4/250
Glarus (Switzerland), 5/256
Glasgow (Scotland), 6/97
Glasgow, University of, 7/12
Glasnost, 5/110, 5/117
Glen Canyon Dam, 6/150
Glen Mor (Scotland), 6/97
Global warming, 1/92, **1/127**
Globalism, 1/2
Glorious Revolution (England), 6/111
Glyptothek museum (Munich), 3/27
Gnassingbé, Faure, 6/41
Gnus, *6/24*
Goa (India), **3/154,** 3/154–155, *3/154–155*
Goat's Hole Cave, 7/50
Gobekli Tepe temple, 7/36, 7/37, *7/37*
Gobi, 2/70–72, 4/150, 4/154, *4/154*
Goethe, Johann Wolfgang von, 3/17, 3/28, 3/39
Golan Heights (Israel), 3/212, 5/269
Gold
    Australia, *1/82, 1/83,* 1/87, 1/89, 7/107
    Brazil, 1/172, *1/176–177*
    Canada, 2/35
    Colombia, 2/98, 2/103, 2/106–107, *2/106–107*

    Congo, *2/117*
    Fiji, 2/216–217
    Mali, 4/108
    Mexico, 4/135, 4/140
    Papua New Guinea, 5/37
    Philippines, 5/69, *5/69*
    Serbia, *5/162–163*
    South Africa, 5/192, 5/196, *5/196*
    Swaziland, 5/240
    United States, 6/128
    Zimbabwe, 6/207, 6/211, 7/101, *7/101*
Gold Chapel (Brazil), 1/172
Gold Coast, 3/42, 3/45, 6/41
Gold Museum (Colombia). *See* Museo del Oro (Colombia)
Gold Rush, 6/128
Golden Age
    England, 6/111
    Greece, 3/58, 3/62, 3/65, 3/66
    Netherlands, 4/202
    Portugal, 5/85
    Spain, 5/199, 5/215
Golden Apples of the Hesperides, 5/216
Golden Gate Bridge (United States), *6/128–129*
Golden Horde, 6/166
Golden Horn, 6/57
Golden Pavilion (Kyoto), *3/269,* 3/280
Golden Rock, 4/63
Golden Terraces (Warsaw), *5/81*
Golding, Bruce, 3/249
Golf, *3/277*
Golfe de la Gonâve, 3/96
Golgotha, 3/217
Golubac Castle, 5/166
Gomera (Canary Islands), 5/218, 5/220
Göncz, Arpád, 3/109
Gondolas, *3/234–235*
Gonds, 3/138
González, Felipe, 5/201
González Macchi, Luis, 5/41
Good Friday, 3/233
Good Friday Agreement, 6/98
Gorbachev, Mikhail S., 4/8, 5/74, 5/110, 5/115, 5/116, 6/75
Gore, Al, *5/246*
Goreme (Turkey), 6/60, 6/61
Gorgas, William C., 5/32
Gorham's Cave, 7/53
Gorillas, *2/115*
Gorkha, Kingdom of, 4/186
Gornostayev, Aleksander, 2/229
Gortys (Crete), 3/69
Goshawks, Northern, *2/74*
Gosier (Guadeloupe), 3/77
Göta Canal (Sweden), 5/249
Götaland Plateau (Sweden), 5/249
Gothic architecture, 2/240, 2/247, 2/252, 3/31, 3/242
Goths, 3/19, 3/34, 4/146, 5/106
Gotland Island, *5/248*
Gouda (Netherlands), 4/204
Goulash, 3/107
Goulash Communism, 3/108

Government Palace (Finland), 2/228
Governor general (office), 2/24
Gowon, Yakubu, 4/238–239
Gozo Island, 4/114
Graham, Martha, 6/158
Grampian Mountains (Scotland), 6/97
Gran Canaria (Canary Islands), 5/219, 5/220
Gran Chaco, 1/54–56, 1/154, 5/42, 5/43
Gran Colombia, 2/99, 2/172, 6/173
Gran Hotel Costa Rica, 2/124
Gran Via (Madrid), *5/210–211*
Granada (Nicaragua), 4/226, *4/226–227*
Granada (Spain), 5/199
Granada, Kingdom of, 2/99, 5/213, 5/214
Granada Cathedral (Nicaragua), *4/226–227*
Grand Anse Beach (Grenada), 3/74, *3/75*
Grand Atlas Mountains, 4/162, *4/162–163*
Grand-Bassam (Côte d'Ivoire), 2/126
Grand Bazaar (Istanbul), 6/63
Grand Canal (China), 2/89
Grand Canal (Venice), *3/242–243*
Grand Canyon (United States), 6/127
Grand Cayman (Cayman Islands), 3/48
Grand Ducal Palace (Luxembourg), 4/81
Grand Duchess Charlotte Bridge (Luxembourg), 4/81
Grand Duchy of Lithuania, 1/130
Grand Duchy of Warsaw, 5/76
Grand Erg Occidental (Sahara), 1/28
Grand Erg Oriental (Sahara), 1/28
Grand Harbour (Malta), 4/114
Grand Mosque (Mauritania), *4/121*
Grand Palace (Bangkok), *6/34–35,* 6/35
Grand' Place (Brussels), *1/140, 1/141*
Grand Prix automobile race, 1/167
Grand Teton National Park (United States), 6/133
Grand Theater of Monte Carlo (Monaco), 4/149
Grand Turk (Greater Antilles), 3/49
Grande Comore (Comoros), 2/108, 2/109
Grande-Terre (Guadeloupe), 3/76, 3/77
Grandes Écoles, 2/235
Granma (newspaper), 2/138
Grapes and wine
    Australia, *1/80*
    Austria, *1/105*
    Bulgaria, *1/185*
    Canary Islands, *5/221*
    Chile, *2/57*
    Croatia, *2/131*
    France, *2/237,* 2/248
    Germany, 3/37
    Greece, *3/55,* 3/60
    Italy, 3/238–239
    Macedonia, 4/83
    Madeira, 1/113
    Portugal, 5/88, 5/90, 5/91, 5/92, *5/92–93*
    South Africa, *5/196*
    Spain, *5/208*

Switzerland, *5/260–261*
Uruguay, *6/164*
Grasslands National Park (Saskatchewan), 2/38
Graubünden canton (Switzerland), 5/260
Gravensteen castle (Belgium), 1/138
Gravettian culture, 7/46–47, *7/46–47,* *7/49*
Graz (Austria), 1/97, 1/103
Great Age of European Discovery, 5/85
Great Artesian Basin (Australia), *1/79*
Great Barrier Reef, 1/69, 1/92–93, *1/92–93*
Great Barrier Reef Marine Park Authority, 1/92
Great Bath, *5/24*
Great Britain (country). *See* England; United Kingdom
Great Britain (island), 6/89, 6/92, 6/94, 6/96
prehistoric human migration, 6/110, 7/50–51, *7/50–51,* 7/53
Great Buddha sculpture (Japan), *3/279*
Great Caucasus Mountains, 3/14
Great Cave (Borneo), 7/23
Great Depression
Argentina, 1/65
Chile, 2/56
Germany, 3/33
Latvia, 4/48
Trinidad and Tobago, 6/46
United States, 6/136
Great Dividing Range (Australia), 1/72
Great Enclosure (Great Zimbabwe), **6/211,** 7/101
Great Escarpment, 1/169
Great European Plain, 5/78
Great Exhibition (London), *6/112*
Great Fall (Guyana), 3/94
Great Falls (Guyana), 3/94
Great Fire (Finland), 2/228
Great Hall of the People (Beijing), 2/83
Great Lakes, 1/4, 7/77
Canada, 2/23, 2/30, 2/31, 2/36
United States, 6/122, 6/123
Great Leap Forward, 2/65
Great Mosque (Córdoba), *5/213*
Great Plains (United States), 6/122, 6/123, 7/77
Great Purge (Soviet Union), 5/114, 5/116, 5/123
Great Rift Valley, 2/208, 3/286, 4/10, 4/90, 6/18, 6/80
Great Smoky Mountains, 6/124
Great Trek (South Africa), 5/192
Great Usutu River, 5/240
Great Wall (China), **2/68,** 2/68–69, *2/68–69*
Great Zimbabwe, 6/210–211, *6/210–211,* **6/211,** *7/99,* 7/100–101, *7/100–101*
Greater Albania, 4/83
Greater Antilles, 2/134, *3/48,* 3/48–49, **3/49,** 4/58
Greater Hinggan Mountains, 2/72

Greater London (England), 6/93
Greater Manchester (England), 6/93
Greater Moravian Empire, 2/144, 2/148, 5/178
Greater Nicobar Island (India), 3/156
Greater Sunda Islands (Indonesia), 3/164
Greater Syria, 5/266, 5/267
Grecian Archipelago, 3/66
Greco, El, *5/214,* 5/215
Greece, 3/50–71, *3/50–71,* **3/53**
Albanian refugees, 1/21
economy, 3/60–61, *3/60–61*
history as nation, 3/52–55, 3/62–63, *3/62–63*
land and climate, 3/54–55, *3/54–55*
people, 3/56–57, *3/56–57*
power in
Cyprus, 2/140–141, 3/52
Macedonia, 4/82, 4/83
Turkey, relations with, 6/68, 6/69
Greece, Ancient, 3/46, 3/62–63, *3/62–63,* **3/65**
Antarctica and, 1/41
civilization, 3/64–65, *3/64–65,* **3/65**
Hellenistic Age, 3/62–63
Italian opera development, 3/237
settlement or power in
Cyprus, 2/140
Damascus, 5/274
Indus River Valley, 5/14
Iraq, 3/187
Israel, 3/212
Libya, 4/67, 4/70
Macedonia, 4/82
Sicily, 3/244
Spain, 5/212
Turkey, 6/68, 6/70
Greek language, 2/142, 3/56, 3/69, 5/267
Greek Orthodox Church, 3/56, *3/56,* 3/57, 3/70–71
Greek Orthodox Monastery of Sumela, 6/70
Greek people
Cyprus, 2/142
Portugal, 5/94
See also Greece
Green Mountain region (Libya), 4/68
Green Mountains, 6/120
Green Plans, 3/238
Green Revolution, 3/135
Green Union party, 2/224
Greenland, 3/72–73, *3/72–73,* **3/73**
prehistoric settlement, 7/75, *7/75*
Greenpeace, *2/33*
Gregory, Lady, 3/206
Grenada, 3/74–75, *3/74–75,* **3/75,** 4/58
Grenadines. *See* St. Vincent and the Grenadines
Grenoble (France), 2/251
Grimaldi family, 4/148
Griot storytellers, 3/88
Gropius, Walter, 3/39
Gros Morne National Park (Newfoundland), 2/38

Gros Piton, 5/134
Gross domestic product, 6/152
Grossglockner, 1/100
Grósz, Károly, 3/109
Grote Markt (Antwerp), *1/137*
Grunewald (Berlin), 3/33
Grunitzky, Nicolas, 6/41
Gruyère cheese, 2/249, *2/249*
Guacharos, *6/177*
Guadalajara (Mexico), 4/134
Guadalcanal (Solomon Islands), 5/185, *5/185*
Guadalquivir River, 5/202
Guadeloupe, 3/76–77, *3/76–77,* **3/77,** 4/58, 4/63
Guam, 3/78–79, *3/78–79,* **3/79,** 5/10, 5/11, 5/12, 7/93
Guanabara Bay, 1/170, *1/170*
Guanches people, 5/220
Guangdong (China), 6/14
Guangzhou (China), 2/64, 2/73
Guantánamo (Cuba), 2/134
Guantánamo Bay (Cuba), 2/132
Guaraní Indians, 5/38, 5/42
Guatavita, Lake, *2/106–107*
Guatemala, 3/80–81, 3/80–85, **3/81,** **3/83**
economy, 3/82–83
history, 3/80–81
land, 3/82–83, *3/83*
people, 3/84–85, *3/84–85*
relations with
Belize, 1/144
Honduras, 3/100
U.S. immigrants, 6/143
Guatemala City (Guatemala), 3/82
Guayaquil (Ecuador), 2/176
Guaymí Indians, 5/30
Guebuza, Armando, 4/171
Gueï, Robert, 2/127
Guernsey (Channel Islands), 6/100
Guernsey cattle, 6/100
Guest workers, 5/265, 6/65
Guevara, Che, 1/154, *2/132–133*
Guggenheim Museum (New York City), *6/158–159*
Guiana Highlands, 1/168, 6/176, 6/179
Guided democracy, 3/162
Guilin (China), 2/59
Guilin Hills, *2/72–73,* 2/73
Guinea, 1/187, 2/128, 3/86–89, *3/86–89,* **3/87**
history, 3/86–87
people and economy, 3/88–89, *3/88–89*
Guinea, Gulf of. *See* Gulf of Guinea
Guinea-Bissau, 2/40, 3/90–91, *3/90–91,* **3/91,** 5/87
Guinea Current, 3/9
Gujarat (India), 3/137
Gujarati language, 3/138
Gül, Abdullah, 6/59
Gulf Cooperation Council, 6/85
Gulf of Aden, 6/194
Gulf of Aqaba, 3/285

Gulf of Bothnia, 2/226, 2/227, 5/248, 5/249
Gulf of Cádiz, 5/202
Gulf of Fethiye, 6/72
Gulf of Finland, 2/204, 2/222, 2/226, 2/228, 5/122
Gulf of Fonseca, 3/102
Gulf of Guinea, 1/146, 2/47, 2/126, 4/240
  Cameroon on, 2/20, 2/21
  Ghana on, 3/40, *3/43*
  São Tomé and Príncipe on, 5/142
  Togo on, 6/40
Gulf of Kotor, *4/157*
Gulf of Mexico, 4/135, 6/124, 6/148, 6/151
Gulf of Oman, 5/8
Gulf of Riga, 2/202
Gulf of St. Lawrence, 2/30
Gulf of Sidra, 4/68
Gulf of Thailand, 6/30, 6/33
Gulf of Tonkin, 6/185
Gulf of Tonkin Resolution, 6/185
Gulf Stream, 2/212, 2/226, 3/118, 3/120, 3/121, 4/250
Gum arabic, 4/120, 4/123, 5/232
GUM department store (Moscow), 5/120, *5/120*
Guoyu language, 6/14
Gupta Empire, 3/128, 3/149, 5/14
Gurgi Mosque (Tripoli), *4/71*
Guri Dam (Venezuela), 6/177
Gurkha (Nepal), 4/185
Gurkhas, 4/188
Gurunsi people, 1/186, 1/187
Gusmao, Xanana, 2/169
Gustav I Vasa, 5/243
Gustav II Adolf, 5/243, 5/246
Gustavus I, 2/228
Gustavus Adolphus. *See* Gustav II Adolf
Gutenberg, Johannes, 3/38
Gutierrez Borbua, Lucio, 2/173
Guyana, 3/92–95, *3/92–95*, **3/93**
  history, 3/92–93
  land, 3/94–95, *3/94–95*
  people, 3/95
  Venezuela, relations with, 6/176
Guzmán Blanco, Antonio, 6/173
Guzmán Reynoso, Abimael, 5/47
Gwaii Haanas National Park (British Columbia), 2/39
Gyanendra, 4/187
Gyndebourne festival, 6/109
Gyoer (Hungary), *3/113*
Gypsies. *See* Roma (people)
Gypsum, 3/248

H

Haakon VII, 4/247
Ha'apai (Tonga), 6/42
Haarlem (Netherlands), 4/204
Habré, Hissene, 2/45
Habsburg family
  Austria, 1/95, 1/98, **1/99,** 1/108
  Croatia, 2/130

Czech Republic, 2/144, 2/148, 2/151
  Germany, 3/20
  Hungary, 3/110–111
  Italy, 3/228
  Switzerland, 5/258
Habyarimana, Juvenal, 5/130
Haciendas, 1/152, 2/172, 2/174, 2/175, 4/134, 4/140
Hadi, Abdel Rabbo Mansour, 6/201
Hadrian, *3/227,* 5/212, 6/73
Hadrian, Gate of, 6/73
Hadrian, Temple of (Turkey), 6/67
Hadrian's Wall (England), *6/110*
Hagåtña (Guam), 3/78
Hagia Sophia (Istanbul), 6/62–63, *6/62–63,* 6/70, 6/71
Hague, The (Netherlands), 2/203, 4/200, *4/201,* 4/203, 4/204–205
Haida Indians, 2/29
Haifa (Israel), 3/209, 3/215
Haiks, *1/28–29*
Haile Selassie I, 2/206, 3/246, 3/251
Hainan Island (China), 2/73
Hainaut (Belgium), 1/142
Haiti, 3/48, 3/96–99, *3/96–99,* **3/97**
  Dominican Republic and, 2/166, 2/167
  economy, *3/98–99,* 3/99
  history, 3/96–97, *3/97*
  land, 3/96, *3/97*
  people, 3/98–99, *3/98–99*
Haitian Creole language, 3/98
Hajibs, 6/55
Hajj, *5/150*
Hajodae (South Korea), 4/22
Hajodae shrine, 4/18–19
Halawa (Syria), *5/272–273*
Haliburton (Ontario), *2/27*
Halicarnassus, 6/73
Halifax (Nova Scotia), 2/37
Hall of Mirrors (Versailles), *2/252–253, 3/20*
Halla-san Mountain, 4/26
Halonen, Tarja, 2/224
Hals, Frans, 4/202
Haltia, Mount, 2/227
Ham, 3/235
Hamams, 6/63
Hamas, 3/213
Hamburg (Germany), **3/30,** 3/30–31
Hammarskjöld, Dag, 5/246, *5/246*
Han dynasty, 2/62, **2/62,** 2/66, 2/69, 2/80
Han people, *2/76–77, 2/76–77*
  Beijing, 2/82
  relations with minorities, 2/78–79
Han River, 2/89, 4/26
Hand-Book for Travellers to Spain
  (Ford), 5/206
Hanga Roa (Easter Island), 2/171
Hangayn Mountains, 4/154
Hanging Gardens of Babylon, 3/187
Han'gul alphabet, 4/31
Hangzhou (China), 2/89
Hani people, *2/78*
Hannibal, 5/212

Hanoi (Vietnam), 6/180, 6/182, *6/183,* 6/186
Hanover Exhibition Center (Germany), *3/36*
Hansa Quarter (Berlin), 3/32
Hanseatic League, 1/138, 1/139, 2/160, 3/30, 3/31, 4/255
Hapsburg family. *See* Habsburg family
Hara-kiri, 3/266, 3/267
Harambee schools, 4/12
Harappa, 5/14, 5/24–25, 7/20, **7/21**
Harare (Zimbabwe), 6/207
Haratine people, 4/122
Hardanger Plateau (Norway), 4/250
Hare Fjord (Iceland), *3/72–73*
Harish Chandra ghat, 3/146
Harold (Saxon king), 6/110
Harold I (king of Norway), 2/213, 4/247
Harold Hårdråde, 4/254
Harratine people, 4/169
Harry (British prince), 4/57
Harz Mountains, 3/24, 3/38
Hašek, Jaroslav, 2/152
Hasina Wajed, 1/121
Hassan II, 4/159, 4/160, *4/161,* 4/162, 6/193
Hassan, Sayyid Muhammad Abdille, 5/186
Hassan Tower (Rabat), 4/166
Hassi Messaoud oil field, *1/31*
Hastings, Battle of, 2/238, 6/110
Hatshepsut, 2/192
Hausa city-states, 4/236
Hausa-Fulani empire, 4/236
Hausa language, 4/235
Hausa people, 4/234, 4/236, 4/237, 4/243
Haut Atlas Mountains, 4/162
Havana (Cuba), *2/132–133,* 2/134, 2/139
  people and economy, 2/136–138, *2/138–139*
Havel, Václav, 2/146, 2/147, *2/149,* 2/153, *2/153*
Havel River, 3/33
Havis Amanda statue, 2/229
Hawaii (United States), 5/11, 6/115, 6/129, *6/129*
  human settlement, 7/94, 7/94–95
Hawr al Hammar Lake, 3/190
Hawr as Saniyah Lake, 3/191
Haya de la Torre, Víctor Raúl, 5/46
Haydarpasa (Turkey), *6/66–67*
Haydn, Joseph, 1/95, 1/106
Hayonim Cave, 7/10, *7/10–11*
Hazara people, 1/15
Heard and McDonald Islands, 1/89
Heathrow Airport, *6/106–107*
Heaven, Temple of (Beijing), 2/82, *2/82*
Hebrew language, 3/215
Hebrew people, 3/209, 3/212, 5/272
Hebron (West Bank), 3/213
Hegira, 5/150
Heian period, 3/268–269
Heidi (Spyri), 5/255
Heimaey Island (Iceland), *3/120*

Heimskringla (Snorri Sturluson), 3/122
Hejaz (Saudi Arabia), 5/148
Hekla, Mount, 3/120
Helgi the Lean, 3/123
Helicopters, *6/183*
Helios (Greek god), *3/69*
Hellenistic Age, 3/62–63
Helmand River, 1/16
Helsinki (Finland), *2/225, 2/228–229, 2/228–229, 2/231*
Helsinki, University of, 2/228
Henley Royal Regatta, *6/108–109*
Henry I (king of Saxony), 3/38
Henry II (king of England), 6/110
Henry II (king of France), 2/242
Henry III (king of Castile), 5/220
Henry III (king of France), 2/243
Henry IV (king of France), 2/244
Henry VII (king of England), 6/111
Henry VIII (king of England), 3/204, 6/111
Henry of Burgundy, 5/94
Henry of Guise, 2/243
Henry the Lion, 3/26
Henry the Navigator, *5/94, 5/95*
Hentiyn Mountains, *4/154*
Heptarchy (England), 6/110
Hera, Temple of (Sicily), 3/244
Heraclius, 5/274
Herat (Afghanistan), *1/15,* 1/18
Hercules, 3/46, 5/216
Herero people, 4/181
Hermit kingdom, 4/23
Hermit people, 2/94
Hermitage (St. Petersburg), *5/116, 5/123*
Hermon, Mount, 5/270
Hernández, José, 1/58–59
Hernández Martínez, Maximiliano, 2/195
Herons
    boat-billed, *6/177*
    night, *6/149*
Herzegovina. See Bosnia-Herzegovina
Herzl, Theodor, 3/212, *3/213*
Heureaux, Ulises (Lilis), 2/167
Heyerdahl, Thor, 5/55
Hezbollah, 3/213, 4/53
Hidalgo y Costilla, Miguel, 4/140
Hideyoshi, 3/266
Hierro (Canary Islands), 5/218, 5/220, *5/221*
High Algarve (Portugal), 5/89
High Atlas Mountains, 4/162
High German, 3/19
High Yemen, 6/194
Highland Games, 6/97, 6/108
Highlands, in Scotland, 6/97
Hill Complex (Great Zimbabwe), 7/101
Hill people of Myanmar, 4/174, 4/175
Hillary, Sir Edmund, 4/188
Hills, 1/4
Himachal Pradesh (India), *3/131*
Himalaya, **4/190–191,** 4/190–197, *4/190–197*
    Bangladesh, 1/118

Bhutan, 1/150–151, *1/151*
    biology, *4/191*
    China, 2/70, 2/74, 2/94
    India, 3/130, *3/131,* 3/158
    myth and legend, 4/190
    Nepal, 4/184
    people, 4/192–193, *4/192–193*
    threats to, 4/194–195, *4/194–195*
    trekking, 4/196–197, *4/196–197*
    wildlife, 4/184
Hindi language, 2/216, 3/95, 3/138, 3/139, 6/47
Hindu Kush range, 1/15, 1/16, 1/18–19, *1/18–19,* 5/18
Hinduism
    Bangladesh, 1/118
    Bhutan, 1/151
    Cambodia, 2/8
    Himalaya people, 4/192
    India, 3/126, *3/138–139*
        culture, 3/140–141, *3/140–141*
        Ganges River, *3/142,* 3/142–143
        Jammu and Kashmir, 3/158
    Indonesia, 3/168–169
    Malaysia, 4/103
    Mauritius, 4/125
    Nepal, 4/188, 4/189
    Singapore, 5/176
    Sri Lanka, 5/226, *5/227*
    United Kingdom, 6/105
    United States, 6/156
Hindustanis, 5/238–239
Hip-hop music, 6/159
Hira, Mount, 5/150
Hirohito, 3/254
Hiroshima (Japan), 3/265, 6/136
Hispania, 5/212
Hispanic Americans, 6/123, 6/126, 6/129, 6/138, 6/142–143, *6/142–143,* **6/143**
Hispaniola, 1/48, 2/166, 2/167, 3/48, 3/96, 4/58
Histadrut, 3/211
Hitler, Adolf, 1/25, 1/98–99, 2/148, 2/204, 2/245, 3/116
    German history and, 3/20, 3/21, 3/26, 3/33
Hittite Empire, 4/50, 6/68
Hmong people, 4/46–47, 6/31, 6/187
Ho Chi Minh, 4/41, 6/181, *6/183,* 6/184
Ho Chi Minh City (Vietnam), 6/180, *6/184, 6/186, 6/187*
Ho Chi Minh Trail, 4/41
Hoa people, 6/180, 6/187
Hoatzins, *3/95*
Hobbies, in United States, 6/157
Hobbits (prehistoric humans), 7/24–25, 7/25
Hofburg (Vienna), 1/108
Hofkirche (Austria), 1/100
Hofmannsthal, Hugo von, 1/108
Hoge Veluwe (Netherlands), 4/201
Hogs, 2/90
Hohensalzburg Castle (Austria), *1/100–101,* 1/101

Hohenzollern family, 3/20, 3/33
Hohle Fels Cave, *7/46, 7/47*
Hokkaido Island (Japan), 3/253, 3/256, 3/272, 3/274
Holder, Eric, 6/145
Holi, *3/138–139*
Holland. See Netherlands, The
Hollande, François, 2/245
Holly trees, 1/55
Hollywood (California), 6/128
Holy Community, 3/70
Holy Roman Empire
    Austria in, 1/98
    Bohemia in, 2/144, 2/148, 2/150
    Danube River and, 5/180
    Germany in, 3/20
    Italy in, 3/229, 3/236
    Liechtenstein in, 4/75
    Slovenia in, 5/182
    Switzerland in, 5/258
Holy Sepulcher, Church of the (Israel), 3/217
Holy Week, *2/206,* 5/206
Home rule (Ireland), 3/205
Homer, 3/64
Homesteaders, 6/122
Hominids, *7/9*
Homo erectus, 7/6, 7/9, 7/34, 7/56–57
Homo floresiensis, 7/24–25, 7/25
Homo sapiens, 7/6, 7/35, 7/37, 7/48, 7/64
Honduras, 3/100–105, *3/100–105,* **3/103**
    economy, 3/104–105, *3/104–105*
    government, 3/103
    history, 3/100–101
    land, 3/102
    people, 3/103
Honecker, Erich, 3/23
Hong Kong (China), 2/61, **2/84,** 2/84–85, *2/84–85,* 4/182
Hong Kong Island, 2/84
Hong Shang, 7/54
Honiara (Solomon Islands), 5/184
Honolulu (Hawaii), 6/129
Honshu Island (Japan), 3/253, 3/256
Hopi Indians, 6/127
Hopper, Edward, 6/158
Hornussen, 5/261
Horse Guards Parade, *6/102*
Horse racing, 3/207
Horses, *3/112–113*
    Australian Outback, *1/78, 1/79*
    gaucho, 1/58–59, *1/58–59*
    Lipizzan, 1/108, *1/109*
Horseshoe Falls (Canada), *2/27*
Horthy, Nicholas, 3/111
Hortobágy national park (Hungary), 3/112
Hospitallers. See Knights Hospitallers of St. John
Hot springs, 3/118
Hôtel de Ville (Brussels), 1/140
Houngans, 3/98
Houphouët-Boigny, Félix, 2/127

House of Commons
  Canada, 2/24
  United Kingdom, 6/90, 6/94, 6/96
House of Lords (United Kingdom), 6/90
House of Representatives (United States),
  6/116
Houston (Texas), 6/147
Hoxha, Enver, 1/20–21
Hoyte, Hugh Desmond, 3/93
Hrabal, Bohumil, 2/152, 2/153
Hradčany (Prague), 2/148, 2/150,
  2/150–151
Hradec Králové (Czech Republic), 2/155
Huancayo (Peru), 5/50, 5/51
Huang He, 2/70, 2/72, **2/73,** 2/89
Huang He Valley, 2/72
Huangpu River, 2/81, 2/89
Huáscar, 5/53
Huayna Capac, 5/52
Hudson, Henry, 2/30, 4/257
Hudson Bay Lowlands (Canada), 2/26
Hugh Capet. See Capet, Hugh
Hugo, Victor, 1/140, 2/252
Huguenots, 2/244
Huis ten Bosch (The Hague), 4/205
Hum region, 1/156
Humala Tasso, Ollanta, 5/47
Human migrations, 7/4–107, 7/4–107
  from Africa, 7/5, 7/6
    Arabian Peninsula route, **7/14,**
      7/14–15, 7/15
    first, 7/10–11, **7/11**
    reasons for, 7/8–9, **7/9**
    second, 7/12, 7/12–13
    slaves to Americas, 7/104, 7/104
    western part to sub-Saharan,
      7/98–101, **7/99,**
      7/100–101
  from Asia
    Australia, **7/26,** 7/26–27, 7/27
    Europe, 7/42, 7/48
    Indonesia and New Guinea, **7/24,**
      7/24–25, 7/24–25
    North America, precolonial, 7/67,
      7/68, **7/68, 7/72**
    Pacific Islands, 7/91, 7/92, **7/93**
    United States, 7/104–105, 7/105
  from Europe
    Americas, 7/102–105, 7/102–105,
      7/103, 7/105
    Australia and New Zealand, **7/106,**
      7/106–107, 7/106–107
  to Asia, 7/5–6, 7/31
    Andaman islanders, evidence from,
      7/16, 7/16–17, 7/16–17
    Arabian Peninsula, **7/14,** 7/14–15,
      7/15
    Arctic region, 7/64–65, 7/64–65,
      7/65
    China, 7/54, 7/56–7/57, 7/56–57,
      7/57
    Egypt, 7/32, **7/32,** 7/32–33, 7/40–
      41, 7/40–41
    India, **7/16,** 7/16–17, 7/19–21,
      7/20–21, **7/21**

    Indonesia and New Guinea, **7/24,**
      7/24–25, 7/24–25
    Iran, 7/14, **7/38,** 7/38–39,
      7/38–39, 7/42
    Japan, 7/60–61, 7/60–61, **7/61**
    Korea, 7/60–61, **7/61**
    Levant and Nile Valley, **7/32,**
      7/32–33, 7/32–33
    Mesopotamia, **7/38,** 7/38–41,
      7/38–41
    Siberia and Northeast Asia, **7/62,**
      7/62–63, 7/62–63
    Southeast Asia, 7/19, 7/22–23,
      7/22–23, **7/23**
    Taiwan, 7/58, 7/58–59, **7/59**
    Turkey, **7/36,** 7/36–37, 7/36–37
  to Australia
    aboriginal, 7/23, **7/26,** 7/26–29,
      7/27–29
    European, 7/106, 7/106–107,
      7/107
  to Europe, 7/42
    Arctic region, 7/64–65, 7/64–65,
      7/65
    Britain, 7/50–51, 7/50–51
    Eastern Europe, 7/44–47, 7/44–47,
      7/45
    Western Europe, **7/48,** 7/48–49,
      7/48–49
  to North America, 7/67, 7/70–71
    Arctic region, 7/74–75, **7/75,** 7/75
    Asian, 7/105, 7/105
    British colonization, 7/104–105,
      7/105
    Clovis culture, **7/68,** 7/68–69,
      7/68–69
    extinctions caused by, 7/72–73, 7/73
    French colonization, 7/104, 7/104
    Portuguese colonization, 7/102–
      103
    pre-Clovis cultures, 7/70–71,
      7/70–71, **7/71**
    Spanish colonization, 7/102–103,
      7/102–103, **7/103**
    transition to agriculture, 7/76–77,
      7/76–77, **7/77**
    Yucatán Peninsula and Central
      America, 7/78, **7/80,**
      7/80–81, 7/80–81
  to Pacific Islands, 7/91
    early voyages, 7/92–93, 7/92–93,
      7/93
    New Zealand, 7/96–97, 7/96–97,
      7/97, 7/106–107, **7/107,**
      7/107
    Polynesian settlement, 7/94–95,
      7/94–95, **7/95**
  to South America, 7/78
    Amazon region, 7/84, 7/84–85,
      7/85
    Andes region, **7/82,** 7/82–83,
      7/82–83
    Monte Verde, **7/86,** 7/86–87, 7/87
    Portuguese colonization, 7/102–
      103, **7/103**

    Spanish colonization, 7/102–103,
      7/103
    Tierra del Fuego, **7/88,** 7/88–89,
      7/88–89
    United States, patterns in, 7/5
    ways of tracking, 7/11
Human Rights Commission (South Africa),
  5/191
Humanists, 3/228
Humboldt Current, 2/176
Humboldt University, 3/32
Hummus, 4/55
Hun Sen, 2/10–11
Hundred Years' War, 2/241, 6/110
Hungarian Empire, 3/110
Hungarian Evangelical Lutheran Church,
  5/164
Hungarian people
  Romania, 5/106–107
  Serbia, 5/155, 5/164
  See also Hungary; Magyars
Hungarian Rhapsodies (Liszt), 3/117
Hungarian Socialist Worker's Party,
  3/109
Hungary, 3/23, 3/106–117, 3/106–117,
  3/109
  economy, 3/112–113, 3/112–113
  history, 3/106, 3/110–111,
    3/110–111, **3/111**
    present-day, 3/108–109
  land, 3/112–113, 3/112–113
    Danube River, 5/180
  people, 3/107–108
  power in
    Bosnia-Herzegovina, 1/156
    Germany, 3/38
    Slovakia, 5/178
    Slovenia, 5/182
  relations with
    Austria, 1/98, 1/103
    Croatia, 2/130
    Soviet Union, 3/111
Hungsu Cave, 7/60
Huns, 3/128, 4/146, 5/14, 5/106, 5/168
Hunter-gatherers. See Hunting
Hunting
  Arctic, 7/65
  Australian Aborigine, 7/29
  Europe, 7/44, 7/45, 7/49
  Maori, 7/96–97
  Middle East, 7/38, 7/39, 7/41
  North America, 2/28–29, 2/28–29,
    7/72–77, 7/73, 7/74, 7/76,
    7/77
Hunyadi, Matthias Corvinus, 3/110,
  3/117
Hunza Valley (Pakistan), 4/195, 5/22–23
Hurling, 3/207
Huron Indians, 2/28
Hurricanes
  by name
    Gilbert, 4/133
    Hugo, 3/77
    Ida, 2/195
    Mitch, 3/101

by place
    Bermuda, 1/148
    Cuba, 2/134
    El Salvador, *2/195*
    Guadeloupe, *3/77*
    Mexico, 4/132
    Nicaragua, 4/228
    Tuamotu Islands, 2/259
Hurt Valley, *4/219*
Hus, John, *2/149*
Husák, Gustáv, 2/146, 2/149
Husayn, Imam, 3/192
Hussein (king of Jordan), 3/283
Hussein, Saddam, 3/186, 3/188, 3/189,
    3/191, 3/199, 4/35–37
Hussein, Sherif, 3/282
Hutongs, 2/82
Hutu people, 1/188–189, 5/130, 5/131,
    5/131
Hyderabad (India), 3/144
Hydroelectric power
    Austria, 1/103, 1/105
    Brazil, 1/176
    Canada, 2/33
    China, 2/89
    Finland, 2/231
    India, 3/133
    Italy, 3/240
    Norway, 4/252
    Portugal, 5/92
    Sweden, 5/253
    Switzerland, 5/262
    United States, 6/150, *6/150*
    Uzbekistan, 6/167
    Venezuela, 6/177
    Zimbabwe, 6/207
Hydroponic farming, 4/36

I

I-Kiribati, 4/16
I Served the King of England (Hrabal),
    2/152
Ibadan (Nigeria), 4/242
Ibadi Islam, 5/8
Ibans, *4/98,* 4/99
Iberian Peninsula, 5/86, 5/199, 5/212
Iberians, 5/94, 5/212
Ibexes, Pyrenean, *1/35*
Ibibio people, 4/243
Ibises, Scarlet, *6/177*
Ibiza Island, 5/216, 5/217, **5/217**
Ibn Abdallah, Idris. *See* Idris ibn Abdallah
Ibn al-Walid, Khalid. *See* Walid, Khalid
    ibn al-
Ibn Saud, Muhammad. *See* Saud,
    Muhammad ibn
Ibn Ziyad, Tariq. *See* Ziyad, Tariq ibn
Ice Age, 2/212, 4/246, 4/250, 5/249
    early humans and, 7/20, 7/33, 7/40
        Arctic, 7/64
        Europe, 7/46, 7/47, 7/50–53
        Japan, 7/60
        Mesoamerica and South America,
          7/78
        Middle East, 7/34, 7/39

    North America, 7/67
    Northeast Asia, 7/62
    megafauna extinctions, 7/72
Ice sheets, Antarctic, 1/41, 1/42, 1/47
Ice shelves, 1/42–43
Ice skating, 4/199, *4/203*
Icebergs, 3/72
    Antarctica, *1/40–41,* 1/41, 1/43, *1/47*
    Chile, *2/52*
Iceland, 3/118–123, *3/118–123,* **3/119**
    economy, 3/119
    history, 3/118–119
        sagas, 3/122–123, *3/122–123*
    land, 3/120–121, *3/120–121*
    people, 3/118
Icelandic Family Sagas, 3/123
Icelandic language, 3/119
Idaho (United States), 6/126, 6/127
Iditarod race, *6/130*
Idris II, 4/166
Idris al Mahdi as-Sanusi, Muhammad,
    4/66, 4/67
Idris ibn Abdallah, 4/158
Idrisid dynasty, 4/158
Ife, *4/244,* 4/245
Ifugao people, 5/64, *5/65*
Igbo people, 4/237, 4/238, 4/243
Iguaçu Falls, 1/51, *1/168–169*
Iguanas, Marine, 2/179, *2/179*
Ijo people, 4/243
IJsselmeer Lake, 4/209
Île-de-France, 2/236
Île de la Gonâve, 3/96
Îles des Saintes (Guadeloupe), 3/76
Iliad (Homer), 3/64
Iliescu, Ion, 5/103
Illinois (United States), 6/122
Illinois River, 6/148
Illyrian people, 1/20, 1/156, 4/32
Imams, 5/146, 6/198
Immigration
    Argentina, 1/56, 1/60
    Australia, 1/85, 1/90, 7/107
    Brazil, 1/165
    France, 2/239
    United Kingdom, 6/91, 6/104–105
    United States, 6/115, 6/138–139,
        6/138–139, 6/146, 7/104–105,
        7/105
    See also Human migrations; Refugees
Immortal Thirty-Three (Uruguay), 6/161
Imperial City (Beijing), 2/82
Imperial Palace (Kyoto), *3/281*
Imperial Palace (Tokyo), 3/270,
    3/270–271, 3/280
Imperial Valley (California), 6/128
Impressionism, 2/253, *2/253*
Inca
    Bolivia, 1/152
    Chile, 2/49
    Ecuador, 2/172
    legend of origin, 5/52
    Peru, 5/44, 5/45, 5/52–53, *5/52–53,*
        5/53, 7/82, 7/83, *7/83*
Incense, 6/198, *6/199*

India, 3/124–159, *3/124–159,* **3/127**
    cities and villages, 3/144–145,
        3/144–145
        Varanasi, 3/146–147, *3/146–147*
    climate, 3/131, **3/131**
        monsoons, 3/132–133, **3/132–**
        133, *3/132–133*
    economy
        agriculture, 3/134–135, *3/134–*
        135, **3/135**
        manufacturing, 3/136–137,
        3/136–137
        service industries, 3/137
    history, 1/18, 5/15
        before independence, *2/128–129,*
        3/128–129, **3/129**
        development of civilization, 7/20,
        7/20–21, **7/21**
        since independence, 3/126–127
    island territories, 3/156–157,
        3/156–157, **3/157**
    land, **3/130,** 3/130–131, *3/130–131*
        Ganges, 3/142–143, **3/142–143**
    people, 3/138–139, *3/138–139*
        city life, 3/145, *3/145*
        culture, 3/140–141, *3/140–141*
        village life, 3/144–145
    population, 3/125, 3/139
    prehistoric human migration, 7/14,
        7/16, 7/16–17, 7/19–21, **7/21,**
        7/42
    Project Tiger, 3/150–151, *3/150–151*
    relations with
        Bhutan, 1/150, 1/151
        China, 3/126, 4/193
        Nepal, 4/186, 4/188
        Pakistan, 3/159, 5/16
    states
        Goa, **3/154,** 3/154–155,
        3/154–155
        Jammu and Kashmir, 3/158–159,
        3/158–159, **3/159**
        Sikkim, **3/152,** 3/152–153,
        3/152–153
    tourist guide, 3/148–149, *3/148–149*
Indian languages, 3/138–139
Indian National Congress, 3/129
Indian Ocean, 1/4
    Bangladesh on, 1/118, 1/119
    India on, 3/130, 3/132
    Maldives in, 4/104
    Mauritius in, 4/124, *4/125*
    Mozambique on, 4/172
    Réunion in, 5/100
    Seychelles in, 5/170
    Tanzania on, 6/20
    See also Tsunamis
Indian Rebellion, 3/129
Indiana (United States), 6/122
Indianapolis (Indiana), 6/123
Indians, American. *See* Native Americans
Indians, Asian
    Bahrain, 1/116
    Fiji, 2/216, 2/220, 2/221
    Guyana, 3/92, 3/93, 3/95

139

Indonesia, 3/161
Jamaica, 3/246
Kuwait, 4/34
Lesser Antilles, 4/60
Malaysia, 4/93, 4/99, 4/102, 4/103
Mauritius, 4/124
Singapore, 5/176, 5/177
South Africa, 5/195
Suriname, 5/238–239
Thailand, 6/31
United Kingdom, 6/104, *6/105*
See also India
Indies, 3/161
Indigirka River, *5/124–125*
Indo-Aryan people, 3/137
Indo-European language, 3/138
Indo-Sikkim Treaty, 3/152
Indochina, 4/40
Indochina War, 6/184
Indonesia, 3/160–173, *3/160–173,*
    3/162–163, 4/207
    art and culture, 3/168–169,
        3/168–169
    climate, 3/164
    economy, 3/170–171, *3/170–171*
    history, 3/161–163
    land, 3/164–165, *3/164–165*
        Mount Toba eruption, 7/13
        rain forest threat, 3/172–173,
            3/172–173, **3/173**
    people, 3/166–167, *3/166–167*
    population, 3/166, **3/167**
    power in
        East Timor, 2/168–169, 3/163,
            5/87
        Madagascar, 4/86
    prehistoric human migration, 7/24,
        7/24, 7/24–25, 7/26
Indonesian people
    Netherlands, 4/202–203
    Suriname, 5/239
    See also Indonesia
Indre River, 2/242
Indris, 4/88, *4/88–89*
Indus River
    India, 3/158
    Nepal, 4/190, 4/191, 4/194
    Pakistan, 5/18, *5/19,* 5/24
    Tibet, 2/94
Indus River Valley, 1/10, 1/18, 5/14,
    5/22
    civilization, 5/14, 5/24–25, *5/24–25,*
        7/20, *7/20–21,* **7/21**
Industrial Revolution, 1/119, 4/252,
    6/89, 6/93, 6/95, 6/112
Industry. *See specific places*
Inés de la Cruz, Sor Juana, 4/145
Inflation
    Argentina, 1/52, 1/65
    Brazil, 1/163, 1/176
    Chile, 2/50
    China, 2/93
Informal economy, 5/11
Ingólfur Arnason, 3/118
Inguwavuma River, 5/240

Initiative, Right of, 5/256
Inn River Valley (Austria), 1/100
Inner Mongolia, 2/71, 2/76
Inner Mongolian Uplands (China), 2/68,
    2/71
Innsbruck (Austria), 1/97, 1/100, *1/101,*
    1/107
Inquilinos, 2/55
Inquisition, 5/214
Institutional Revolutionary Party
    (Mexico), 4/128–129
Interior, U.S. Department of the, 6/133
International Court of Justice, 4/180,
    4/204
International Geophysical Year, 1/46
International Labor Organization, 5/265
International Polar Year 2007-2008,
    1/47
International Union for Conservation of
    Nature and Natural Resources,
    6/24
Inthanon Mountain, 6/30
Inti, 5/53, 7/83
Intramuros (Manila), 5/66
Inuit, 2/23, 2/28–29, 3/73, 7/74, 7/75
    See also Paleo-Eskimos
Inuit-Inupiac language, 2/28
Inuk, 7/75, *7/75*
Ioannina River Valley, 3/54
Ionian Islands (Greece), 3/63
Ionian Sea, 3/61, 3/66, 3/224
Ios (Greece), 3/66
Iowa (United States), 6/122
Ipanema beach (Rio de Janeiro), 1/161,
    1/170, *1/170*
IRA. *See* Irish Republican Army
Iraklion (Crete), 3/68
Iran, 3/174–185, *3/174–185,* **3/177**
    art and architecture, 3/184–185,
        3/184–185
    economy, 3/180–181, *3/180–181,*
        3/181
    government, 3/176
    history, 3/175–177, 3/182–183,
        3/182–183, **3/183**
    land and climate, 3/180, *3/180–181*
    people, 3/176, 3/178–179,
        3/178–179
        Kurds, 3/198
    prehistoric migrations, 7/14, **7/38,**
        7/38–39, *7/38–39,* 7/42
Iran-Iraq War, 3/177, 3/186, 3/188,
    3/192, 3/193, 3/194, 3/199, 4/36
Iranian people
    Bahrain, 1/116
    Kuwait, 4/34
    See also Iran
Iraq, 3/186–199, *3/186–199,* **3/189**
    economy, 3/194–195, *3/194–195,*
        3/195
    history
        early, 3/186–187, 3/196–197,
            3/196–197, **3/197**
        modern, 3/188–189, *3/188–189*
        land, 3/190–191, *3/190–191*

    people, 3/192–193, *3/192–193*
        Kurds, 3/198–199, *3/198–199*
    prehistoric human migration, 7/38–39
    relations with
        Kuwait, 4/34–35
        Saudi Arabia, 5/145
        United States, 6/136, 6/137
Iraq-Iran War. *See* Iran-Iraq War
Iraq War (2003-2011), 3/186, 3/189,
    3/193, 3/195, 3/199, 6/136, 6/137
Irazú volcano, 2/125
Irbil (Iraq), 3/198
Ireland, 3/200–206, *3/200–206,* **3/203,**
    6/98
    government, 3/202
    history, 3/204–205, *3/204–205,*
        6/110
    land, 3/201
    people, 3/201, 3/206–207,
        3/206–207
Irhil m' Goun peak, 4/169
Irian Jaya. *See* Papua (Indonesia)
Irish Citizen Army, 3/204
Irish Derby, 3/207
Irish Free State, 3/205, 6/98
Irish Grand National, 3/207
Irish language, 3/202–203
Irish people
    Australia, 1/85
    Canary Islands, 5/221
    United States, 6/138
    See also Ireland; Northern Ireland
Irish Republican Army, 3/205, 6/98
Irish Sea, 6/99, 6/100
Irkutsk (Siberia), 5/127
Iron Age, 5/216
Iron and steel
    Austria, 1/103, *1/103,* 1/104
    Bantu, 7/98
    Belgium, 1/142
    Brazil, 1/176
    Canada, 2/35
    Gabon, 3/11
    Germany, 3/36-37
    Luxembourg, 4/80
    Mauritania, 4/123
    Norway, 4/252
    Spain, 5/209
    Sweden, 5/252
    United Kingdom, 6/106
    United States, 6/123, *6/150–151*
    Zimbabwe, 6/207
Iron Gate, 5/166–167, 5/180
Iroquois Indians, 6/140
Iroquois League, 2/28
Irrawaddy Delta, 4/174
Irrawaddy River, 4/174, 4/175
Irrawaddy River Valley, 4/174, 4/175
Irrigation
    Afghanistan, 1/17
    Australia, 1/81
    Bulgaria, *1/183*
    Canary Islands, 5/221
    China, 2/90
    Egypt, 2/184, 2/185, *2/185*

Inca, *5/52–53*
Iraq, 3/190, *3/191*
Italy, *3/239*
Madeira, 1/113
Pakistan, *5/19*, 5/22
Philippines, 5/64
Saudi Arabia, 5/149
Spain, 5/212
Sudan, 5/232
Sumer, 3/196
Syria, 5/271, *5/272*
Yemen, 6/197
Irving, Washington, 6/158
Isabella I, 3/116, 5/214, *5/215*
Isar River, 3/27
Isfahan (Iran), *3/183*, *5/150*
Ishikari Plain (Japan), 3/256
IsiNdebele language, 6/207
Iskür River, 1/182
Isla de la Juventud (Cuba), 2/134
Islam
    Afghanistan, 1/11–15
    African art, 4/112, *4/112–113*
    Albania, 1/22
    Algeria, 1/24–26, 1/29
    art of, *3/184–185*, *3/184–185*
    Azerbaijan, 1/111
    Bahrain, 1/116
    Bangladesh, 1/118, *1/121*, 1/124, *1/124*
    Bosnia-Herzegovina, 1/156, 1/157
    Brunei, 1/179, *1/179*
    Cameroon, *2/16*
    Chad, 2/44–45, 2/46
    China, 2/79
    Comoros, 2/109
    Côte d'Ivoire, 2/128
    Djibouti, 2/162
    Egypt, 2/181, 2/186–187, 2/192
        Al-Azhar University and, 2/188–189
        architecture, 2/190
        fundamentalism, 2/182–183, 2/188–189, 2/193
    Ethiopia, 1/208
    Gambia, 3/12
    growth and spread of, 5/150–151, *5/150–151*
    Guinea, 3/86, 3/88
    Himalayan people, 4/192
    India, 3/126, 3/129, 3/141
    Indonesia, 3/166, *3/167*, 3/168–169
    Iran, 3/175, 3/176, 3/179, 3/182–183
        arts, *3/184–185*, *3/184–185*
    Iraq, 3/192
    Israel, 3/209
    Jerusalem's role, 3/216, 3/217
    Jordan, 3/282, 3/286
    Kosovo, 4/32
    Kuwait, 4/34
    Kyrgyzstan, 4/38, 4/39
    Libya, 4/67, 4/70
    Madagascar, 4/86
    Malaysia, 4/98, 4/103
    Mali, 4/108
    Mauritania, 4/122

Mauritius, 4/125
Morocco, 4/158, 4/160, 4/164, 4/165, 4/166
Mozambique, 4/173
Niger, 4/232
Nigeria, 4/236, 4/239, *4/242*, 4/243
Oman, 5/8
Pakistan, 5/14, 5/15, 5/20, *5/20–21*, 5/21
Philippines, 5/58, 5/63, 5/64, 5/65
Portugal, 5/94
Red Crescent Societies, 5/265
Russia, 5/117
Saudi Arabia, 5/144, 5/146–147
    origins and growth, 5/150–151, **5/151**
Singapore, 5/176
Somalia, 5/186
Spain, 5/213, 5/214
Sri Lanka, 5/226, 5/227, *5/227*
Sudan, 5/228–231, 5/234
Syria, 5/267, 5/272, 5/274
Tajikistan, 6/16–17
Togo, 6/40
Tunisia, 6/48, 6/51, 6/54, 6/55
Turkey, 6/57, 6/58, *6/69*, 6/71
Turkmenistan, 6/74
Uganda, 6/81
United Arab Emirates, 6/85
United States, 6/156
Uzbekistan, 6/166, 6/167
Yemen, 6/195, 6/198, 6/201
See also Shi`ite Muslims; Sunni Muslims
Islamic Empire, 2/192, 3/187, 5/151, **5/151**
Islamic Salvation Army, 1/27
Islamic Salvation Front, 1/26, 1/27
Islands, Bay of (New Zealand), *4/221*
Islandwana, Battle of, 7/99
Islas Malvinas, 2/210
Isle of Anglesey, 6/94
Isle of Man, 6/100–101, **6/101**, *6/101*
Isle of Pines, 4/212
Isle of Wight (England), 6/92
Ismail (ruler of Egypt), 2/193
Ismailis, 4/54
Israel, 3/208–219, *3/208–219*, **3/211**, **3/212**, 3/214–215, *3/214–215*
    agriculture, 3/218–219, *3/218–219*
    economy, 3/211
    government, 3/210
    history, **3/212**, 3/212–213, *3/212–213*
    land and climate, 3/209
    people, 3/214–215, *3/214–215*
        kibbutz life, 3/218–219, *3/218–219*
    population, 3/214
    relations with
        Egypt, 2/193
        Jordan, 3/282–283
        Lebanon, 4/53
        Syria, 5/269
    Star of David, 5/265
Issa people, 2/162–163

Issoufu, Mahamadou, 4/233
Issyk-Kul, Lake, 4/38
Istanbul (Turkey), 6/57, 6/62–63, *6/62–63*, 6/65
    historic sites, 6/70
    history, 6/63, 6/68
    industry, 6/66
Isthmus of Panama, 5/32
Isthmus of Suez, 2/193
Istiqlal Party, 4/159, 4/160
Itagon (Philippines), 5/69
Italian language, 3/232, 5/141
    dialects, 2/238, 2/254, 3/232
    Libya, 4/70
    Switzerland, 5/260, *5/261*
Italian people
    Canada, 2/23
    France, 2/239
    Puerto Rico, 5/96
    United States, 6/138, *6/139*
    Uruguay, 6/160, 6/165
    See also Italy
Italian Somaliland, 5/186
Italy, 3/220–245, *3/220–245*, **3/223**
    Albanian refugees, 1/21
    arts, 3/236–237, *3/236–237*
    economy, 3/222
        agriculture, 3/238–239, *3/238–239*
        industry, 3/240–241, *3/240–241*, **3/241**
    foods, 3/234–235, *3/234–235*
    government, 3/222
    history
        early, 3/226–227, *3/226–227*, **3/227**
        modern, 3/222–223, 3/228–229, *3/228–229*, **3/229**
    land, 3/224–225, *3/224–225*, **3/225**
    people, 3/232–233, *3/232–233*
    power in
        Albania, 1/20–21
        Corsica, 2/255
        Eritrea, 2/202
        Libya, 4/67
        Monaco, 4/148
        Somalia, 5/186
    relations with
        Germany, 3/26
        San Marino, 5/140, 5/141
        Vatican City, 6/170–171
Itasca, Lake, 6/148
Itching Parrot, The (Fernández de Lizardi), 4/145
Itsekiri people, 4/243
Iturbide, Agustín de, 4/140
Iu Mien, 4/46
Ivan I (prince of Moscow), 5/120
Ivan III, the Great, 5/112
Ivan IV, the Terrible, 5/112, *5/112*
Ivory, 2/126, 6/24
Ivory Coast. *See* Côte d'Ivoire
Iwo Jima (Japan), 3/258
Ixtacihuatl, 4/130
Izalbal, Lake, 3/82
Izgara, 6/73

Izmir (Turkey), *6/65, 6/66, 6/67*
Izu islands (Japan), 3/258

J

Jabal ad Duruz range, 5/270
Jabal al Qara mountains, 5/8
Jabal an Nusayriyah Mountains, 5/267, 5/270
Jabir, Sheik, 4/35
Jackman, Hugh, *1/87*
Jackson, Jesse, 6/145
Jacobites, 4/54
Jacobzoon May, Jan, 4/257
Jade Mountain, 6/10
Jadwiga, 5/76
Jagan, Cheddi B., 3/92–93
Jagan, Janet, 3/93
Jagdeo, Bharrat, 3/93
Jagiełło, Władysław, 4/76, 5/76
Jagiełłonian dynasty (Poland), 5/76
Jaguars, 2/125, *2/125*, *6/177*
Jahan, Shah, 3/148, 3/149
Jainism, 3/141
Jakarta (Indonesia), 3/162, 3/164, 3/167, 3/171
Jakovo (Serbia), *5/161*
Jalasayi ghat, 3/146
Jamaa Mosque (Kenya), *4/13*
Jamaica, 1/129, 3/48, 3/246–251, *3/246–251*, **3/249**, 4/58
economy, 3/248–249
government, 3/249
Grenada, relations with, 3/75
history, 3/246–247, 3/249
land, 3/248, *3/249*
people, 3/246, *3/246–247*
reggae music, 3/250–251, *3/250–251*
Jamaican Labor Party, 3/249
Jamaican people, in Costa Rica, 2/120, *2/121*
Jambiyas, *6/200, 6/201*
James (apostle), 5/205
James I (king of Aragon), 5/217
James II (king of England), 3/204
James VI (king of Scotland), 6/111
Jamestown (Virginia), 6/134, 7/104, *7/105*
Jammeh, Yahya, 3/13
Jammu and Kashmir (India), 3/126, 3/158–159, *3/158–159*, **3/159**, 4/193
Jamuna River, 1/118, 1/122
Jan Mayen Island (Norway), 4/250, 4/257
Jansz, Willem, 1/86
Japan, 3/252–281, *3/252–281*, **3/255**
climate, 3/257
economy, 3/254
agriculture, 3/260–261, *3/260–261*
fishing, 3/258, 3/261, *3/261*
industry, 3/262–263, *3/262–263*
gardens, 3/280–281, *3/280–281*
history, 3/264–265, *3/264–265*, **3/265**
samurai, 3/266–267, *3/266–267*

human settlement, 7/60–61, *7/60–61*, **7/61**
land, 3/256–257, *3/256–257*
island groups, 3/258–259, *3/258–259*, **3/259**
people, 3/272–273, *3/272–273*
ethnic and social groups, 3/274–275, *3/274–275*
samurai, 3/267
popular culture, 3/276–277, *3/276–277*
population, 3/253, 3/270, **3/275**
power in
China, 2/63, 2/64
Federated States of Micronesia, 2/215
Guam, 3/79
Kiribati, 4/17
Korea, 4/23, 4/27
Laos, 4/40–41
Malaysia, 4/95
Manchuria, 3/265
Mariana Islands, 3/79
Marshall Islands, 4/116
Myanmar, 4/179
Nauru, 4/183
Pacific Islands, 5/12
Palau, 5/26–27
Papua New Guinea, 5/34–35
Philippines, 5/66, 5/70–71
Russia, 5/114
Singapore, 5/174
Taiwan, 6/14
Thailand, 6/37
Vietnam, 6/181
relations with
China, 2/63, 2/64–65, 3/253
Germany, 3/265
Italy, 3/265
Peru, 5/47
religion, 3/278–279, *3/278–279*
Japan Communist Party, 3/254
Japan Current, 3/257
Japan Socialist Party, 3/255
Japanese Alps, 3/256
Jarbah Island, 6/54
Jarl, Birger, 5/250
Jaruzelski, Wojciech, 5/74
Jasper National Park (Alberta), 2/39
Java (Indonesia), 3/161, 3/166
disasters, 3/163
economy, 3/170
land, 3/164, *3/165*
Java man, 3/161
Jawahar Tunnel, 3/159
Jawara, David, 3/13
Jawara tribe, 7/16
Jay Pritzker Pavilion (Chicago), *6/158–159*
Jazz, 2/152, 6/139, 6/158, 6/159
Jean, Michaëlle, *2/31*
Jebali, Hamadi, 6/50
Jebel Toubkal peak, 4/169
Jebusite people, 3/216
Jefferson, Thomas, 6/135, *6/135*

Jellabas, 4/165
Jena (Germany), 3/39
Jenney, William Le Baron, 6/158
Jerba Island, 6/53
Jerez (Spain), *5/206–207*
Jermuk Waterfall and Spa (Armenia), *1/67*
Jersey (Channel Islands), 6/100
Jerusalem (Israel), **3/216**, 3/216–217, *3/216–217*
history, 3/212
urban life, 3/215
world religions and, 3/209, 3/216
Jervis Bay, 1/89
Jesuits, 5/38
Jesus Christ, 3/217, 3/233
Jet stream, 3/132
Jews
Belarus, 1/131
Czech Republic, 2/150
Israel, 3/209, 3/214–215, *3/214–215*
Jerusalem's importance, 3/216
Netherlands, 4/207
Netherlands Antilles, 4/211
Poland, 5/80
Russia, 5/116, 5/117
Spain, 5/214
Syria, 5/272
Tunisia, 6/54
United Kingdom, 6/104
United States, 6/156
World War II and, 3/21, 3/212, *3/212*, 4/207, 5/246
Jharkland (India), 3/137
Jhelum River, 3/158, *3/159*, 5/18, 5/22
Jiangsu Province (China), 2/81
Jiao Zhi kingdom, 6/180
Jidai Matsuri festival, *3/266–267*
Jihad, 3/86, 5/151
Jim, Russell, *6/141*
Jiménez de Quesada, Gonzalo, 2/99, 2/106
Joan of Arc, 2/241, *2/241*
Jodhpur (India), *3/149*
Jogaila. *See* Jagiełło, Władysław
Jóhanna Sigurdardóttir, 2/119
Johannesburg (South Africa), 5/192, 5/193, 5/196
John (king of England), 6/110
John (ruler of Portugal), 1/164
John VI (king of Portugal), 5/95
Johnson, Lyndon B., 6/185
Johnson-Sirleaf, Ellen, 4/65
Jola people, 3/12–13
Jomon people, 7/56, 7/60–61, 7/61
Jonathan, Goodluck, 4/239
Jonathan, Leabua, 4/57
Jordan, 3/282–287, *3/282–287*, **3/285**
economy, 3/284–285
government, 3/284
history, 3/282–283
land, 3/286, *3/286–287*
people, 3/286–287, *3/286–287*
relations with
Israel, 3/212, 3/213, 3/282–283
Syria, 5/269

Jordan, Michael, 6/145
Jordan, River, 3/209, 3/282, 3/285
Jordan River Valley, 3/286
Joropo, 6/179
Jos Plateau, 4/240, 4/240, 4/244
Jose I (king of Portugal), 5/84, 5/90
José Marti International Airport, 2/138
Jost van Dyke Island, 6/188
Jostedal Glacier, 4/250
Jotunheimen range, 4/250
Joyce, James, 3/201
Joyner-Kersee, Jackie, 6/145
Juan Carlos I, 5/200, 5/215
Juan Fernández Islands (Chile), 2/53
Juárez (Mexico), 4/134
Juárez, Benito, 4/140
Juba (South Sudan), 5/236
Judah, 3/209
Judaism. See Jews
Judo, 3/276
Julian Alps, 5/183
Julian calendar, 3/71
Julius II (pope), 3/236
Junggar Basin (China), 2/71
Juno and the Paycock (O'Casey), 3/206
Juntas, 2/195
Jura canton (Switzerland), 5/260
Jura Mountains, 2/236, 5/255,
    5/256–257
Justice, Palais de (Belgium), 1/135,
    1/141
Justinian I, 6/62
Jute, 1/118, 1/120, 1/123
Jutes, 6/91, 6/104, 6/110
Jutland Peninsula, 2/156, 2/158–159

K

K2 (mountain), 3/158, 4/190, 5/18
Ka Lae (Hawaii), 6/118
Kabakas, 6/78
Kabalega Falls National Park, 6/80,
    6/80–81
Kabalo (Congo), 2/114
Kabbah, Ahmed Tejan, 5/173
Kabila, Joseph, 2/111
Kabila, Laurent, 2/111
Kabuki theater, 3/276
Kabul (Afghanistan), 1/13
Kachin people, 4/174, 4/178
Kaczynski, Lech, 5/76
Kádár, János, 3/108
Kadavu Island, 2/219
Kadima party, 3/210
Kaffiyyehs, 3/287
Kafirs, 1/15
Kagame, Paul, 5/131
Kaieteur Fall (Guyana), 3/94
Kailas, Mount, 4/193
Kairouan (Tunisia), 6/48
Kaiser Wilhelm Memorial Church (Berlin),
    3/32
Kaisersaal of Frankfurt Cathedral
    (Frankfurt), 3/28
Kakadu National Park (Australia), 7/28
Kalahari Desert, 1/158, 1/159, 4/180

Kalash people, 5/18, 5/21
Kalenjin people, 4/14
Kalevala, 2/222
Kalgoorlie (Australia), 1/79
Kali Gandaki River, 4/191
Kali River, 1/124–125
Kalidasa, 3/128
Kalimnos (Greece), 3/66
Kamakura (Japan), 3/279
Kamarang River, 3/94
Kamba people, 4/14
Kamchatka Peninsula, 5/125
Kamchiya River, 1/183
Kamehameha I, 7/94, 7/95
Kami gods, 3/269, 3/278
Kamikaze suicide missions, 3/267
Kammu, 3/268
Kampala (Uganda), 6/79, 6/79
Kampongs, 3/167, 4/93, 5/176
Kampuchea. See Cambodia
Kamsar (Guinea), 3/89
Kan, Naoto, 3/255
Kana characters, 3/268
Kanada language, 3/138–139
Kanchenjunga, Mount, 3/130, 4/190,
    4/193
Kanchipuram (India), 3/149
Kanda Town (Japan), 3/262
Kandy (Sri Lanka), 5/226–227, 5/227
Kanem kingdom, 2/44–45, 4/236
Kangaroos, 1/74, 1/74–75, 1/79, 1/91
Kankan (Guinea), 3/88
Kansas (United States), 6/122
Kansas City (Kansas), 6/123
Kanto Plain (Japan), 3/256, 3/270
Kanuri people, 4/234, 4/243
Kapil, 3/143
Kapuas River, 3/164
Karachi (Pakistan), 5/20
Karakoram Highway, 4/195
Karakoram Range, 2/70, 3/158, 4/190,
    5/18
Karakul sheep, 6/17, 6/75
Karakum Canal, 6/74, 6/75
Karakum Desert, 6/74
Karamanli mosque (Libya), 4/70–71
Karamojong people, 6/81, 6/81
Karanga people, 6/210, 7/100
Karaoke music, 3/277
Karaouiyine University, 4/166
Karate, 3/276
Karavostasi (Greece), 3/70
Karbala (Iraq), 3/192
Karelia, 2/223, 2/230
Karen people, 4/174, 4/178, 6/31
Kariba Gorge hydroelectric complex
    (Zimbabwe), 6/207
Karimov, Islam, 6/167
Karkonosze Range, 5/79
Karlovy Vary (Czech Republic), 2/155
Karlsbad. See Karlovy Vary (Czech
    Republic)
Karlskirche (Vienna), 1/109
Karolinska Institute, 5/247
Karpathos (Greece), 3/66

Kars (Turkey), 6/70
Karst, 3/77
Karukera, 3/77
Karyas (Greece), 3/70
Karzai, Hamid, 1/13
Kashifs, 5/228, 5/229
Kashmir. See Jammu and Kashmir (India)
Kashmiri language, 3/138
Kat. See Khat
Kata Tjuta (Australia), 1/89
Katanas, 3/266
Katchall island (India), 3/156
Kathmandu (Nepal), 4/184, 4/185,
    4/188–189, 4/193–194
Kathmandu Valley, 4/189, 4/192
Katowice (Poland), 5/83
Kaunda, Kenneth, 6/202, 6/203
Kauri trees, 4/225
Kava, 2/221
Kavango people, 4/181
Kawahla people, 5/235
Kawaiko-chan singers, 3/277
Kawasaki (Japan), 3/270, 3/272
Kawashima, Kiko, 3/265
Kawm Umbu (Egypt), 2/187
Kayagum zither, 4/31
Kayokyoku songs, 3/277
Kazakh people, 2/70, 2/78, 2/78, 2/79,
    4/9, 4/154, 5/116, 6/74
Kazakh Soviet Socialist Republic, 4/8
Kazakhstan, 4/8–9, 4/8–9, 4/9
Kazan (Russia), 5/118
Keita, Modibo, 4/109
Kelantan River, 4/97
"Kelp Highway" migration theory, 7/87
Kemal, Mustafa. See Atatürk, Kemal
Kemerovo (Siberia), 5/127
Kemijoki River, 2/227
Kempenland (Belgium), 1/135
Kenneth I MacAlpin, 6/110
Kennewick Man, 7/71
Kent State University, 6/185
Kentucky (United States), 6/124, 6/125
Kenya, 4/10–15, 4/10–15, 4/13, 7/99
    economy, 4/12, 4/15, 4/15
    government, 4/12–13
    land and wildlife, 4/10/4/10–11
    people, 4/14–15, 4/14–15
Kenya, Mount, 4/10
Kenya African Democratic Union, 4/11
Kenya African National Union, 4/11
Kenya African Union, 4/11
Kenyatta, Jomo, 4/11
Kerala (India), 3/141
Kerekou, Mathieu, 1/146
Keren (Eritrea), 2/202
Kermanshah (Iran), 3/198, 3/199
Key, John, 4/217
Khabarovsk (Siberia), 5/127, 5/128,
    6/129
Khajuraho (India), 3/134–135
Khalid (king of Saudi Arabia), 5/145
Khalkha Mongolian language, 4/154
Khalkidhiki Peninsula, 3/70
Khamenei, Ali, 3/177, 3/183

143

Khammouan Plateau, 4/44
Khamseh people, 3/178
Khan, Yahya. *See* Yahya Khan, Agha Mohammad
Khan (title), 1/14
Khanates, 4/8, 4/38
Khania (Crete), 3/69
Kharif, 3/135
Khartoum (Sudan), 5/232, *5/232–233,* 5/233–235
Khartoum North (Sudan), 5/234
Khasis, 3/138
Khat, 2/163, 6/197
Khin people, 6/180
Khios (Greece), 3/66
Khirokitia (Cyprus), 2/140
Khiva (Uzbekistan), 6/166
Khlamati River, 1/122
Khmer kingdom, 2/8–9, 2/14–15, *2/14–15*
Khmer people, 2/8, 6/31, 6/36, 6/180, 6/187
Khmer Rouge, 2/9–13, 6/182
Khmu people, 4/46
Khoikhoi people, 5/189, 5/192
Khomeini, Ruhollah, 3/175, 3/176–177, 3/183, *3/183*
Khorat Plateau, 6/30, *6/30*
Khosrau II, *3/183*
Khrushchev, Nikita S., 5/115
Khuzistan Plain (Iran), 3/180
Khyber Pakhtunkhwa (Pakistan), 5/15, 5/16, *5/18–19,* 5/20
Khyber Pass, 1/18–19, *1/18–19*
Kibaki, Mwai, 4/13
Kibbutzes, 3/218–219, *3/218–219*
Kidd, Captain, 4/86
Kiel, Treaty of, 4/247
Kieslowski, Krzysztof, 5/73
Kiev (Ukraine), 5/112, 6/82
Kievan Rus, 1/130, 5/112, 6/82
Kigali (Burundi), 1/189
Kikongo language, 2/112
Kikuyu people, 4/11, 4/14, 7/99
Kilchurn Castle (Scotland), 6/96
Kilimanjaro, 6/18, 6/21, *6/21*
Kim Dae-jung, 4/25
Kim Il Sung, 4/18, 4/21
Kim Jong-Il, 4/18, 4/25
Kim Jong-un, 4/18
Kimchi, *2/26*
Kimonos, 3/272
Kinabalu, Mount, 4/96–97, 4/97
Kinabatangan River, 4/97
King, Martin Luther, Jr., 6/145
King Bird Jaguar, *7/81*
Kingdom of... *See under key word, as in* Granada, Kingdom of
Kingship, Age of, 7/57
Kingston (Jamaica), 3/249, 3/250, *3/250*
Kingstown (St. Vincent and the Grenadines), 5/136, *5/136*
Kinh people, 6/186
Kinshasa (Congo), 2/110, 2/114
Kirchberg (Luxembourg), 4/81

Kirchner, Néstor, 1/53, 1/65
Kirghiz people, *1/15*
Kiribati, 4/16–17, **4/17,** *4/17,* 4/182, 5/10, 6/77
Kirkuk (Iraq), *3/194–195,* 3/195, 3/198, 3/199
Kirov, Sergey, 5/123
Kirov Theater (St. Petersburg), *5/116*
Kirtinasha River, 1/122
Kiruna (Sweden), 5/252, *5/252*
Kiswahili language, 4/14
Kites, *4/98, 4/103*
Kiwis, 4/224, *4/224*
Kizilbash people, 1/15
Kjølen Mountains, 4/250
Klang River, *4/95*
Klar River, *5/252*
Klaus, Vaclav, 2/147
Klee, Paul, 5/255
Kluane National Park (Yukon), 2/39
Klyuchevskaya volcano, 5/125
Knesset, 3/210, 3/219
Knights Hospitallers of St. John, 2/143, *3/69,* 4/115
Knorrs, *3/122–123*
Knossos (Crete), 3/68–69
Koalas, 1/74, *1/74,* 1/91
Kobe (Japan), 3/255, 3/274
Kobuk River, 7/74
Kodiak Island, *6/130*
Koguryo, 4/22, 4/30
Kohl, Helmut, 3/18, 3/23
Koizumi, Junichiro, *3/265*
Kola nuts, 5/172
Kolelas, Bernard, 2/117
Kölen Mountains, 5/248
Kolingba, André, 2/43
Kolkata (India), 1/119, 3/136, 3/144, 3/148
Kölln (Germany), 3/33
Kolo dance, 5/165
Kolossi Castle (Cyprus), 2/143
Komati River, 5/240
Kon-Tiki, 5/55
Konare, Alpha Oumar, 4/109
Kong Le, 4/41
Kongo kingdom, 1/39
Kongo people, 1/39, 2/117
Konya (Turkey), 6/71
Kookaburras, 1/74
Kopet-Dag Mountains, 6/74
Kora, 4/112
Koran. *See* Qur'ān
Korcë (Albania), 1/23
Korea, 3/264, 4/22–23
  art and culture, 4/30–31, *4/30–31*
  human settlement, 7/60–61, **7/61**
  See also Korea, North; Korea, South
Korea, North, 4/18–21, *4/18–21,* **4/19**
  climate, 4/20
  history, 4/18–19
  Japanese loyalty, 3/274
  land, 4/20, *4/20–21*
  people and economy, 4/20–21, 4/21

  relations with
    China, 4/18
    Korea, South, 4/18–19, 4/22–23, 4/25
    Soviet Union, 4/18
Korea, South, 4/22–31, *4/22–31,* **4/25**
  culture and art, 4/30–31, *4/30–31*
  economy, 4/28–29, *4/28–29*
  history
    early, 4/22–23
    recent, 4/24–25
  land and climate, 4/26–27, *4/26–27*
  North Korea, relations with, 4/18–19, 4/22–23, 4/25
  people, 4/26–27, *4/27*
Korea Strait, 4/22, 4/26
Korean language, 3/274
Korean Peninsula, 4/19, 4/22, 4/26–27
  human settlement, 7/60–61, **7/61**
Korean people, in Japan, 3/272, 3/274
Korean War, 4/19, 4/21, 4/23, 4/24, 5/246
Korean Workers' Party, 4/21
Koro Island, 2/219
Koror (Palau), 5/26, 5/27
Koryak Mountains, 5/125
Koryo, 4/23
Kos (Greece), 3/66
Kosciuszko, Tadeusz, 5/76
Kosher, 3/215
Kosovo, 4/32–33, *4/32–33,* **4/33,** 5/155, 5/182
  Serbia, relations with, 5/157, 5/160–161
Kosovo Liberation Army, 4/33, 5/157, 5/160
Kosovo Polje, Battle of, 4/32, 4/157, 5/157, 5/158
Kosovska Mitrovica (Kosovo), *4/33*
Kosrae Island, 2/214, 2/215
Kossuth, Louis, 3/110
Kostenki site, 7/44–46
Koštunica, Vojislav, 5/161
Kota Kinabalu (Malaysia), 4/93
Kota Tampan site, 7/23
Kotor, Gulf of, *4/157*
Koudougou (Burkina Faso), 1/187
Kouklia (Cyprus), 2/143
Kountché, Seyni, 4/233
Kourion (Cyprus), 2/143
Kourou (French Guiana), 2/256
Kouroussa (Guinea), 3/88
Kowloon (Hong Kong), 2/84
Kowloon Peninsula, 2/84, 2/85
Kozloduy (Bulgaria), 1/185
Kpelle people, 4/64
Kragujevac (Serbia), 5/165
Krak des Chevaliers fort (Syria), 5/267
Krakatau (Indonesia), *3/165*
Kraków (Poland), *5/77,* 5/82
Krapina Neandertals, Museum of the, *7/34–35*
Krasnoyarsk (Siberia), 5/127
Kremlin (Moscow), 5/120, *5/120–121,* 5/121

Krill, 1/44, *1/45*
Krim, Abd al-, 4/159
Krio language, 5/172
Kristiansand (Norway), 4/254
Kristiansten fortress (Norway), 4/255
Krona, 5/244
Kru people, 2/128
Kruger, Paulus, *5/192, 5/193*
Kshatriyas, 3/140
Kuala Lumpur (Malaysia), 4/93, *4/95,*
    4/96, 4/99
Kubitschek, Juscelino, 1/165
Kublai Khan, 2/62, *2/63,* 2/69, 4, 4/150,
    4/178, 151
Kuching (Malaysia), 4/93, *4/102–103*
Kuenga (Siberia), 5/129
Kufuor, John A., 3/42, 3/45
Kujala Kadphises, 1/10
Kumasi (Ghana), 3/42
Kunlun Range, 2/70
Kuomintang, 2/64
Kuopio (Finland), *2/226–227*
Kurdish language, 3/192, 3/198, 5/272
Kurdish people
    Armenia, 1/67
    Iran, 3/178
    Iraq, 3/188, 3/189, 3/191, 3/192,
        3/192–193, 3/193, *3/198–199,*
        3/198–199, 4/54
    Syria, 5/272
    Turkey, 6/59
Kurdistan, *3/190–191,* 3/191,
    3/198–199, **3/199**
Kurfürstendamm (Berlin), 3/32
Kuril Islands (Japan), 3/258
Kurosawa, Akira, 3/276
Kuršiu Marios lagoon, 4/76
Kushan Empire, 1/10–11, 3/128, 5/14,
    5/228, 6/166
Kuwait, 3/186, 4/34–37, **4/35,** *4/35–37*
    economy, 4/36–37, *4/36–37*
    history, 3/188, 3/195, 4/34–35, 6/136
    Syria, relations with, 5/269
    See also Persian Gulf War of 1991
Kuwait (city), 4/34
Kuwait Oil Company, 4/36
Kuznetsk Basin, 5/124
Kwa-Zulu Natal (South Africa), 5/192,
    5/195
Kwahu Plateau, 3/40
Kwajalein atoll (Marshall Islands), 4/117
Kwekwe (Zimbabwe), 6/207
Kyffhäuser Mountains, 3/38
Kyoto (Japan), 3/264, 3/268–269, *3/268–*
    269, 3/274, 3/280, *3/280 281*
Kyrenia (Cyprus), *1/143*
Kyrenia Mountains, 2/142
Kyrgyz people, 4/39
Kyrgyzstan, 4/38–39, *4/38–39,* **4/39**
Kyushu Island (Japan), 3/253, 3/256
Kyz Kalasy tower (Azerbaijan), 1/110
Kyzylkum Desert, 6/166

L

Labor Party (Barbados), 1/129

Labor Party (Israel), 3/210, 3/213
Labour Party (Ireland), 3/202
Labour Party (New Zealand), 4/217
Labour Party (United Kingdom), 6/107
Labuan (Malaysia), 4/93, 4/96
Labyrinth, 3/68
Lacandón Indians, *4/130*
Laccadive Islands (India), **3/157**
Lace, 1/139, *1/140,* 1/143, *1/167*
La Cima (Peru), 5/51
La Condamine (Monaco), 4/149
Ladakh (India), 3/159, 4/193
La Désirade (Guadeloupe), 3/76
Ladies Day (United Kingdom), 6/108,
    6/109
Ladin language, 3/232
Ladinos, 2/198, 3/84–85
Ladysmith (South Africa), *5/193*
Laetoli site, *7/8*
Lagoa das Sete Cidades (Azores), *1/112*
Lagoa do Fogo (Azores), *1/112*
Lagos (Nigeria), 4/236, 4/237, *4/240,*
    4/242
Lagos Escobar, Ricardo, 2/51
Lagting, 2/213
La Guajira Peninsula, 2/104
Laguna de Bay (Philippines), *5/68–69*
Lake Albert, 6/80
Lake Argentino, *1/55*
Lake Assal, *2/163*
Lake Baikal, 5/125, *5/125–126, 7/62–63*
Lake Balaton, 3/112, *3/112–113*
Lake Chad, 2/20, 2/44, 2/46, *2/47,*
    4/234
Lake Chilwa, *4/91*
Lake Constance, 1/100, 1/107, 5/257
Lake Country (Chile), *2/52, 2/53, 2/54*
Lake District (England), 6/92, *6/92*
Lake Edward, 6/80
Lake Enriquillo, 2/166
Lake Eyre, *1/73*
Lake Garda, *3/224–225*
Lake Geneva, 5/257
Lake George, 6/80
Lake Guatavita, *2/106–107*
Lake Issyk-Kul, 4/38
Lake Itasca, 6/148
Lake Izalbal, 3/82
Lake Lanao, 5/64
Lake Mälaren, 5/249, 5/250
Lake Mamry, 5/78
Lake Managua, 4/228
Lake Mansarowar, 4/193
Lake Maracaibo, 6/172, 6/176, 6/179
Lake Mead, 6/133
Lake Mungo, 7/26, *7/27*
Lake Nicaragua, 4/228, *4/229*
Lake Nios, 2/21, **2/21**
Lake Nyasa, 4/90, 6/21
Lake Ontario, 2/36
Lake Poopó, 5/54
Lake Siecha, *2/106*
Lake Siljan, *5/248–249,* 5/249
Lake Sniardwy, 5/78
Lake Tana, *2/207, 2/208*

Lake Tanganyika, 1/188, 6/18
Lake Texcoco, 4/138, 4/140
Lake Titicaca, 1/152, 5/44, **5/54,**
    5/54–55, 5/54–55
Lake Toba, *3/164*
Lake Turkana, *4/15*
Lake Valencia, 6/177
Lake Van, *6/60–61,* 6/61
Lake Vänern, 5/249
Lake Vättern, 5/249
Lake Victoria, 4/11, 6/21, 6/25, 6/80
Lake Volta, 3/40, 3/41
Lake Wörther, 1/101, *1/101*
Lake Xochimilco, *4/138–139*
Lake Ypacaraí, *5/42–43*
Lakeba Island, 2/219
Lakes, 1/4
Lakhon dance, *6/35*
Lakshadweep (India), 3/156–157,
    3/156–157, **3/157**
La Libertad (El Salvador), 2/199
La Macarena National Park (Colombia),
    2/103
Lamaism, 1/150, 1/151, 2/94–97, 4/151,
    4/155
Lamaseries, 2/96
Lambaréné (Gabon), 3/8
Lamet people, 4/46
Lan Xang kingdom, 4/40
Lanao, Lake, 5/64
Land of the Midnight Sun, 2/222, 4/246
Land's End (England), 6/92
Landsgumeinde (Switzerland), 5/256,
    5/259
Landtags (Austria), 1/95
Lang Rong Rien Cave, 7/23
La Nueva Isabela (Dominican Republic),
    2/167
Lanzarote (Canary Islands), 5/218, 5/219,
    5/220, *5/220–221,* 5/221
Lao Lum people, 4/46
Lao people, 4/40, 4/46
Lao People's Revolutionary Party, 4/42,
    4/43
Lao Sung people, 4/46
Lao Thoeng people, 4/46
Laocoön (statue), 6/171
Laos, 4/40–47, *4/40–47,* **4/43**
    Chinese culture, 2/79
    climate, 4/44
    economy, 4/44–45, *4/45*
    history
        early, 4/40–41
        recent, 4/42–43
    land, 4/44, *4/44–45*
    people, 4/46–47, *4/46–47*
La Palma (Canary Islands), 5/218, *5/218,*
    5/219, 5/220
La Paz (Bolivia), 1/152, *1/152–153,*
    1/155
Lapilli, 5/221
Lapis lazuli, 1/17
Lapita Culture, 7/91
    pottery, *7/92, 7/93*
Lapland (Norway), 4/246

La Plata (Argentina), *1/56–57*
    Viceroyalty of, 1/64–65, 5/39, 6/160
Lapps. *See* Sami
La Quebrada cliffs, 4/142–143
La Rioja (Spain), *5/208*
La Savane square (Fort-de-France), 4/119
Lascaux Cave (France), 2/240
La Serena (Chile), 2/49
Las Palmas (Canary Islands), 5/218
Lassi people, 5/20
Last Judgment, The (Michelangelo), 6/171
Last Night at the Proms, 6/109
Las Vegas (Nevada), *6/126–127*
Latakia (Syria), 5/271
Lateran, Treaty of the, 6/171
Latifah, Queen, 6/145
Latin language, 2/240, 3/34
    Ireland, 3/204
    Portugal, 5/94
    Romania, 5/106
    Spain, 5/212
    Syria, 5/267
Latvia, 4/48–49, *4/48–49,* **4/49**
Latvian people, 4/48, 4/49
Laurentian Mountains, 2/38
Laurier, Wilfred, 2/31
Lausanne (Switzerland), 5/261
Lausanne, Treaty of, 3/199, 6/198
Lava-lavas, 5/138
Lavaux vineyards, *5/260–261*
Laven people, 4/46
Lavender, *2/249*
Law and Justice party (Poland), 5/75
LDP (Japan), 3/254
Le Corbusier, 5/255
Le family, 6/180–181
Lead, *1/63*
League of Nations
    Finland and, 2/223
    Marshall Islands and, 4/116
    Nauru and, 4/182
    Papua New Guinea and, 5/34
    Sweden and, 5/246
    Switzerland and, 5/265
    Syria and, 5/267
    Yap Islands and, 2/215
Leather-Stocking Tales, The (Cooper),
    6/158
Lebanon, 4/50–55, *4/50–55,* **4/53**
    government, 4/52
    history, 3/213
        early, 4/50–51
        recent, 4/51–53
    people, 4/54–55, *4/54–55*
    Syria, relations with, 5/269
Lebanon, Mount, 4/51
Le Clos-Luce house, 2/242
Lee Myung-bak, 4/25
Lee Teng-hui, 6/13, 6/15
Leeds (England), 6/93
Leeward Islands, 4/58, 4/62–63,
    4/62–63, **4/63,** 6/188, 6/190
Left Bank (Paris), 2/239, 2/246
"Legend of Sleepy Hollow, The" (Irving),
    6/158

Legislative Palace (Tegucigalpa), *3/103*
Legislative Yuan, 6/12
Legong dance, *3/169*
Leiden (Netherlands), 4/204
Leidseplein (Amsterdam), *4/205*
Lempa River, 2/194, 2/197
Lemurs, 4/88–89, *4/88–89*
    ring-tailed, 4/88, *4/88–89*
Lemus, José María, 2/195
Lena River, 5/118, 5/125
Lene Hara Cave, 7/24
Lenin, Vladimir I., 4/170, 5/114, *5/115,*
    5/122
Lenin shipyard (Gdańsk), 5/77
Leningrad (Russia). *See* St. Petersburg
    (Russia)
León (Nicaragua), 4/226
Leonardo da Vinci, 2/242, 2/243, 2/247,
    3/221, *3/229, 3/236, 3/237,* 6/171
Leopards, 2/74, *4/185*
Leopold I (king of Belgium), 1/134
Leopold II (king of Belgium), 2/110
Lepchas, 3/153
Lérida (Spain), 5/202
Lesbos (Greece), 3/66
Lesotho, 4/56–57, *4/56–57,* **4/57**
Lesotho Bank, 4/56
Lesotho Congress for Democracy, 4/57
Lesser Antilles, 4/58, 4/58–63,
    4/58–63, **4/63**
    See also specific islands and countries
Lesser Nicobar Island (India), 3/156
Lesser Sunda Islands (Indonesia), 3/164
Letsie III, 4/57
Letzburgesch language, 4/80
Levant region, 7/10, 7/14, **7/17,** **7/32,**
    7/32–33, *7/32–33,* 7/42
Lhasa (Tibet), 2/95, 2/96, *2/96–97*
Li Peng, 2/60
Li River, 2/59
Li-wu River, *6/10–11*
Liang Bua cave, 7/25
Liba (Poland), 5/78
Liberal-Democratic Party (Japan), 3/254
Liberal Party (Colombia), 2/99, 2/100
Liberation Front of Occupied South
    Yemen, 6/199
Liberation Tigers of Tamil Eelam, 5/225
Liberation Tower (Kuwait), 4/35
Liberté, Egalité, Fraternité, 2/234
Liberty, Age of (Sweden), 5/243
Libreville (Gabon), 3/9, *3/9,* 3/11
Libuše, 2/150
Libya, 1/24, 4/66–73, *4/66–73,* **4/67**
    government, 4/66
    history, 4/66–67
    land and economy, 4/68–69, *4/68–69*
    people, 4/70–71, *4/70–71*
    power in
        Chad, 2/45
        Western Sahara, 6/193
Libyan Arab Airlines, 4/69
Libyan Desert, 2/181, 4/68, 5/232
Liechtenstein, 1/100, 4/74–75, *4/74–75,*
    4/75

Liechtenstein, Johann-Adam, 4/75
Liège (Belgium), 1/133
Lignite, 1/23, 1/105, 3/61
Ligurian dialect, 3/232
Ligurian Sea, 3/224
Lihir Island, 5/37
Likud bloc, 3/210
Lilies, Turks-cap, *1/35*
Lim Fiord (Denmark), 2/158
Lima (Peru), 5/48–49, 5/50
Limassol (Cyprus), *2/142,* 2/143
Limnos (Greece), 3/66, *3/66*
Limpopo River, 6/206
Lincoln, Abraham, *6/117,* 6/133, 6/135
Lincoln Memorial, *6/117*
Linden (Guyana), 3/95
Line Islands, 4/16, 4/17
Line of Demarcation, 1/164
Lingala language, 2/112
Linz (Austria), 1/97, 1/102, *1/104,* 1/107
Lion Gate, *3/62*
Lion Rock (Sri Lanka), *5/226*
Lions, *6/24–25*
Lisbôa da Costa, António Francisco,
    1/172
Lisbon (Portugal), 2/86, *5/84–85, 5/85,*
    5/87, 5/88, 5/90
Lissouba, Pascal, 2/117
Lisu people, 6/30
Liszt, Franz, 3/39, *3/110,* 3/117
Literature
    China, 2/66, 2/67
    France, 2/252
    Ireland, 3/206, *3/207*
    Italy, 3/237
    Japan, 3/268
    Korea, 4/31
    Mexico, 4/145
    Serbia, 5/165
    Spain, 5/215
    Switzerland, 5/255
    United States, 6/158
Lithuania, 4/76–77, *4/76–77,* **4/77**
    power in
        Belarus, 1/130
        Latvia, 4/48
        Ukraine, 6/82
Little Caucasus Mountains, 3/14
Little Cayman (Cayman Islands), 3/48
Little Mermaid, The (statue), *2/161*
Little Wild Goose Pagoda (China), 2/80
Liverpool (England), 6/93
Living Buddha, 4/151
Livingstone, David, 4/91, 6/206
Livonia, 4/48
Lizard Point (England), 6/92
Lizards, 2/178, *2/179*
Llaima Volcano, *2/52*
Llamas, *5/48*
Llanberis (Wales), *6/94*
Llaneros, 6/176
Llanos, 2/102, 6/176
Loa River, *2/53*
Lobi people, 1/186, 1/187
Lobito (Angola), *1/38*

Loch Awe, *6/96*
Lochs (Scotland), 6/97
Locks (canal), 5/32, *5/33*
Locomotives. *See* Railroads
Loess, 2/72
Loess Hills (China), *2/71*
Lofoten Islands (Norway), *4/250*, *4/251*
Logan, Mount, 2/39
Loire River, 2/242, 2/243
Loire Valley (France), 2/233, 2/242–243, *2/242–243*, **2/243**, 2/248
Lom (Bulgaria), *1/184*
Lomami River, 2/114
Lombard dialect, 3/232
Lombards, 3/227
Lomé (Togo), *6/41*
Lomwe people, 4/90
Lon Nol, 2/9
London (England), 6/102–103, *6/102–103*, **6/103**
    economy, 6/93, *6/93*, 6/106, *6/107*
    festivals, 6/108–109, *6/108–109*
    history, 6/102, *6/113*
London, City of, 6/102–103
London, University of, 6/93
London Missionary Society, 2/118
London Stock Exchange, *6/107*
Long houses, 3/166
Long Island (New York), *6/156*
Long March, 2/64, **2/64**
Long Mountains, 4/250
Long Range Mountains, 2/38
Longshan culture, 7/57
Longyear, John Munro, 4/257
López, Carlos Antonio, 5/39
López, Francisco Solano, 5/39
López Arellano, Osvaldo, 3/101
Lord's Resistance Army, 6/79
Lorenzo the Magnificent, *3/229*, 3/236
Lorraine, Duke of, *1/98–99*
Los Angeles (California), 6/128, 6/129, 6/142
Los Glaciares National Park, *1/54–55*
Los Chorros (El Salvador), 2/199
Lost-wax process, 4/245
Louangphrabang (Laos), 4/40
Lough Neagh, 6/98
Loughs, 6/98
Louis XIV (king of France), 2/244
Louis XVI (king of France), 2/244, *2/244*, 2/247
Louis Napoleon. *See* Napoleon III
Louisiade Archipelago (Papua New Guinea), 5/36
Louisiana (United States), 6/124
Louisiana Purchase, 6/135
Louvre (France), 2/247, *2/247*
Low Countries, 4/202
Low German, 3/19
Lower Algarve (Portugal), 5/89
Lower Austria, 1/101–101, *1/102*, *1/103*
Loya jirga, 1/13
Loyalty Islands, 4/212
LRA (Uganda), 6/79
Lu people, 4/46

Lua people, 6/31
Lualaba River, 2/114
Luanda (Angola), *1/37*, 1/38
Luanda-Chokwe people, 1/39
Luangwa River, *6/204*
Lucayo Indians, 1/114
Luganda language, 6/81
Lugo, Fernando, 5/41
Luhya people, 4/14
Lukashenko, Aleksandr, 1/131
Lula da Silva, Luiz Inácio, *1/163*, *1/165*
Lully, Jean Baptiste, 2/252
Lumber. *See* Forests and forest products
Lüneburger Heide (Germany), *3/25*
Luo people, 4/14
Lurs people, 3/178
Lusaka (Zambia), 6/205
Lusignan family, 2/140
Lusitania (ancient Rome), 5/94
Lusty Man (carving), *6/99*
Luther, Martin, 3/20, *3/21*, 3/38
Lutheran Cathedral (Finland), 2/228, *2/228–229*
Lutheranism
    Denmark, 2/156
    Finland, 2/222
    Germany, 3/19
    Sweden, 5/243
Luvua River, 2/114
Luxembourg (city), 4/78–81, *4/80*, *4/81*
Luxembourg (country), 1/133, 4/78–81, *4/78–81*, **4/79**
    history, 4/76–77
    people and economy, 4/80–81, *4/80–81*
Luzon (Philippines), 5/60, 5/61, 5/63, 5/64, 5/66, *5/68*, *5/69*
Ly family, 6/180
Lydian people, 6/68
Lyon (France), 2/251

M

M. V. Lomonosov Moscow State University, 5/121
Ma Ying-jeou, 6/13
Maasai people, 4/14, *4/14*, 6/22, *6/22–23*, 6/24
Maasai Steppe (Tanzania), 6/21
Macapagal-Arroyo, Gloria, 5/71
MacArthur, Douglas, 5/71
Macau (city, Macau), 2/86
Macau (region, China), 2/61, 2/86–87, *2/86–87*, **2/87**, 5/87
Macauley, Vincent, 7/12
Macdonald, John A., 2/31, *2/31*
Macedonia, 2/130, 3/62, 4/82–83, *4/82–83*, **4/83**
    Serbia, relations with, 5/159, 5/160
    Taiwan conquest, 6/16
Macedonia-Thrace (Greece), 3/54
Macedonian language, 4/82
Machiavelli, Niccolò, 3/237
Machu Picchu, 5/44, **7/82**, *7/82*
Macías Nguema, Francisco, 2/201
Mackenzie, William, 2/31

Macumba, *1/166*
Madagascar, 4/84–89, *4/84–89*, **4/87**, 7/92
    government, 4/86
    history, 4/86–87
    wildlife, 4/88–89, *4/88–89*
Madan, 3/191
Madeira Island, 1/113
Madeira Islands, 1/112, 1/113, **1/113**, *1/113*
    Portugal, relations with, 5/87, 5/95
Madhya Pradesh (India), 3/137
Mádl, Ferenc, 3/109
Madras (India). *See* Chennai (India)
Madrid (Spain), 5/199, *5/205*, 5/210–211, *5/210–211*
    economy, 5/209
    history, 5/201
Madrid, Treaty of, 1/164
Madrid Protocol, 1/46–47
Madurese people, 3/166
Maelstrom Current, 4/251
Maeterlinck, Maurice, 1/133
Mafia, 3/223, 3/275
Magdalena River, 2/102
Magdalenian Culture, 7/49
Magellan, Ferdinand, 2/53, 3/79, 5/57, 5/70, 7/88, *7/88*
Magerit fortress (Spain), 5/210
Maghreb, 1/24, 4/66, 4/166, 6/48
Magi, 3/233
Magic Lantern Theater, 2/153
Magna Carta, 6/90, 6/110
Magnesite, 1/105
Magpies, Yellow-billed, *4/185*
Magyar language, 3/106, 3/117
Magyars, 2/148, 3/106, 3/110, 3/116, 5/178
Mahabad (Iran), 3/198
Mahabharata, 3/154
Maharaja Sagar, 3/143
Maharaja's Palace (India), *3/149*
Mahathir bin Mohamad, 4/95
Mahavelona (Madagascar), 4/85
Mahé (Seychelles), 5/170
Mahendra, 4/186
Mahler, Gustav, 1/106
Mahoganies, Philippine, 5/68
Mahouts, 6/32
Mahuad, Jamil, 2/173
Maiden's Tower (Azerbaijan), 1/110
Maidu Indians, 6/140
Maiduguri (Nigeria), *4/242*
Main-Danube Canal, 5/181
Main River, 3/24, 3/28
Maïnassara, Ibrahim Baré, 4/233
Maine (ship), 2/132
Maine (United States), 6/120, 6/132
Mainz (Germany), 3/35, 3/38
Maire, Jakob le, 6/44
Maisons de la Culture, 2/252
Maize. *See* Corn
Majardah River, 6/52
Majardah Valley, *6/52*
Majdanpek (Serbia), *5/162–163*

Maji Maji rebellion, 6/19
Majlis, 3/176, 3/177
Major, John, 6/107
Majorca Island, 5/203, 5/216, *5/216–217*, **5/217**
Majuro (Marshall Islands), *4/117*
Makalu, Mount, 4/190
Makarios III, 2/140–141
Makati (Philippines), *5/66–67*
Makhosetive, 5/241
Makonde people, 6/22
Makonnen, Ras Tafari. *See* Haile Selassie I
Makrani people, 5/20
Makua-Lomwe people, 4/173
Malá Strana (Prague), 2/150, *2/150–151*
Malabo (Equatorial Guinea), 2/200
Malacca, Strait of, 4/96, 4/102
Malagasy language, 4/86
Malaita Island, 5/185
Mälaren, Lake, 5/249, 5/250
Malaria, 4/94, 4/108, 5/32, 5/225
Malate Church (Manila), 2/59
Malawi, 4/90–91, **4/91**, *4/91*
Malawi, Kingdom of, 4/91
Malay language. *See* Bahasa Malaysia language
Malay Peninsula, 4/93, 5/174, 6/30
Malay people
    Brunei, 1/179
    Indonesia, 3/166
    Malaysia, 4/93, 4/98, *4/98*
    Philippines, 5/62
    Singapore, 5/176, 5/177
    Sri Lanka, 5/226
    Thailand, 6/31
Malaya, 4/94
Malaya Siya site, 7/62
Malayalam language, 3/139, 3/156
Malaysia, 1/178, 4/92–103, *4/92–103*, **4/94**
    climate, 4/96
    economy, 4/95, 4/100–101, *4/100–101*
    government, 4/95
    history, 4/94–95, 4/102–103, *4/102–103*
    land, 4/96–97, *4/96–97*
    people, 4/93, 4/98–99, *4/98–99*
    population, 4/93
    prehistoric migration, 7/23
    Singapore in, 5/174
Maldives, 4/104–107, **4/105**, *4/106*
    coral reef life, 4/106–107, *4/106–107*
    history, 4/104
    land and climate, 4/104–105
    people, 4/105
Male (Maldives), **4/105**
Mali, 1/187, 2/128, 4/108–113, **4/109**, *4/110–113*, 4/232
    arts, 4/112–113, *4/112–113*
    history, 4/108–109
    land, 4/110
    people and economy, **4/110**, *4/110–111*, *4/110–111*
Mali Empire, 3/13, 3/86, 4/108, 4/120, 5/152

Mali People's Democratic Union, 4/108
Maliki, Nouri Kamel al-, 3/189
Malinke people, 3/86, 3/88, 4/110
Mall of America, 6/156
Mall of the Emirates (Dubai), *6/86–87*
Mallaig (Scotland), 6/97
Mallows, Carolina, 6/149
Malta, 4/114–115, *4/114–115*, **4/115**
Malta Culture, 7/62–63, *7/63*
Malta Island, 4/114, 115
Maltese language, 4/114
Malula (Syria), *5/272–273*
Mama Ocllo, 5/52
Mamluks, 1/66, 2/192, 4/51, 5/267
Mammoths, 7/47–49, 7/62, 7/64, *7/64–65*, *7/65*, 7/72, *7/73*
Mamontovaya Kurya site, 7/64
Mamry, Lake, 5/78
Mana Island, *2/218–219*
Managua (Nicaragua), 4/229, 4/230
Managua, Lake, 4/228
Manama (Bahrain), *1/117*
Manas chronicle, 4/39
Manaus (Brazil), *1/165*
Manchester (England), 6/93
Manchu dynasty, 6/14
Manchuria, 2/64, 2/72, 3/264, 5/129
    power in
        China, 2/62
        Mongolia, 4/151
Manchurian Plain, 2/72
Manchus
    China, 2/62, 2/64, 2/85
    Korea, 4/23
    Mongolia, 4/151
Manco Capac, 5/52
Mandarin Chinese language, 2/76, 2/82, 2/96, 5/176, 6/14
Mandé people, 1/186, 1/187, 2/128
Mandela, Nelson, 5/189, 5/190–191, *5/193*, 5/195, *5/195*, 5/247
Mandinka people, 3/12, 3/13, 3/90, 4/110
Manet, Édouard, 2/253
Manganese, 3/11
Mangroves, 2/219, 3/86, 5/36–37
Manikarnika ghat, 3/146
Manila (Philippines), 5/59, 5/61, 5/66–67, *5/66–67*, **5/67**, 5/69
Manila Bay, 5/66
Manipur (India), 3/126
Manitoba (Canada), 2/26, *2/32–33*, 2/39
Manjako people, 3/90
Manley, Michael, 3/249
Mann, Thomas, 3/17
Manneken-Pis (statue, Brussels), 1/140
Mannerheim, Carl Gustaf, *2/225*
Mansa Musa, 4/108
Mansarowar, Lake, 4/193
Manufacturing. *See specific places*
Manukas, 4/224
Manx language, 6/101
Manzanillo (Mexico), 4/143

Mao Zedong, 2/64, 2/65, *2/65*, 2/79, *2/80*, 2/82
Maoists, in Nepal, 4/187
Maori, 2/118, 4/215, 4/220, **4/222**, *4/222–223*, *4/222–223*
    British colonization and, 7/107, *7/107*
    New Zealand settlement by, 7/96–97, *7/96–97*
Maori language, 4/220, 4/223
Mapuche Indians, 2/48–49, 2/54
Maputo (Mozambique), *4/171*, 4/173
Maquis, 3/55
Marabouts, 4/111
Maracaibo (Venezuela), *6/175*, 6/177
Maracaibo, Lake, 6/172, 6/176, 6/179
Maracaibo Basin, 6/176
Maranao people, 5/65
Marathas, 3/129, 3/154
Marathi language, 3/138
Marble, *3/241*
Marchfeld, Battle of, 1/98
Marco Polo. *See* Polo, Marco
Marcos, Ferdinand E., 5/71, *5/71*
Marcos, Imelda, 5/71
Margaret (queen of Denmark and Norway), 2/156, 4/247, 5/243
Margaret Island, 3/114
Margherita Peak, 6/80
Maria Theresa, 1/98, *1/98*
Mariachis, 4/145
Mariana Islands, 3/78–79, *3/78–79*, **3/79**, 5/11, 5/12, 7/93
Mariana Trench, *3/78*
Mariánsk Lázně (Czech Republic), 2/147
Marib Dam (Yemen), *6/198*
Marie Antoinette, 2/247
Marie-Galante (Guadeloupe), 3/76
Marienbad (Czech Republic), *2/147*
Marignano (Italy), 5/258
Marigot (St. Martin), 4/62
Marimbas, *4/144–145*, 4/145
Marines, U.S., in Nicaragua, 4/226
Marinus, St., 5/140
Maritime Provinces (Canada). *See* Atlantic Provinces (Canada)
Marka people, 1/186, 1/187
Market Square (Helsinki), 2/229
Markets of Trajan, 3/231
Markina Gora site, 7/45
Marley, Bob, 3/250–251, *3/251*
Marley, David "Ziggy," 3/250, 3/251
Marmara (Turkey), 6/66
Marmara, Sea of, 6/62, 6/63, 6/69
Marmaris (Turkey), 6/72
Marmas tribe, 1/124
Maromokotro, 4/84
Maronite Church, 4/52, 4/54
Maronite people, 2/142
Maroons, 2/257, 3/247, 5/239, *5/239*
Marquesas Islands, 2/258, 7/94–95
Marrakech (Morocco), 4/166, *4/166–167*, 4/168
Marriage of Figaro, The (Mozart), 1/106
Marsamxette harbor (Malta), 4/114
Marsh Arabs, 3/191, *3/197*

Marshall, John, 4/116
Marshall Islands, 4/116–117, *4/116–117*, **4/117**, 5/10, 5/11, 5/12–13, 7/93
Marsupials, 1/74, 1/75, *7/29*
Martelly, Michel, 3/97
Martí, José, 2/132
Martial arts, 3/276
Martín Fierro (Hernández), 1/58–59
Martinelli, Ricardo, 5/29
Martinique, 4/58, 4/60, 4/118, *4/118–119*, **4/119**
Marx, Karl, 1/37, 4/170
Marx-Engels Platz (Berlin), 3/32
Marxism. *See* Communism
Mary (queen of England), 6/111, 6/112
Mary, Virgin, 5/207
Maryland (United States), 6/124, 6/134
Marzouki, Moncef, 6/50
Masai people. *See* Maasai people
Masaryk, Tomáš G., 2/148, 2/152, *2/152*
Mashona people. *See* Shona people
Masjid-e-Imam mosque, 3/185
Masks, 4/113, *4/135*
Mason-Dixon Line, 6/124
Massachusetts (United States), 6/120
Massif Central (France), 2/237, 2/248, 2/249
Massinissa, 1/24
Masurian Lake District (Poland), 5/78–79
Masvingo (Zimbabwe), 6/210, 7/100
Matabele people. *See* Ndebele people
Matadi (Congo), 2/114
Matcharak Lake, 7/74–75
Maté, 1/55, 1/59, *5/38–39*, *6/160*
Matejko, Jan, 5/73
Mathematics, 2/181, 3/100
Matmata hills (Tunisia), *6/55*
Matmatah (Tunisia), *6/52–53*
Matsu Islands, 6/10
Matsuri festivals, 3/278, *3/278*
Matthias of Arras, 2/151
Matucana (Peru), 5/51
Mau Mau, 4/11
Maurice (prince of Nassau), 4/125
Mauritania, 4/120–123, **4/121**, *4/121–123*
   history, 4/120–121
   people and economy, 4/122–123, *4/122–123*
   Western Sahara, power in, 6/193
Mauritius, 4/124–125, *4/125*, **4/125**
Mauritius Island, 4/124
Mauritshuis Museum, 4/205
Maurya, Chandragupta, 3/128
Mauryan Empire, 3/128, 3/154, 5/14
Mausoleum of Halicarnassus, 6/73
Mausolus, 6/73
Max Planck Institute for Evolutionary Anthropology, 7/34–35, 7/45
Maximilian I, 1/100
May Day, 2/229, *3/34*
Maya
   Belize, 1/145, *1/145*
   Guatemala, 3/80, *3/80–81*, 3/82, 3/84, *3/84*

Honduras, 3/100, *3/100–101*
Mexico, 2/194, *4/135*, 4/140, *4/142*, 4/144
   rise of, 7/81, *7/81*
Maya Mountains, 1/144
Mayan language, 1/145, 3/84, 4/130, 4/137
Mayflower Compact, 6/134
Mayombé Escarpment, 2/116, *2/117*
Mayotte (Comoros), 2/108
Mazandarani people, 3/178
Mazar-e Sharif (Afghanistan), *1/10–11*, 1/11, *1/14–15*
Mazaruni River, 3/94
Mazatlán (Mexico), 4/143
Mazowiecki, Tadeusz, 5/75
Mba, Leon, 3/9
Mbandaka (Congo), 2/114
Mbanza, 1/39
M'Bochi people, 2/117
Mbundu people, 1/39
Mbuti people, 2/110, 2/112
McKinley, Mount, 6/130
McKinley, William, 2/132
McLean Lake (Saskatchewan), *2/34*
McMurdo Station, *1/46–47*
MDC (Zimbabwe), 6/209
Mead, Lake, 6/133
Meat industry, 1/57, *1/57*, 1/59, 1/62
Mecca (Saudi Arabia), 3/282, 5/21, 5/144, *5/150*, 5/150–151, 5/213
Mechelen (Belgium), *1/139*
Mecit, Abdul, 6/70
Medellín (Colombia), 2/100
Media kingdom, 3/182
Medici, Lorenzo de', 3/236
Medici family, 3/236
Médicis, Catherine de, 2/242
Medina (Saudi Arabia), 5/144, 5/150, 5/151
Medinas, 4/166
Meditation, 3/279
Mediterranean Sea
   Algeria on, 1/26, 1/28
   Armenian empire on, 1/66
   Balearic Islands in, 5/216, *5/216–217*
   Corsica on, 2/254, *2/255*
   Cyprus on, 2/140
   France on, 2/248
   Gibraltar on, 3/46
   Greece on, 3/55, *3/55*
   human migration and, 7/10, *7/32*, *7/40*
   Israel on, 3/209, *3/215*
   Italy on, 3/224
   Lebanon on, 4/50
   Libya on, 4/68
   Malta on, 4/114
   Monaco on, 4/148
   Montenegro on, 4/156, *4/157*
   Morocco on, 4/158
   Roman conquest, 3/226
   Sicily on, 3/244
   Spain on, 5/199
   Syria on, 5/266, 5/267, 5/270
   Tunisia on, 6/48

Turkey on, 6/57, 6/67, 6/72–73, *6/72–73*
   Venetian control, 3/242
Medvedev, Dmitry, 5/111
Meech Lake Accord, 2/32
Meeting houses, Maori, *4/222*
Megafauna, 7/28–29, *7/29*, 7/72–73, *7/73*
Megalopolises, 6/121
Meghna River, 1/118, 1/122
Mehmet II, 6/68, *6/69*, 6/70
Meiggs, Henry, 5/50
Meiji period, 3/262, 3/264
Meiji Shrine (Tokyo), 3/270
Mejato atoll, 4/117
Meknès (Morocco), 4/166, *4/166*, 4/168
Mekong Basin, 4/40, 4/44
Mekong Delta (Vietnam), 6/186
Mekong River
   Cambodia, 2/8, 2/12, *2/12–13*
   Laos, 4/40, 4/43, 4/44, 4/46
   Thailand, 6/30
   Tibet, *2/94*
   Vietnam, 6/183, 6/186, *6/186*
Melaka (Malaysia), 3/168, 4/96, 4/99, 4/102
Melanesia, 5/10, 5/12, 5/184
   Austronesian settlement, 7/92–93
Melanesians
   Fiji, 2/216, 2/220
   genetic studies, 7/92–93
   Nauru, 4/182
   New Caledonia, 4/212
   Pacific Islands, 5/10
   Papua New Guinea, 5/34
   Solomon Islands, 5/184, *5/185*
   Vanuatu, 6/168, 6/169
Melbourne (Australia), 1/69, 1/72, *1/77*, 1/83, 1/84–85, *1/85*, 1/89
Melekeok (Palau), 5/27
Melk (Austria), *1/103*
Mella, Ramón, 2/167
Mellars, Paul, 7/14
Melons, *5/38–39*
Melting pot concept, 1/166, 6/139
Melville, Herman, 6/9
Memorial Gardens (Melbourne), *1/85*
Menai Strait, 6/94
Mendaña, Alvaro de, 5/185, 6/77
Mende people, 5/172
Mendez, Aparicio, 6/163
Mendoza (Argentina), 1/63
Mendoza, Pedro de, 1/62, 1/64
Menelik II, 2/206
Menem, Carlos Saúl, 1/52–53, 1/65
Menes, 2/192, 7/41
Mengistu Haile-Mariam, 2/206
Meninas, Las (Velázquez), *5/210–211*
Mennonites, 5/42
Menzel, Jiri, 2/152, 2/153
Mer de Glace glacier, 2/237
Mérida Range, 6/176
Merina kingdom, 4/86
Merkel, Angela, 3/18, *3/23*
Merovingian dynasty, 2/240–241

Merseyside (England), 6/93
Mertz Glacier, *1/47*
Mesa Central (Mexico), 4/130
Mesa del Norte (Mexico), 4/130
Mesa Verde National Park (United States), *6/126–127*
Mesaoria Plain, 2/142
Meseta (Spain), 5/88, 5/199, 5/202, 5/204
Meseta Central (Costa Rica), 2/122
Mesha, 3/282
Mesoamerica. *See* Central America; Mexico
Mesolithic Period, 7/40
Mesopotamia, 3/186, 3/187, 3/196–197, *3/196–197,* **3/197**
prehistoric human migration, **7/38,** *7/38–39,* 7/38–41
Mesopotamia region (Argentina), 1/53, 1/54, *1/54*
Messiaen, Olivier, 2/253
Messina, Strait of, 3/244
Mestizos
Angola, 1/38
Argentina, 1/56, 1/59
Belize, 1/145
Bolivia, 1/152
Chile, 2/54, 2/55
Colombia, 2/104
Costa Rica, 2/120
Ecuador, 2/174, *2/177*
El Salvador, 2/198
Honduras, 3/103
Mexico, 4/127, 4/136, 4/140
Nicaragua, 4/228–229
Paraguay, 5/38, *5/39, 5/43*
Peru, 5/45, 5/46
United States, 6/142
Uruguay, 6/165
Venezuela, *6/178*
Metal-working industry, 2/231
Metallurgy. *See* Gold; Iron and steel; Metal-working industry
Methanol, 6/47
Methodists, 2/220, 6/44–45
Métis, 2/31
Metropolitan areas, 6/147
Metternich, Klemens von, 1/98, *1/99*
Mexican Americans, 6/129, 6/138, 6/139, 6/142, 6/143
Mexican Revolution of 1910, 4/134, 4/141, 4/144, 4/145
Mexico, 4/126–145, *4/126–145,* **4/129**
art and culture, 4/144–145, *4/144–145*
climate, 4/131
economy, 4/127, 4/134–135, *4/134–135*
tourism, 4/142–143, *4/142–143*
government, 4/128
history, 4/128–129, 4/140–141, *4/140–141,* **4/141**
human migrations, 7/78, **7/80,** 7/80–81, *7/80–81,* 7/102, *7/102–103*
land, 4/130–131, *4/130–131*

natural forces, 4/132–133, *4/132–133,* **4/133**
people, 4/136–137, *4/136–137*
population, **4/136,** 4/137, 4/139
relations with
Canada, 2/34
Nicaragua, 4/226
United States, 6/126, 6/135
Mexico, Gulf of, 4/135, 6/124, 6/148, 6/151
Mexico, Plateau of, 4/130
Mexico, Valley of, 4/130–131
Mexico City (Mexico), 4/130, *4/130–131,* 4/138, *4/138–139,* **4/139**
earthquake, 4/132, *4/132, 4/133*
economy, 4/134, *4/134*
origins, 7/102
tourism, 4/142
Miami (Florida), 2/138, 6/143
Miami Beach (Florida), *6/125*
Michelangelo, 3/221, 3/236, 3/237, 6/171
Michener, James, 6/9
Michigan (United States), 6/122, 6/123, 6/133
Mickiewicz, Adam, 5/73
Micronesia, 3/78, 3/79, *3/79,* 4/116, 5/10–11, 5/26, 7/93
Micronesia, Federated States of, 5/10, 5/13, 5/27
Micronesians, 2/214, 2/215, 4/16, 4/116, 4/182, 5/10
Mid-Atlantic Ridge, 1/112
Middle Ages, 1/138–139, 2/241, 2/252
Middle Atlas Mountains, 4/162, 4/168–169, *4/168–169*
Middle East, 1/4
See also Human migrations; *and specific countries*
Middle Kingdom (Egypt), 2/192
"Middle Way" of Sweden, 5/244
Midges, 1/44
Midlands (England), 6/92
Mieszko I, 5/76
Migration, Animal, 6/24–25
Migration, Human. *See* Human migrations
Mikonos (Greece), 3/66, *3/66–67*
Mikulov (Czech Republic), *2/154–155*
Milan (Italy), 3/228, *3/229,* 3/234, 3/236
Millennium Bridge (London), *6/103*
Miller, Arthur, 6/159
Miller, Gifford, 7/28–29
Miller, Portia Simpson, 3/249
Millet, 7/56
Mills, John Atta, 3/42
Milos (Greece), 3/66
Milošević, Slobodan, *5/160,* 5/160–161, 5/169
Milwaukee (Wisconsin), 6/123
Min River, 2/73
Mina language, 6/40
Minamoto family, 3/264, 3/269
Minamoto Yoritomo, 3/264, *3/266–267*
Minangkabau people, 3/166
Minarets, 6/62

Minas Gerais (Brazil), 1/172, 1/176
Mindanao Island (Philippines), 5/60, 5/61, 5/64, *5/64–65,* 5/68
Mindelo (Cape Verde), *2/41*
Mindoro (Philippines), 5/61
Ming dynasty, 2/62, 2/69, 2/80–83
Minho River, 5/88
Minicoy Island (India), **3/157**
Mining
Albania, 1/23
Algeria, 1/30
Angola, 1/38, *1/38–39*
Antarctica, 1/46
Australia, 1/71, 1/79, 1/82–83, *1/82–83,* 1/87, 1/92
Austria, 1/102, *1/102,* 1/105
Belgium, 1/142
Bolivia, 1/153, 1/154
Botswana, 1/159
Brazil, 1/172, *1/176–177*
Canada, 2/34, 2/35
Chile, 2/50, 2/56, *2/56*
Colombia, 2/103
Congo region, *2/112, 2/117*
Gabon, 3/11
Greece, 3/61
Guinea, 3/89
Indonesia, 3/170
Jamaica, 3/248, *3/249*
Kyrgyzstan, 4/39
Laos, 4/44
Luxembourg, 4/80
Mauritania, *4/122,* 4/123
Mexico, 4/135
Morocco, 4/162
Nauru, 4/182, 4/183
New Caledonia, 4/213
New Zealand, 4/218
Nigeria, 4/241
Norway, 4/252, 4/257
Papua New Guinea, 5/35
Philippines, 5/69, *5/69*
Poland, *5/83*
Portugal, 5/92
South Africa, 5/196, *5/196*
Sudan, 5/233
Swaziland, 5/241
Thailand, 6/33
United Kingdom, 6/108
United States, 6/127
Zambia, 6/203, *6/203*
Zimbabwe, 6/207, *6/207*
See also specific types
Minneapolis (Minnesota), 6/149
Minnesota (United States), 6/122, 6/123
Minnesota Woman fossil, 7/69
Minnon Chinese dialect, 6/14
Minoan culture, 3/62, 3/68
Minorca Island, 5/203, 5/216, *5/216,* **5/217**
Minos, 3/68
Palace of, *3/68–69*
Minotaur, 3/68
Minsk (Belarus), 1/130
Mirococha (Peru), 5/51

Miskito Indians, 3/103, 4/226
Misratah (Libya), 4/69
Missionaries
  Fiji, 2/220
  Gabon, 3/9
  New Zealand, 7/107
  Tonga, 6/44
  Uganda, 6/81
Mississippi (United States), 6/124
Mississippi River, 6/123, 6/123,
    6/124–125, 6/148, 6/148–149,
    6/148–149
Mississippian Culture, 7/76–77, 7/76–77
Missouri (United States), 6/122, 6/123
Missouri River, 6/148
Mistras (Greece), 3/54
Mitochondrial Adam and Eve, 7/12–13
Mitochondrial DNA. See Genetic studies
Mitrovica (Kosovo), 4/33
Mixed member proportional voting
    system, 4/217
Mixtec language, 4/137
Mizo people, 3/138
Mladec Cave, 7/44
Moabites, 3/282
Moai statues, 2/170, 2/170–171, 2/171,
    7/95, 7/95
Moas, 4/222, 4/224, 4/225, 7/96, 7/97
Mobutu, Joseph Désiré, 2/110, 2/111
Moche Indians, 5/44
Mochico Indians. See Moche Indians
Mocho Mountains, 3/248
Model Parliament, 6/110
Modern Art, Museum of (Brussels), 1/141
Modoc Indians, 6/140
Mogadishu (Somalia), 5/187, 5/187
Mogul Empire. See Mughal Empire
Mohammad (prophet). See Muhammad
    (prophet)
Mohammed V (king of Morocco), 4/159,
    4/166
Mohammed VI (king of Morocco), 4/159,
    4/161
Mohawk Indians, 6/141
Mohéli (Comoros), 2/108
Mohenjo-Daro, 5/14, 5/24–25
Moi, Daniel arap, 4/13
Moldavia, 4/146, 5/104, 5/104, 5/105
Moldavian Carpathian Mountains, 5/104
Moldavian Popular Front, 4/146
Moldova, 4/146–147, 4/147, 4/147
Moldovan language, 4/147
Moldoveanu, Mount, 5/104
Molière, 2/252
Moluccas (Indonesia), 3/104–105, 7/92
Mombasa (Kenya), 4/11
Møn Island (Denmark), 2/158, 2/158–159
Mon-Khmer language, 4/46
Mon people, 4/174, 4/178, 6/31, 6/36
Mona Lisa (Leonardo da Vinci), 2/247
Monaco, 4/148, 4/148–149, 4/149
Monaco (town, Monaco), 4/148, 4/149
Monarchy, Absolute, 2/234
Monastery of St. Gregorius (Greece),
    3/70–71

Monastiraki Square (Athens), 3/59
Mondego River, 5/88
Monet, Claude, 2/253
Mongkut. See Rama IV
Mongol Empire, 4/150–151
  power in
    Afghanistan, 1/11, 1/15
    China, 2/62, 2/69
    Georgia, 3/14
    Iran, 3/182
    Iraq, 3/187, 3/197
    Kazakhstan, 4/8
    Kyrgyzstan, 4/38
    Moldova, 4/146
    Myanmar, 4/178
    Russia, 5/112, 5/120
    Siberia, 5/126
    Tajikistan, 6/16
    Turkey, 6/68
    Turkmenistan, 6/74
    Ukraine, 6/82
    Uzbekistan, 6/166
Mongol people, 2/78, 2/78–79, 4/150,
    4/154, 5/127
Mongolia, 4/150–155, 4/150–155,
    4/153
  government, 4/153
  history, 4/150–153
  land, 4/154–155
  people, 4/154–155, 4/154–155
Mongolian Border Uplands (China), 2/68,
    2/72
Mongolian language, 4/154
Mongolian People's Revolutionary
    Party, 4/152, 4/153, 4/155
Monkeys, 3/151
  black spider, 6/177
  capuchin, 6/177
  golden, 2/74
  red howler, 6/177
  squirrel, 2/125, 2/125
Monk's Mound, 7/76–77
Monks of Mount Athos, 3/70–71, 3/71
Monrovia (Liberia), 4/65, 4/65
Møns Klint (Denmark), 2/158–159
Mons region (Belgium), 1/137
Monsoons
  Bangladesh, 1/118, 1/126, 1/126–127
  Cambodia, 2/12
  India, 3/131, 3/132–133, 3/132–133,
    3/132–133
  Japan, 3/257
  Korea, 4/20, 4/26
  Malaysia, 4/96
  Myanmar, 4/175
  Pakistan, 5/18
  Sri Lanka, 5/223
Mont Blanc, 2/237, 2/237
Mont-Bouet Market (Libreville), 3/9
Montaigne, Michel de, 2/252
Montana (United States), 6/126, 6/127
Monte Carlo (Monaco), 4/148–149, 4/149
Monte Verde site, 7/70, 7/86, 7/86–87,
    7/87
Montefeltro, Duke Federico da, 3/236

Montego Bay (Jamaica), 3/247, 3/249,
    3/250, 3/251
Montelbaans Tower (Amsterdam), 4/206
Montenegrin language, 4/156
Montenegrins, in Serbia, 5/164
Montenegro, 2/130, 4/33, 4/156–157,
    4/156–157, 4/157
  Serbia, relations with, 5/155, 5/157,
    5/159, 5/161
Monterrey (Mexico), 4/134
Montevideo (Uruguay), 6/160, 6/161,
    6/162, 6/165
Montezuma II, 4/140, 7/102–103
Montgomery (Alabama), 6/145
Monti, Mario, 3/323
Montmartre (Paris), 2/232–233, 2/233
Montreal (Quebec), 2/23, 2/27, 2/30,
    2/34, 2/36
Montserrat, 4/58, 4/63
Monument to the Two Congresses
    (Buenos Aires), 1/61
Moon landing, 6/136
Moore, Henry, 4/244
Moorish Gardens (Damascus), 5/274–275
Moors (land), 6/92
Moors (people)
  Gibraltar, 3/46
  Mali, 4/111
  Mauritania, 4/120, 4/122
  Morocco, 4/158, 4/166
  Portugal, 5/90
  Spain, 5/199, 5/210, 5/213, 5/214,
    5/217
  Sri Lanka, 5/226
Moose, 2/39, 2/227
Mora, Juan Rafael, 2/123
Morača River, 4/156
Moraines, 2/158, 2/226
Morales, Evo, 1/155
Morant Bay Rebellion, 3/247
Morava River, 5/167
Morava River Valley, 2/155
Moravia (Czech Republic), 2/144, 2/148,
    2/155
Moravian Lowlands, 2/155
Moravian people, 2/145
Mordvin people, 5/116
Mormons, 6/156
Moro, Aldo, 3/222
Moro Naba (title), 1/187
Moro National Liberation Front, 5/65
Moro people, 5/64–65
Moroccan people, 3/232
Moroccan people, in France, 2/239
Morocco, 1/25, 4/109, 4/158–169,
    4/158–169, 4/161
  economy, 4/162–163, 4/162–163
    communication, 4/161
    tourism, 4/168–169, 4/168–169
    transportation, 4/161
  education, 4/161
  government, 4/160
  history, 4/158–161
  human fossils, 7/6
  land, 4/162

people, 4/164–165, *4/164–165*
power in
Niger, 4/232
Portugal, 4/158
Spain, 4/158
Western Sahara, 6/193
royal cities, 4/166–167, *4/166–167*
Moroni (Comoros), 2/109
Morphou Plain, 2/142
Morris dancers, *6/108*
Morrison, Toni, 6/145
Mortadella sausage, 3/235
Moscoso, Mireya, 5/29
Moscow (Russia), *5/116, 5/117,* 5/120–121, *5/120–121,* **5/121**
history, *5/114,* 5/120–121
Volga River and, 5/118
Moscow River, 5/120
Moselle River, 3/35, 3/37, 4/81
Moshavim, 3/219
Moshavim shitufi, 3/219
Moshi-Dagomba people, 3/40, 3/42
Moshoeshoe, 4/57
Moshoeshoe II, Mototlehi, 4/57
Mosquitia (Honduras), 3/102
Mosquito Coast, 3/102, 4/228
Mosquito Indians. *See* Miskito Indians
Mossi people, 1/186–187
Mostar (Bosnia-Herzegovina), *1/157*
Mosul (Iraq), 3/190
Motilones, 6/179
Motion pictures
Cuba, 2/139
Czechoslovakia, *2/152,* 2/153
India, *3/136–137*
Japan, 3/276
Poland, 5/73
United States, 6/158, 6/159
Motorcycles, *3/148–149, 3/166*
Mouhot, Henri, 2/15
Mound builders, 7/76, *7/76–77*
Mount Agel, 4/148
Mount Ararat, *6/61*
Mount Aso, *3/256–257*
Mount Athos, 3/70–71, *3/70–71,* **3/71**
Mount Bia, 4/44
Mount Bromo, *3/165*
Mount Cameroon, 2/20
Mount Cook. *See* Aoraki/Mount Cook
Mount Elgon, 6/80
Mount Erebus, *1/42, 1/43*
Mount Etna, 3/224, 3/225, 3/244
Mount Everest, 2/70, 2/94, 4/184, 4/188, 4/190, *4/190*
name, 4/193
pollution, *4/195*
trekking, 4/196
Mount Fuji, 3/253, 3/256, 3/280
Mount Gimie, 5/134
Mount Haltia, 2/227
Mount Hekla, 3/120
Mount Hermon, 5/270
Mount Hira, 5/150
Mount Kailas, 4/193

Mount Kanchenjunga, 3/130, 4/190, 4/193
Mount Kenya, 4/10
Mount Kilimanjaro. *See* Kilimanjaro
Mount Kinabalu, *4/96–97, 4/97*
Mount Lebanon, 4/51
Mount Logan, 2/39
Mount Makalu, 4/190
Mount McKinley, 6/130
Mount Moldoveanu, *5/104*
Mount Nemrut, 6/70
Mount Olympus, 2/142, *2/142*
Mount Pelée, 4/118
Mount Pinatubo, 5/58–60
Mount St. Catherine, 3/74
Mount Scenery, 4/62–63
Mount Soufrière, 3/77, 5/136
Mount Titano, 5/140
Mount Toba, *7/12, 7/13, 7/13*
Mount Vesuvius, 3/224
Mountains, 1/4
Mourne Mountains, 6/99
Mouscron (Belgium), *1/142*
Mouse Island (Greece), *3/54–55*
Mousterian industry, 7/34, 7/44, 7/64
Mouvement Populaire (Morocco), 4/160
Movement for Democracy and Justice in Chad, 2/45
Movement for Democratic Change (Zimbabwe), 6/209
Movement for Multiparty Democracy (Zambia), 6/202
Movies. *See* Motion pictures
Movimiento a Socialismo (Bolivia), 1/155
Movimiento Nacionalista Revolucionario (Bolivia), 1/154
Moyen Atlas Mountains, 4/162
Mozambique, 4/170–173, *4/170–173,* **4/171,** 5/240, 6/210
economy and people, 4/172–173, *4/172–173*
history, 4/170–171, 7/99
land, 4/172
relations with
Portugal, 5/87
Zimbabwe, 4/170
Mozambique National Resistance. *See* Renamo
Mozart, Wolfgang Amadeus, 1/95, *1/99,* 1/101, 1/106, *1/106,* 1/107
Mozarteum (Austria), *1/106*
MPLA (Angola), 1/36–37
MPRP (Mongolia), 4/152, 4/153, 4/155
Mros tribe, 1/124
MRTA (Peru), 5/47
Mswati III, 5/241
Mu Us Desert, 2/70, 2/71
Mubarak, Hosni, 2/183, 2/193
Muchinga Mountains, 6/204
Muezzins, 2/190, *5/233*
Mugabe, Robert, 6/208, 6/209
Mughal Empire
Afghanistan, 1/11
Bangladesh, 1/119
India, 3/125, 3/128, 3/148, 3/158

Indus River Valley, 5/15
Muhajirs, 5/20
Muhammad (prophet), 2/188, 3/217, 4/158, 4/166, 5/144, 5/150–151, 6/63
Muhammad Ali, 2/192–193, *2/193*
Muir Glacier, *6/131*
Mujahideen, 1/12–13
Mujib, Sheik, 1/120–121
Mujica, José, 6/163
Mulai Idris, Mosque of, 4/166
Mulattoes, 1/114, *1/166,* 1/167, 3/98
Mulay Ismail, 4/168
Mulhacén mountain, *5/202*
Muluzi, Bakili, 4/91
Mumbai (India), 3/127, *3/138–139,* 3/144, *3/145,* 3/148, 3/149
economy, 3/136, *3/136–137*
floods, *3/132, 3/133*
Mun River, 6/30
Mungo, Lake, *7/26, 7/27*
Mungo Man and Woman, 7/26–27, *7/27*
Munich (Germany), 3/26–27, *3/26–27,* **3/27**
Munich Agreement, 2/148
Munkholm Island (Norway), 4/255
Munt Plein (Amsterdam), 4/206
Muong people, 6/180, 6/187
Muong Swa, 4/40
Mur River, 1/103
Murals, *6/142*
Murasaki Shikibu, 3/268
Murgab River, 6/74
Murillo Square (Bolivia), *1/155*
Muroroa atoll (French Polynesia), 2/259
Murray River, *1/72, 7/29*
Musandam Peninsula, 5/8
Muscat (Oman), 5/8, *5/9*
Museo del Oro (Colombia), 2/98, 2/106, *2/107*
Museum of Decorative Arts (Copenhagen), 2/161
Museum of Fine Arts (Brussels), 1/141
Museum of Modern Art (Brussels), 1/141
Museum of Natural History (Brussels), 1/141
Museum of Natural History (Vienna), *1/108*
Museum of the Krapina Neandertals, 7/34–35
Museumplein (Amsterdam), 4/206
Museveni, Yoweri, 6/79
Musgrave Ranges, *1/78*
Musharraf, Pervez, 5/17
Mushrooms, *5/117*
Music
by group
Arab, *6/87*
Australian Aborigine, *1/77*
Roma, 3/116, 3/117
West Indian, 3/250–251
by place
Africa, 4/112
Australia, 1/91

Austria, 1/101, 1/106–107, *1/106–107*
Brazil, 1/171
Caribbean, 4/60, *4/60*
China, 2/67
Colombia, 1/104
Côte d'Ivoire, *2/129*
Germany, 3/17
Indonesia, *3/168–169*, 3/169
Ireland, 3/201, 3/206, *3/206–207*
Italy, 3/237
Japan, 3/277
Korea, *4/30*, 4/31
Mexico, *4/144–145*, 4/145
Portugal, 5/90
Scotland, *6/96*
Serbia, 5/165
United States, 6/158
Wales, 6/94, 6/95
by type
bagpipe, *6/96*
Baroque, 2/252, 3/237
calypso, 3/250, 4/61, *4/61*
choral, 6/94, 6/95
gamelan, *3/168–169*, 3/169
opera, *1/106*, 2/67, 3/237
popular, 6/159
prehistoric, *7/46*
reggae, 3/250–251, *3/250–251*
salsa, *2/136*, 3/250
samba, 1/171
See also Drums
Musical comedy, 6/158
Muslims. *See* Islam
Muspellsheim, 3/122
Mussolini, Benito, 3/222, 3/229, 4/82
Mustique (St. Vincent and the Grenadines), *5/137*
Mutations, 7/11, 7/13, 7/31
Mutesa II, Sir Edward, 6/78
Mutharika, Bingu wa, 4/91
Mutsuhito, *3/264*, *3/265*, 3/270
Muzorewa, Abel T., 6/208
Mwami (title), 1/188
Mwami Mutara III, 5/130
Mwanamutapa Empire, 6/210, **6/211,** 7/100
Mwanawasa, Levy, 6/202
Mwinyi, Ali Hassan, 6/19
Myanmar, 4/174–179, *4/174–179*, **4/177**
Ayutthaya kingdom and, 6/36
Bangladesh history, 1/124
Chinese culture, 2/79
economy, 4/175
history
early, 4/178–179, *4/178–179*
recent, 4/176–177
land and people, 4/174–175, *4/174–175*
Mycenae (Greece), 3/62
Mycenaean culture, 3/62, *3/63*
Myohyang Mountains, *4/20–21*
Myrdal, Alva R., 5/246–247
Myrrh, *6/198*, 6/199

N

Nabataean people, 3/282, 5/144
Nadir Shah, 3/182
NAFTA, 2/34
Naga people, 4/174, 5/226
Nagasaki (Japan), 3/265, 6/136
Naginatas, 3/266
Nago (Japan), *3/259*
Nagorno-Karabakh (Azerbaijan), 1/67, 1/111
Nagoya (Japan), 3/272
Naha (Japan), 3/258
Nahanni National Park (Northwest Territories), 2/39
Nahua Indians, 2/120, 2/198
Nahuatl language, 4/137
Najd (Saudi Arabia), 5/148
Najib Razak, 4/95
Nakhichevan (Azerbaijan), 1/111
Naktong River, 4/26
Nam Viet kingdom, 6/180
Nama people, 4/181
Námafjall (Iceland), *3/121*
Namaste, 3/148
Nambung National Park (Australia), *1/72*
Namgyal, Penchu, 3/152
Namibia, 2/111, 4/180–181, *4/180–181*, **4/181,** 7/99
Nan Madol ruins, 2/215
Nanak, 3/141
Nanjing (China), 2/80, 2/80–81, 2/89
Nanjing, Treaty of, 2/62
Nantes (France), 2/242
Nanyang Technological Institute, 5/177
Naples (Italy), 3/234
Naples, Bay of, 3/224
Napo River, 2/106
Napoleon I, 2/172, 2/233, *2/244,* 2/244–247, 3/21, 4/79
campaigns
Britain, 6/112
Corsica, 2/255
Egypt, 2/193
Italy, 3/228
Liechtenstein, 4/75
Malta, 4/115
Poland, 5/76, 5/80
Portugal, 5/95
Russia, 5/121
Spain, 2/172, 5/215
Sweden, 5/243
Gibraltar and, 3/46
Napoleon III, 2/244–245, 2/256
Napoleonic Wars
Austria, 1/98
Britain, 6/112
Denmark, 2/160
Germany, 3/20–21
Iceland, 3/119
Poland, 5/76
Scandinavia, 4/247
Sweden, 5/243, 5/246
Naricanda (India), *3/145*
Narmer. *See* Menes

Narrows (St. Kitts and Nevis), 5/132
Nassau (Bahamas), *1/115*
Nassau, House of, 4/78
Nasser, Gamal Abdel, 2/193
National Action Party (Mexico), 4/129
National Archaeological Museum (Greece), 3/59
National Assembly of People's Power (Cuba), 2/135
National Assembly of Wales, 6/94
National Autonomous University of Mexico, 4/138, 4/144
National Cathedral (Mexico), 4/144
National Center for the Performing Arts (Beijing), 2/83
National Coalition Party, 2/224
National Council (Switzerland), 5/256
National Council of Maubere Resistance, 2/168
National Council of Timorese Resistance, 2/168
National Democratic Congress (Ghana), 3/42, 3/45
National Democratic Party (Egypt), 2/182
National Democratic Party (Mongolia), 4/153
National Democratic Rally (Algeria), 1/26
National Front (Colombia), 2/100
National Guard (Nicaragua), 4/226, 4/231
National Hydrocarbon Agency (Italy), 3/240
National Insurance Act (Norway), 4/248
National League for Democracy (Myanmar), 4/176–177
National Liberation Army (Colombia), 2/100, 2/101
National Liberation Army (Morocco), 4/159
National Liberation Front (Albania), 1/20–21
National Liberation Front (Algeria), 1/25, 1/26
National Liberation Front (Yemen), 6/199
National Mall (Washington, D.C.), *6/138*
National Monument (Costa Rica), 2/124
National Museum (Copenhagen), 2/161
National Museum of Anthropology (Mexico), 4/144
National Museum of Budapest (Hungary), *3/111*
National Palace (Mexico City), 4/138
National Palace Museum (Taiwan), 6/15
National Park Service, 6/133
National parks
Canada, 2/38, **2/38,** 2/39
Costa Rica, **2/124,** 2/124–125
United States, **6/132,** 6/132–133, *6/132–133*
National Party (New Zealand), 4/217
National Party (South Africa), 5/193
National Resistance Council (Uganda), 6/79
National Resistance Movement (Uganda), 6/79

National Salvation Front (Romania), 5/102–103
National Socialist Party. *See* Nazism
National Theater (San José), 2/124
National Theatre (Mexico City), 4/142
National Union for the Total Independence of Angola, 1/36–37
National University of Singapore, 5/177
Nationalism, 2/223
Nationalist Party (China), 2/64
Nationalist Party (Taiwan), 6/12, 6/13, 6/15
Nationalrat (Austria), 1/96
Nations, Palace of, 5/265
Native Americans
  Americas, Precolonial, 7/67, 7/78
    Amazon region, 7/84–85, *7/84–85,* **7/85**
    Andes region, **7/82,** 7/82–83, *7/82–83*
    animal extinctions and, 7/72–73, *7/73*
    Arctic, 7/74–75, **7/75,** *7/75*
    Central America, 7/80–81
    Clovis, 7/68–69, *7/68–69*
    European migrations and, 7/102, *7/102–103*
    Mexico, **7/80,** 7/80–81, *7/80–81*
    pre-Clovis, 7/70–71, *7/70–71,* **7/71**
    Tierra del Fuego, **7/88,** 7/88–89, *7/88–89*
    transition to agriculture, 7/76–77, *7/76–77,* **7/77**
  Argentina, 1/51, 1/56
  Bahamas, 1/114
  Bolivia, 1/152, *1/153,* 1/154
  Brazil, *1/164,* 1/166–167
  Canada, 2/23, 2/28–29, *2/28–29*
  Chile, 2/48–49, 2/54
  Colombia, 2/102, 2/104, *2/104*
  Costa Rica, 2/120
  Cuba, 2/132
  Ecuador, 2/172, 2/174, *2/176–177*
  El Salvador, 2/194, *2/194,* 2/198
  French Guiana, 2/257
  Guatemala, 3/84, *3/84,* 3/85
  Guyana, 3/94
  Mexico, 4/127, 4/136–137, 4/140
    arts, 4/144, 4/145
  Mississippi River, 6/148
  Nicaragua, 4/229, 4/231
  Panama, 5/30
  Paraguay, 5/38
  Peru, 5/44–45, *5/48,* 5/49
  United States, 6/122, 6/127, 6/129, 6/133, 6/138, 6/140–141, *6/140–141,* **6/141**
    Alaska, 6/131
    arts, 6/158
    history, 6/134, 6/136, 6/140–141
    population, 6/141
  Uruguay, 6/160
  Venezuela, 6/172, 6/178, 6/179, *6/179*

See also Paleo-Indians; *and specific groups*
NATO. *See* North Atlantic Treaty Organization
Natural gas
  Afghanistan, 1/17
  Algeria, 1/30
  Australia, 1/82
  Bangladesh, 1/121
  Bolivia, 1/154
  Brunei, 1/179
  Canada, 2/33, 2/35
  Indonesia, 3/170
  Iran, 3/180, *3/181*
  Italy, 3/240
  Malaysia, 4/93, 4/100
  Norway, 4/252
  Thailand, 6/33
  United Arab Emirates, 6/87
  United Kingdom, 6/106
  United States, 6/150
  Uzbekistan, 6/167
Natural History, Museum of (Brussels), 1/141
Natural History, Museum of (Vienna), *1/108*
Natural selection, *2/178*
Naturalism, 2/252–253
Nature reserves, 2/75
Nauru, 4/182–183, **4/183,** *4/183,* 5/10, 5/12
Nauruan language, 4/182
Navajo Indians, 6/127, 6/140, *6/140,* *6/141*
Navarre, Kingdom of, 5/214
Naxos (Greece), 3/66
Naypyidaw (Myanmar), 4/177
Nazarbayev, Nursultan, 4/8–9
Nazaré (Portugal), *5/90*
Nazarenos, 5/206, *5/206*
Nazareth (Israel), 3/209
Nazca Indians, 5/44
Nazism
  Austria, 1/98–99
  Belarus, 1/131
  Germany, 3/18, 3/21, 3/26, 3/33
  Hungary, 3/211, 3/212, 5/80–81
  Italy, 3/229
  Roma and, 3/116
Ndebele people, *5/194,* 6/207, 6/209
N'Djamena (Chad), 2/47
Ndlovukazi, 5/241
NDP (Greece), 3/52, 3/53
Ne Win, 4/176
Neander Valley (Germany), 7/34
Neandertals, 7/6, 7/17
  Arctic, 7/64, 7/65
  Europe, 7/44, 7/48, 7/49, 7/52–53, *7/52–53*
  Middle East, 7/33, 7/34–35, *7/34–35,* 7/39
Nebraska (United States), 6/122, *6/122*
Nebuchadnezzar II, 3/187
Necas, Petr, 2/147
Neckar River, 3/24

Negev Desert, 3/209
Negril (Jamaica), 3/249
Negrito people, 7/13
  Andaman Islands, *7/16,* 7/16–17, 7/19, 7/89
  Philippines, 5/62, 5/64, 7/58, 7/59, 7/92
  Taiwan, *7/58,* 7/58–59, **7/59**
Negros Island (Philippines), 5/69
Nehru, Jawaharlal, *3/129*
Nelson's Column (Barbados), 1/128
Nemanja, Stefan, 5/158
Nemrut, Mount, 6/70
Neo-Confucianism, 2/62
Neo-Destour Party, 6/49
Neolithic period, 2/140
Nepal, 4/184–197, *4/184–197,* **4/187**
  history, 4/186–187
  land, 4/184–185, *4/184–185*
  people, 4/188–189, *4/188–189*
  population density, 4/194
  Sikkim, power in, 3/152
  See also Himalaya
Nepalese people, 1/151, 3/153
  See also Nepal
Nepali language, 1/151, 3/153, 4/189
Nerevta River, *1/157*
Nervión River, 5/205
Nessaea obrinus, *2/125*
Netanyahu, Benjamin, 3/213
Netherlands, Kingdom of the, 4/200, 4/210
Netherlands, The, 4/198–209, *4/198–209,* **4/201**
  Australia exploration, 1/86, 7/106
  cities, 4/204–205, *4/204–205*
  government, 4/200
  history, 4/202
  land, 4/199
  people, 4/202–203, *4/202–203*
  population, 4/200, 4/202–203
  power in
    Belgium, 1/134
    Ghana, 3/44
    Guyana, 3/92
    Indonesia, 3/161, 3/162
    Lesser Antilles, 4/59, 4/60
    Luxembourg, 4/79
    Malaysia, 4/94, 4/102
    Maldives, 4/104
    Mauritania, 4/120
    Mauritius, 4/125
    Myanmar, 4/178
    Netherlands Antilles, 4/210
    Nigeria, 4/236
    Senegal, 5/153
    South Africa, 5/189, 5/192, 7/99
    Sri Lanka, 5/224
    Suriname, 5/238
    Trinidad and Tobago, 6/46
  relations with
    Belgium, 1/133, 1/134
    Kuwait, 4/36
    Saudi Arabia, 5/145

sea, battle with, 4/199, 4/208–209, *4/208–209*, **4/209**
town and country, 4/202–203
Netherlands Antilles, The, 4/62, 4/200, 4/210–211, *4/210–211*, **4/211**
Neuschwanstein Castle (Germany), *3/16–17*, 3/17
Neusiedler Lake, 1/103
Neutrality
Austria, 1/95, 1/99
Denmark, 2/157
Finland, 2/223, 2/224
Liechtenstein, 4/75
Norway, 4/247
Sweden, 5/246
Switzerland, 5/255, 5/258, 5/264
Neva River, 5/122, *5/123*
Nevada (United States), 6/126
Nevado del Ruiz volcano, 2/98
Neves, Tancredo de Almeida, 1/163
Nevis. *See* St. Kitts and Nevis
New Amsterdam, 5/238, 6/120
New Bridge (Bratislava), *5/180–181*
New Britain Island, 5/36
New Brunswick (Canada), 2/31, 2/32
New Caledonia, 4/212–213, *4/212–213*, **4/213**, 5/10, 5/11
New Caledonia (island), 4/212
New Clean Government Party, 3/254
New Deal, 6/136
New Delhi (India), 3/148–149
New Democracy Party (Greece), 3/52, 3/53
New Economic Policy (Malaysia), 4/101
New Economic Policy (Russia), 5/114
New England (United States), 6/120–121
New France, 2/30
New Guinea, 1/86, 3/161, 5/10, 5/12, 5/34–36
human settlement, 7/17, 7/24, **7/24**, *7/24–25*, 7/26, 7/92
New Hampshire (United States), 6/120
New Hebrides, 6/169
New Island (Falkland Islands), *2/210–211*
New Jersey (United States), 6/120
New Kingdom (Egypt), 2/192
New Kingdom of Granada, 2/99
New Komeito party, 3/254
New Mexico (United States), 6/126, 6/127
New National Party (Grenada), 3/75
New Order (Indonesia), 3/162
New Orleans (Louisiana), 6/124
New Patriotic Party (Ghana), 3/42, 3/45
New Serbia party, 5/156
New South Wales (Australia), 1/70, 1/84, 1/88
economy, 1/80, *1/80*, *1/82*, 1/83, *1/83*
history, 1/86, 7/106, 7/107
New Spain, 7/102
New Territories (Hong Kong), 2/84, 2/85
New towns (Singapore), 5/176
New York (United States), 6/120, 6/121
New York City (New York), 6/121, *6/121*, *6/141*
arts, 6/158

ethnicity, *6/138*, *6/139*
Hispanic Americans, 6/143
terrorist attack, 6/137, *6/137*
urban life, 6/146, *6/146*
New Zealand, 4/214–225, *4/214–225*, **4/217**, 5/10, 5/11
economy, 4/215, 4/218–219, *4/218–219*
government, 4/216, 4/217
history, 4/216–217
aboriginal settlement, 7/96–97, *7/96–97*, **7/97**
British settlement, 7/106–107, *7/107*
land and climate, 4/215
people, 4/215, 4/220–221, *4/220–221*
population, 4/216
power in
Cook Islands, 2/119
Nauru, 4/182, 4/183
Pacific Islands, 5/12
Samoa, 5/139
wildlife, 4/224–225, *4/224–225*
See also Maori
New Zealand Wars, 4/223
Newcastle (Australia), 1/88
Newcastle upon Tyne (England), *6/92–93*, 6/93, 6/106
Newfoundland and Labrador (Canada), 2/22, 2/30, 2/38
Newport (Wales), 6/95
Ngalop people, 1/151
Ngo Dinh Diem, 6/184
Ngoni people, 4/90, 4/91
Nguni people, 6/210, 7/100
Nguyen Ahn, 6/181
Nguyen family, 6/181
Ngwane II, 5/240
Ngwenyama, 5/241
Niagara Falls, *2/26*
Niah Caves (Borneo), 7/23
Niamey (Niger), 4/235
Niari Valley, 2/116
Nicaragua, 4/226–231, *4/226–231*, **4/229**
economy, 4/226
government, 4/229
history, 4/226–227
crises, 4/230–231, *4/230–231*
Honduras, relations with, 3/101
land and climate, 4/228, *4/229*
people, 4/228–229
Nicaragua, Lake, 4/228, *4/229*
Nicarao, 4/226
Nicarao Indians, 4/226
Nicholas (king of Montenegro), 4/157
Nicholas II (czar of Russia), 2/223, 2/228, 5/109, 5/113, 5/114
Nickel, 2/35, 4/213
Nicosia (Cyprus), 2/141, 2/143
Nidelv River, *4/255*
Nieuwe Kerk (Amsterdam), 4/206
Nieuwe Kerk (Delft), 4/205
Niflheim, 3/122

Niger, 4/112, 4/232–235, *4/232–235*, **4/233**
history, 4/232–233
land, 4/234, *4/235*
people, 4/234–235, *4/234–235*
Niger-Benue River, 4/240
Niger Delta, 4/236, 4/240
Niger Progressive Party, 4/233
Niger River, 4/110, *4/111*, 4/232, 4/234, *4/235*, 4/244
Nigeria, 4/236–245, *4/236–245*, **4/238**, 7/98
art and culture, 4/244–245, *4/244–245*
economy, 4/240–241, 4/241
government, 4/238
history
early, 4/236–237
recent, 4/238–239
land, 4/240, *4/240*
people, 4/242–243, *4/242–243*
relations with
Cameroon, 2/19
Ghana, 3/45
Nikko (Japan), 3/279
Nikko National Park (Tokyo), *3/256–257*
Nile Delta, *2/184–185*, 2/192
Nile River, 1/4, **7/32**
Egypt, 2/181, 2/184, 2/185, 2/190
Ethiopia, 2/208
Sudan, 5/228, 5/232, *5/233*, 5/234
See also Albert Nile River; Blue Nile River; Victoria Nile River; White Nile River
Nile River Valley, 2/186, 5/233
prehistoric human migration, 7/10, 7/14, 7/15, 7/32–33, **7/33**, 7/40, *7/40–41*
Nilgiri Hills (India), *3/130*, 3/131
Nimeiri, Gaafar, 5/229, 5/230
Nîmes (France), 2/240
Ninety-Five Theses (Luther), 3/38
Ninety Mile Desert, 1/81
Nios, Lake, 2/21, **2/21**
Nippon. *See* Japan
Nirvana, 6/38
Niš (Serbia), 5/163, 5/165
Niue, 5/11
Niyazov, Saparmurad A., 6/75
Nizhniy Novgorod (Russia), *5/117*
Njal Thorgeirsson, 3/123
Nkrumah, Kwame, 3/44, 3/45, *3/45*
NLD (Myanmar), 4/176–177
Nobel, Alfred, 5/246, 5/247
Nobel Prizes, *1/123*, *1/133*, *2/168*, 4/177, *4/255*, *5/190*, 5/247
Noda, Yoshihiko, 3/255
Noh plays, 3/276
Nok civilization, 4/244, *4/244*
Nokia, *2/231*
Nomads
Afghanistan, 1/14, 1/16
Algeria, 1/24, *1/29*
Chad, *2/46–47*
China, 2/78

Djibouti, 2/162, 2/163
Egypt, 2/187
Iran, *3/179*
Iraq, 3/191
Jordan, 3/287
Kenya, 4/10, 4/14, *4/15*
Kyrgyzstan, 4/39
Libya, *4/68*, 4/70
Mali, 4/109, 4/111
Mongolia, 4/150, 4/152, *4/154*, 4/155
Morocco, 4/164, *4/165*
Niger, 4/232, 4/235
Roma, 3/116–117, *3/116–117*
Saudi Arabia, 5/144
Sikkim, 3/149
Somalia, 5/186
Sudan, 5/234
Tibet, 2/95
Turkmenistan, 6/74
Venezuela, 6/178
Western Sahara, 6/192
Yemen, 6/200, *6/200*
Nonviolent disobedience, 3/129
Noord-Holland (Netherlands), 4/203
Noordwijk resort (Netherlands), 4/199
Norfolk Island, 1/89
Noriega, Manuel, 5/28–29
Norman Cathedral (Sicily), 3/244
Norman Island, 6/189
Normandy (France), 2/238, 2/240, 2/249
Normans
    Canary Islands, 5/221
    Ireland, 3/204
    Italy, 3/232
    Sicily, 3/245
    United Kingdom, 6/91, 6/104
Norodom Ranariddh, 2/10–11
Norodom Sihamoni, 2/11
Norodom Sihanouk, 2/9, 2/11
Norrmalm (Stockholm), 5/251
Norse mythology, 3/122–123, *3/122–123*
Norse people, 2/213, 2/238
    See also Vikings
Norsk Hydro, 4/252
Norte Chico (Chile), 2/52
North Albanian Alps, 1/22
North America, 1/4
    See also Human migrations; *and specific*
      countries
North American Free Trade Agreement,
    2/34
North American Plate, 3/120, *4/132*
North Atlantic Current, 4/250
North Atlantic Treaty Organization
    Afghanistan in, 1/13
    Belgium as headquarters, 1/133, 1/140
    Belgrade bombing, 5/169
    Bermuda, defense for, 1/149
    Czech Republic in, 2/147
    Estonia in, 2/205
    Greece in, 3/52
    Kosovo peace plan, 4/33, 5/157,
      5/160–161
    Latvia in, 4/49

Lithuania in, 4/77
Luxembourg in, 4/79
Macedonia and, 4/83
Poland in, 5/75
Romania in, 5/103
Russia and, 5/111
Slovakia in, 5/178
Slovenia in, 5/183
Turkey in, 6/57
North Carolina (United States),
    6/124–125
North Dakota (United States), 6/122
North East Land (Norway), 4/256
North Island (New Zealand), 4/215,
    4/224
North Korea. *See* Korea, North
North Sea
    Belgium on, 1/133, 1/135, 1/138,
      1/142
    Czech Republic and, 2/154
    Danube River and, 5/181
    Denmark on, 2/158, *2/158*, 2/159
    Germany on, 3/24
    Netherlands on, 4/199, 4/208–209
    Norway on, 4/252, *4/252–253*
    United Kingdom on, 6/89, 6/92, 6/97,
      6/106
North Vietnam. *See* Vietnam
North Yemen. *See* Yemen (Sanaa)
Northern Desert (Chile), 2/52, 2/54
Northern Ireland, 3/201, 3/205, **6/98,**
    6/98–99, *6/98–99*
    history, 6/98, 6/112
Northern Ireland Assembly, 6/98
Northern Marianas, 3/78–79, *3/78–79,*
    3/79, 5/10
Northern Mountains (Spain), 5/202
Northern Mountains (Thailand), 6/30
Northern Rhodesia, 6/202, 6/208
Northern Territories (Japan), 3/258
Northern Territory (Australia), 1/70,
    1/88, *1/89,* 1/91
Northwest Passage, 3/72, 4/254
Northwest Territories (Canada), 2/29,
    2/39
Norway, 4/246–257, *4/246–257,* **4/249**
    cities, **4/254,** 4/254–255, *4/254–255*
    Denmark, relations with, 2/156
    economy, 4/252–253, *4/252–253*
    government, 4/248
    history, 4/246–247
    land, 4/250–251, *4/250–251*
    people, 4/249
    power in
      Faroe Islands, 2/213
      Greenland, 3/72
      Iceland, 3/118–119
    Svalbard, 4/256–257, *4/256–257,*
      4/257
    Sweden, relations with, 5/242–243
Norwegian language, 4/248
Norwegian Nobel Committee, 5/247
Norwegian people, in U.S., 6/123
Nossa Senhora do Pilar, Church of
    (Brazil), *1/164*

Notre Dame Cathedral (France), 2/241,
    2/247, 2/252
Notre Dame Cathedral (Luxembourg),
    4/81
Notre-Dame-des-Victoires Church
    (Quebec), 2/36
Notting Hill Carnival, 6/108, *6/108*
Nouakchott (Mauritania), *4/121,*
    4/122–123
Nouméa (New Caledonia), 4/212
Nova Scotia (Canada), 2/31, 2/37, 2/38
Nové Mĕsto (Prague), 2/150
Novi Sad (Serbia), 5/165
Novokuznetsk (Siberia), 5/127
Novosibirsk (Siberia), 5/127
NRM (Uganda), 6/79
NSF (Romania), 5/102–103
Nuba Mountains, 5/232
Nubian Desert, 5/232
Nubian people, *2/186, 2/187,* 5/228,
    5/234
Nuclear families, 6/157
Nuclear power
    Belarus, 1/131
    France, *2/250*
    India, *3/137*
    Japan, 3/255
    Spain, *5/208–209*
    Switzerland, 5/262
    United Kingdom, 6/106
    United States, 6/150
Nuclear weapons and testing
    Antarctica, 1/44
    French Polynesia, 2/259
    Iran, 3/177
    Kazakhstan, 4/9
    Korea, North, 4/18, 4/19, 4/25
    Marshall Islands, 4/117
    Ukraine, 6/83
    World War II, 3/265, 6/136
Nuer people, 5/229, 5/236
Nueva San Salvador (El Salvador), 2/199
Nuevitas (Cuba), 2/134
Nujoma, Sam, 4/180
Nuku'alofa (Tonga), 6/42
Numidia, 1/24
Nunavut (Canada), 2/29
Nung people, 6/180, 6/187
Nupe people, 4/243
Nurek Dam, 6/17
Nuremberg (Germany), 3/39, *3/39*
Nuristani people, 1/15
Nutrias, *6/149*
Nuuk (Greenland), 3/72
Nuwara Eliya (Sri Lanka), 5/222
Nyamwezi people, 6/18, 6/22
Nyanja people, 4/90–91
Nyasa, Lake, 4/90, 6/21
Nyasaland, Protectorate of, 4/91
Nyerere, Julius, 6/19
Nyhavn Canal (Denmark), *2/160*
Nynorsk Norwegian language, 4/248
Nyoman River, 1/130

O

Oahu (Hawaii), 6/129
Oases, 4/73, 6/52–53
Ob River, 5/124
Obama, Barack, 2/135, 6/137, 6/137, 6/145
Obasanjo, Olusegun, 4/239
Obelisk of Luxor (France), 2/247
Obiang Nguema Mbasogo, Teodoro, 2/201
Obote, Milton, 6/78
Òbuda (Hungary), 3/114
O'Casey, Sean, 3/206
Ocean Drive (Miami Beach), 6/125
Ocean Island. See Banaba Island
Oceania. See Pacific Islands
Oceanographic Museum (Monaco), 4/149
Oceans, 1/4
Ocelots, 6/177
Ocho Rios (Jamaica), 3/249
Ocmulgee National Monument (Georgia), 6/132
O'Connell, Daniel, 3/205
Octavian. See Augustus
October Revolution, 3/14, 5/114, 5/123, 5/123, 6/74
Odeon (Athens), 3/58
Odeon (Turkey), 6/73
Oder River, 2/154, 3/24, 5/76, 5/79
Odinga, Raila, 4/13
Odisha (India), 3/136, 3/137
Odoacer, 3/227
Oduduwa, 4/245
Odyssey (Homer), 3/64
Ofu Island (American Samoa), 1/32
Ogaden (Ethiopia), 2/206, 5/187
Ogbomosho (Nigeria), 4/242
Oghuz people, 6/74
Ogooué River, 3/8, 3/9, 3/10, 3/11
Ogotai, 4/150
Ogun, 4/245
O'Higgins, Bernardo, 2/49
Ohio (United States), 6/122
Ohio River, 6/122, 6/148, 6/149
Oil industry. See Petroleum
Oilbirds, 6/177
Ojukwu, Odumegwu, 4/238–239
Okapis, 2/115
Okavango River, 1/159
Oker River, 3/24
Okhotsk (Siberia), 5/126
Okinawa (Japan), 3/258, 7/61
Oklahoma (United States), 6/126
Okoumé trees, 2/200, 3/10
Oktoberfest (Germany), 3/27, 3/27, 3/35
Olav I, 4/247
Olav II, 4/247, 4/255
Old Basilica of Constantine (Vatican City), 6/170
Old Bridge (Bosnia-Herzegovina), 1/157
Old City (Beijing), 2/82, 2/82–83, 2/83
Old City (Jerusalem), 3/216, 3/217, 3/217
Old Havana, 2/133

Old Kingdom (Egypt), 2/192
Old Norse language, 3/118
Old State House (Boston), 6/120
Old Town (Stockholm), 5/250–251
Old Town (Warsaw), 5/81
Olduvai Gorge, 6/20
Olgas (Australia), 1/89
Olinda (Brazil), 1/172
Olive oil, 2/234
Olives, 3/55, 3/61, 3/225, 3/239, 6/53
Olmecs, 7/81, 7/81
Olmert, Ehud, 3/213
Olosega Island (American Samoa), 1/32
Ölü Deniz lagoon, 6/72
Olympic Games, 2/37, 2/61, 2/83, 3/64
Olympic National Park (Washington), 6/128
Olympic Stadium (Berlin), 3/33
Olympic Village (Munich), 3/27
Olympio, Sylvanus, 6/41
Olympus, Mount, 2/142, 2/142
Oman, 5/8–9, 5/9, 5/9
Oman, Gulf of, 5/8
Omar (leader of Mecca), 5/150
Omar Ali Saifuddin Mosque (Brunei), 1/179
Ombudsmen, 4/220
Omdurman (Sudan), 5/234
Omdurman, Battle of, 5/229
Omertà code, 3/244
Omonia Square (Athens), 2/58
Omsk (Siberia), 5/127
Omyéné people, 3/8
On-Ogurs, 3/106
Ondol heating system, 4/27
Onion Portage site, 7/74–75
Ontario (Canada), 2/23, 2/32, 2/34, 2/36–37
   early, 2/31
   land, 2/27
   wilderness, 2/39
Ontario, Lake, 2/36
Opals, 1/82, 1/83
OPEC. See Organization of the Petroleum Exporting Countries
Open-Door Policy, 2/63
Opera, 1/106, 1/165, 3/28–29, 3/237
   Austria, 1/106–108
   Beijing, 2/67
Opera House of Manaus (Brazil), 1/165
Operation Bootstrap (Puerto Rico), 5/96, 5/97
Opium, 1/16, 2/62, 4/44, 4/47
Opium War, 2/62
Opossums, Mouse, 2/125
Orange Free State (South Africa), 5/192
Orange Revolution, 6/83
Oraons, 3/138
Orapa (Botswana), 1/159
Orbán, Viktor, 3/109
Orchids, Grass pink, 6/149
Orchirbat, Punsalmaagiyn, 4/153
Ore Mountains, 2/154, 3/34
Oregon (United States), 6/128, 6/132
Oregon Country, 6/128

Oregon Trail, 6/128
Orellana, Francisco de, 2/106
Organization of the Petroleum Exporting Countries, 1/30, 3/11, 3/194, 4/36, 4/37
Oriental Pearl Tower (Shanghai), 2/81
Oriente (Ecuador), 2/176, 2/177
Origin of Species, The (Darwin), 2/178
Orinoco River, 2/102, 6/176
Oriya language, 3/138
Oromifa language, 2/208
Oromo people, 2/208
Orontes River, 5/271
Orozco, José, 4/144
Ortega, Daniel, 4/229, 4/230, 4/231
Osaka (Japan), 3/272, 3/274
Oslo (Norway), 4/249, 4/254, 4/255
Oslo Fiord (Norway), 4/254
Ostend (Belgium), 1/138, 1/138
Östermalm (Sweden), 5/251
Ostrava (Czech Republic), 2/155
Ota Dokan, 3/270
Otakar II, 2/150
Otavalo (Ecuador), 2/172
Otavalo Indians, 2/176–177
Othello's Tower, 2/143
Otomi language, 4/137
Ottawa (Ontario), 2/24, 2/36, 2/36–37
Otters, 6/149
   giant, 3/95
Otto (king of Greece), 3/63
Otto I (king of Germany), 1/98, 3/20, 3/38
Ottoman Empire, 5/157, 6/62
   historic sites, 6/70
   history, 6/68, 6/68–69, 6/69
   Malta campaign, 4/115
   power in
      Albania, 1/20
      Algeria, 1/24
      Armenia, 1/66–67
      Austria, 1/98, 1/98–99
      Bosnia-Herzegovina, 1/156
      Bulgaria, 1/180, 1/181
      Crete, 3/68
      Croatia, 2/130
      Cyprus, 2/140
      Egypt, 2/192
      Georgia, 3/14
      Greece, 3/51, 3/59, 3/63
      Hungary, 3/110
      Iraq, 3/187
      Jordan, 3/282
      Kosovo, 4/32
      Lebanon, 4/51
      Libya, 4/67
      Macedonia, 4/82
      Mesopotamia, 3/197
      Moldova, 4/146
      Montenegro, 4/157
      Palestine, 3/212
      Saudi Arabia, 5/144, 5/151
      Serbia, 5/155, 5/156, 5/158, 5/164, 5/169
      Syria, 5/267, 5/274
      Tunisia, 6/48

157

Yemen, 6/195, 6/198
See also Turkey
Ou River, *4/47*
Ouaddai kingdom, 2/45
Ouagadougou (Burkina Faso), 1/187, 2/129
Ouattara, Alassane, 2/127
Oude Kerk (Delft), 4/205
Ould Daddah, Mokhtar, 4/121
Oum er Rbia River, 4/168
Our Lady of Peace church (Côte d'Ivoire), 2/128
Ouro Prêto (Brazil), *1/164, 1/172*
Ousmane, Mahamane, 4/233
Out of Africa theory, 7/6, **7/11,** 7/12
Outback, Australian, 1/78–79, *1/78–79*
Ovambo people, 4/180–181
Ovamboland, 4/181
Oviedo Silva, Lino, 5/41
Ovimbundu people, 1/39
Owendo (Gabon), 3/11
Owls
collared Scops, *2/75*
snowy, *2/38*
Oxen, *4/185, 5/23*
Oxford University, 6/93, *6/110–111*
Oy people, 4/46
Oyashio Current, 3/257
Ozone layer, 1/46, *1/46*

P

Pachacuti, 5/52
Pachinko, 3/276, *3/277*
Pacific, War of the (1879-1883), 1/153
Pacific Barrier radar system, 4/117
Pacific Coast Ranges, 6/130
Pacific Islands, 3/264, **5/10–11,** *5/10–13, 5/10–13*
economy, 5/11
history, 5/12–13, *5/12–13,* **5/13**
See also Human migrations
Pacific Islands, Trust Territory of the, 5/12
Pacific Islands Forum, 2/221
Pacific Ocean, 1/4
American Samoa in, 1/32
Austronesian migration, 7/92–93, **7/93**
Bolivia on, 1/153
Canada on, 2/37
Chile on, 2/52
Colombia on, 2/98
Cook Islands in, 2/118
Costa Rica on, 2/122
depth, 5/60
Easter Island in, 2/170
Ecuador on, 2/174
El Salvador on, 2/199
Federated States of Micronesia in, 2/214
Fiji in, 2/216
French Polynesia in, 2/258
Galapagos in, 2/178
Hawaii in, 6/129
hurricanes, 4/133
Japan on, 2/257, 3/256

Kiribati in, 4/16
Malaysia on, 4/96
Mariana Trench and, *3/78*
Marshall Islands in, 4/116
Mexico on, 4/142–143
Nauru in, 4/182
New Zealand in, 4/215
Nicaragua on, 4/228, 4/229
Pacific Islands in, 5/10, 5/11
Panama on, 5/28, 5/32–33
Papua New Guinea in, 5/34
Peru on, 5/48
Philippines in, 5/57, 5/60
Ring of Fire and, 1/43, 4/132, 5/36
Samoa in, 5/138
Solomon Islands in, 5/184
Tahiti in, 6/8
Tonga in, 6/42, *6/44–45*
Tuvalu in, 6/76
United States on, 6/128, 6/151
Vanuatu in, 6/168
Pacific Rim National Park (Canada), 2/39
Paddies, Rice, 3/170, *3/171, 4/28, 4/45*
Padma River, 1/118
Paekche, 4/22, 4/30
Paektu-san Mountain, 4/20
Paestum (Italy), *3/224–225*
Páez, José Antonio, 6/173
Pagan (Myanmar), 4/178
Pago Pago (American Samoa), 1/32, *1/32–33*
Pahang River, 4/97
Pahlavi, Mohammad Reza, 3/175, 3/182–183, *3/183*
Pahlavi, Reza Shah, 3/175, 3/182
PAICV (Cape Verde), 2/40
PAIGC (Africa), 2/40, 3/90, 3/91
Paihia (New Zealand), *4/221*
Painted Desert, 6/127
Painting. *See* Art
Pakehas, 4/223
Pakhto language. *See* Pashto language
Pakhtun people. *See* Pashtun people
Pakistan, 5/14–25, *5/14–25,* **5/17**
history
early, 5/14–15, **5/24,** 5/24–25, *5/24–25,* 7/20
recent, 5/16–17
land, 5/18–19, *5/18–19*
people, 5/20–21, *5/20–21*
population, 5/20
relations with
Afghanistan, 1/12, 1/14, 1/19
Bangladesh, 1/118, 1/120, 1/124, 1/126
India, 3/126–127, 3/159, 5/16
Pakistan People's Party, 5/16, 5/17
Pakistani people
Bahrain, 1/116
Kuwait, 4/34
Malaysia, 4/99
Mozambique, 4/173
United Kingdom, 6/104
See also Pakistan
Palace (Munich), 3/26

Palace of Culture and Science (Warsaw), *5/80*
Palace of Fine Arts (Mexico City), 4/138, *4/142, 4/143*
Palace of Minos (Crete), *3/68*
Palace of Nations, 5/265
Palace of Sans Souci, 3/39
Palace of the Parliament (Bucharest), *5/103*
Palace of the Popes (France), *2/252*
Palace of the Prince (Monaco), *4/149*
Palais de Justice (Belgium), *1/135, 1/141*
Palatine Hill (Rome), 3/230
Palau, 2/214, 5/10, 5/13, 5/26–27, *5/27,* **5/27**
Palau Islands, 5/13
Palenque ruins, *4/142*
Paleo-Eskimos, 7/74
Paleo-Indians
Amazon region, 7/84–85, *7/84–85,* **7/85**
animal extinctions and, 7/72–73, *7/73*
Arctic settlement, 7/74–75, *7/74–75*
North American route, **7/72**
South America, **7/86,** 7/86–87, *7/87*
Tierra del Fuego and, 7/89
transition to agriculture, 7/76–77, *7/76–77,* **7/77**
Yucatán Peninsula and Central America, **7/80,** *7/80,* 7/80–81, *7/81*
See also Clovis culture
Paleolithic Period
Asia, 7/19, 7/36–38, 7/60
Britain, 7/50–51, *7/50–51*
Europe, 7/49
Mesoamerica, 7/78
See also Paleo-Indians
Palermo (Italy), 3/223, 3/244
Palestine, 3/209, 3/212, 3/282–283, 5/266, 5/267
division of, 5/268
Palestine Liberation Organization, 3/209, 3/212, 3/213, 3/283, 4/52, 4/53, 4/67
Palestinians
Israel, 3/209, 3/213, *3/214,* 3/215
Jordan, 3/283, 3/285, *3/285,* 3/286
Kuwait, 4/34
Lebanon, 4/52–54
Palikir (Federated States of Micronesia), 2/214
Palm oil, 1/146, 1/147, *3/41,* 4/237, 4/241, *5/131*
Palma (Majorca), 5/216, *5/216–217*
Palmettos, *6/149*
Palmyra (Syria), 5/266
Pamir Mountains, 6/166
Pamirs, 2/70, 4/190, 6/16, *6/17*
Pampas, 1/51, 1/54, 1/55, 1/62, 1/63
gauchos of, 1/58–59, *1/58–59*
Pamplona (Spain), *5/206–207,* 5/207
PAN (Mexico), 4/129
Pan-African Congress (South Africa), 5/193

158

Panama, 5/28–33, *5/28–33,* **5/29**
  government, 5/28
  history, 2/99, 5/28–29
  land, 5/28
  people and economy, 5/30–31,
    5/30–31
Panama, Isthmus of, 5/32
Panama Canal, 5/28, *5/29, 5/32,*
    5/32–33, **5/33**
  building of, 5/29, 5/32
  people and economy, 5/30–31
Panama Canal Zone, 4/29, 5/30
Panama City (Panama), 5/31, *5/31*
Panama hats, *2/175*
Panay Island (Philippines), 5/61
Panchayat system of Nepal, 4/186,
    4/187
Panchen Lama, 2/97
Pancho Indians, 2/198
Pandas
  giant, 2/74, *2/74–75,* 2/75, **2/75,**
    2/82
  red, *2/74*
Pangeboche (Nepal), *4/193*
Panhandle (Alaska), 6/131
Panhellenic Socialist Movement, 3/53
Pannonian Plains, 2/130, 5/155, 5/163,
    5/166, 5/167
Pannonian Sea, 5/166
Pans (drums), *4/60, 4/61*
Pantheon, *3/226–227*
Pantufo (São Tomé and Príncipe), *5/143*
Panungs, 6/31
Papadopoulos, George, 3/52
Papal States, 3/227, 6/170–171
Papandreou, George, 3/52, *3/63*
Papeete (Tahiti), 2/258, *6/8–9, 6/9*
Paper production, 5/253
Papermaking, 2/231, *2/231*
Paphos (Cyprus), 2/143
Papiamento language, 4/60
Papineau, Louis, 2/31
Papua (Indonesia), 3/163, 3/166, 5/11
  land, 3/164, 3/165
Papua New Guinea, 5/10, 5/11, 5/12,
    5/34–37, *5/34–37,* **5/35**
  fossils, 7/24
  history, 5/34–35
  land and economy, 5/36–37, *5/36–37*
  people, 5/34
  Solomon Islands and, 5/184
Papuans, 7/24–26, 7/92
  See also Papua New Guinea
Pará (Brazil), *1/176–177*
Paraguay, 5/38–43, *5/38–43,* **5/41**
  early history, 5/38–41
  land and economy, 5/42–43, *5/42–43*
  relations with
    Argentina, 5/39
    Bolivia, 1/154
    Brazil, 5/39
    Uruguay, 5/39
    with Bolivia, 5/42
Paraguay River, 5/38, 5/42, 5/43
Parakeets, 1/74

Paraná (Brazil), *1/168,* 1/169, 1/174
Paraná River, 1/54, 1/176, 5/38, 5/43
Paria Peninsula (Venezuela), 6/172
Paricutín (Mexico), 4/133
Paricutín (volcano), 4/133
Paris (France), 2/233, *2/239,* 2/246,
    2/246, 2/246–247, *2/246–247*
  arts, 2/246–247, 2/252
  history, 2/241, *2/244,* 2/246–247
  industry, 2/251
  railroad hub, 2/251
Paris, Treaty of (1763), 7/104
Paris Basin, 2/236, 2/246, 2/251
Park Chung Hee, 4/24, 4/28
Parks, Rosa, 6/145
Parler, Peter, 2/151
Parliament
  Canada, 2/24–25
  France, 2/234
  Ireland, 3/202
  Jamaica, 3/249
  New Zealand, 4/217
  Scottish, 6/96
  United Kingdom, 3/204, 3/205, 6/90,
    6/103, 6/110, 6/111
Parliament, Palace of the (Bucharest),
    5/103
Parliament Building (Austria), *1/97,*
    1/109
Parliament Building (Budapest), *2/109,*
    3/114
Parliament Hill (Ottawa), 2/36, *2/36–37*
Parliamentary monarchy, 5/200
Paros (Greece), 3/66
Parrots, 1/74
  Gaiah, *1/75*
Parthenon (Greece), 3/58, *3/58–59, 3/65*
Parthians, 3/182, 5/14
Parti Démocratique de Guinée, 3/86
Parti Québécois, 2/32, *2/32*
Partido Acción Nacional (Mexico), 4/129
Partido Revolucionario Institucional
    (Mexico), 4/128–129
Partisans (Yugoslavia), 5/158–159, *5/159*
Partnership for Peace, 5/103
Parvati, 4/193
Paseo, 5/204
Paseo de la Reforma (Mexico City), 4/138
Pashto language, 1/14, 5/20
Pashtun people, 1/13, 1/14, 1/19, *1/19,*
    5/20
Pashtunwali code, 1/14
Pasig River, 5/66, 5/67, *5/68–69*
PASOK (Greece), 3/53
Pass of Llanberis (Wales), *6/94*
Pasta, 3/234, *3/235,* 3/238
Patagonia (Argentina), *1/54–55,* 1/54–56,
    1/63, 1/65
Patan (Nepal), 4/185
Patassé, Ange-Félix, 2/43
Pathet Lao movement, 4/41, 4/47
Patmos (Greece), 3/66
Patois language, 4/60
Patrai (Greece), 3/61
Patrick, St., 3/201, 3/204, *3/205*

Patriotic Alliance for Change, 5/41
Patterson, Percival J., 3/249
Paul (regent of Yugoslavia), 5/158
Paul, Saint, the Apostle, 4/114
Pause-pulse migration, 7/91
Pavlovsk Palace (St. Petersburg), 5/122
Pax Romana, 3/227
Payton, Walter, 6/145
Paz, Octavio, 4/145
Paz Estenssoro, Víctor, 1/154
Peace, and Sweden, 5/246–247
Peace of Westphalia, 3/20
Peace Palace (The Hague), 4/204
Peaches, *3/218*
Peanuts, 3/11, *4/241*
Pearl Harbor (Hawaii), 3/264–265, 5/12,
    5/70, 6/136
Pearl of the Antilles, 2/134
Pearl River. *See* Zhu Jiang
Pearls, *3/273*
Peasants, Russian, 5/109, 5/112
Peat, 2/26, 3/207, *3/207*
Peat bogs, 5/78
Pedro I (ruler of Brazil), 1/164
Pedro II (ruler of Brazil), 1/164–165,
    1/176
Pegnitz River, *3/39*
Peking (China). *See* Beijing (China)
Pelé, *1/165*
Pelée, Mount, 4/118
Peleliu Island (Palau), 5/26
Peloponnesian War, 3/62
Peloponnesus (Greece), 3/54, *3/54,* 3/62
Pemba (Tanzania), 6/19, 6/21
Pembrokeshire Coast National Park,
    6/94
Pencak silat, 3/167
Penguins, 1/45, *2/210–211*
  Adélie, 1/45
  emperor, 1/45, *1/45*
  yellow-eyed, *4/224*
Peninsular Malaysia, 4/93, 4/95, *4/96,*
    4/96–97
Peninsular War, 7/103
Penn, William, 6/120
Pennines range, 6/92
Pennsylvania (United States), 6/120,
    6/121, 6/133, 6/134
Pentagon Building, 1/13, 6/137
People's Action Party (Singapore), 5/175
People's Assembly (Egypt), 2/181
People's Communist Party of Cuba,
    2/139
Peoples' Friendship University of
    Russia, 5/121
People's Front (Latvia), 4/48
People's National Congress (Guyana),
    3/93
People's National Party (Jamaica), 3/249
People's Power Party (Thailand), 6/29
People's Progressive Party (Guyana),
    3/92, 3/93
Pepin of Herstal, 2/240
Pepin the Short, 2/240, 3/227
Peppers, *1/174*

Perahera festival, 5/227
Perak River, 4/97
Perast (Montenegro), 4/157
Perch, 2/39
Peres, Shimon, 3/213
Pérez, Carlos Andrés, 6/174
Perfume, 5/101, 6/198, 6/199
Pergamum (Turkey), 6/68, 6/70
Perge (Turkey), 6/73
Pericles, 3/58, 3/65
Perija Mountains, 6/179
Perito Moreno glacier, 1/54–55
Permafrost, 6/130
Permanent Court of Arbitration, 4/204
Pernambuco (Brazil), 1/167
Perón, Eva, 1/65
Perón, Isabel, 1/52, 1/65
Perón, Juan, 1/52, 1/65, 1/65
Perpetual Covenant (Switzerland), 5/258
Perrault, Charles, 2/242
Perry, Matthew C., 3/264
Persepolis, 3/184
Persia, 3/182
Persian Empire
    Islam, spread of, 5/151
    power in
        Afghanistan, 1/10, 1/11
        Armenia, 1/66
        Azerbaijan, 1/111
        Greece, 3/65
        Iran, 3/175
        Iraq, 3/187
        Israel, 3/212
        Lebanon, 4/50
        Mesopotamia, 3/196
        Taiwan, 6/16
        Turkey, 6/68
Persian Gulf
    Bahrain on, 1/116
    Iran on, 3/175
    Iraq on, 3/186
    Kuwait on, 4/34, 4/36
    oil formation, 4/36, 4/37
    Oman on, 5/8
    Saudi Arabia on, 5/144
    Syria on, 5/266
Persian Gulf War of 1991
    Iraq and, 3/186, 3/188, 3/191, 3/193, 3/195, 3/199
    Kuwait and, 4/34–35, 4/37
    Saudi Arabia and, 5/145
    Syria and, 5/269
    United Arab Emirates and, 6/85
    United States and, 6/136
Persian language, 3/182
Persian people, 3/178
Perth (Australia), 1/73, 1/86, 1/89, 1/89
Peru, 5/44–55, 5/44–55, 5/47
    Central Railroad, 5/50–51, 5/50–51, 5/51
    early civilization, 7/82, 7/82–83, 7/82–83
    Easter Island raids, 2/171

    history
        early, 5/44–45, 5/52–53, 5/52–53, **5/53**
        recent, 5/46–47
    land and economy, 5/48–49, 5/48–49
        Lake Titicaca, **5/54,** 5/54–55, 5/54–55
Peru Current, 2/176
Pescadores islands, 6/10
Peshawar (Pakistan), 5/20–21
Peso (Argentina), 1/53
Pest (Hungary), 3/114, 3/115
Pestera cu Oase site, 7/44, 7/44–45
Peter I (ruler of Serbia), 5/158
Peter I, the Great, 5/112, 5/112, 5/120, 5/122, 5/123, 5/243
Peter II (ruler of Yugoslavia), 5/158–159
Peter IV (king of Aragon), 5/217
Peter, Saint (Biblical figure), 6/170
Peter, Saint, of Mount Athos, 3/70
Peter Island (British Virgin Islands), 6/188–189
Peterhof Palace (St. Petersburg), 5/122, 5/123
Petit Piton, 5/134, 5/135
Petite-Terre (Guadeloupe), 3/76
Petra (Jordan), 3/282
Petrarch, 2/150
Petrified Savanna (Martinique), 4/118
Petrobras (oil company), 1/176
Petrograd (Russia). See St. Petersburg (Russia)
Petroleum
    Algeria, 1/30, 1/30–31
    Angola, 1/38
    Australia, 1/92
    Azerbaijan, 1/110
    Bahrain, 1/116, 1/117
    Bangladesh, 1/121
    Brazil, 1/174, 1/176
    Brunei, 1/179
    Bulgaria, 1/184
    Cameroon, 2/20, 2/21
    Canada, 2/33, 2/35, 2/35, 2/37
    Chad, 2/44, 2/47
    Congo, 2/113, 2/116
    Ecuador, 2/177
    Equatorial Guinea, 2/200
    formation of, 4/36, 4/37
    Gabon, 3/11
    Indonesia, 3/170
    Iran, 3/175, 3/177, **3/181**
    Iraq, 3/188, 3/189, 3/190, 3/194–195, **3/195**
    Italy, 3/240, 3/241
    Kuwait, 4/34, 4/36–37, 4/37
    Libya, 4/66, 4/67, 4/68, 4/69
    Malaysia, 4/93, 4/100, 4/100
    Mexico, 4/135
    Netherlands Antilles, 4/211
    Nigeria, 4/240, 4/241, 4/241
    Norway, 4/252, 4/252–253
    Oman, 5/9
    Pacific Islands, 5/11
    Qatar, 5/98

    Russia, 5/124
    Saudi Arabia, 5/145
        oil fields, 5/148, 5/148–149
        social benefits, 5/147
    Sudan, South, 5/233, 5/236
    Syria, 5/271
    Trinidad and Tobago, 6/47, 6/47
    Tunisia, 6/53
    United Arab Emirates, 6/85–87, 6/87
    United Kingdom, 6/97, 6/106
    United States, 6/126, 6/131, 6/150
    Uzbekistan, 6/167
    Venezuela, 6/172, 6/173, 6/174, 6/175, 6/176, 6/177
    Yemen, 6/196
Petronas Twin Towers, 4/93
Petrusse River, 4/81
Peul people, 3/88
Pharaohs, 2/181
Pharmaceuticals, 5/263
Pheasants, 4/185
Phetsarat, 4/41
Philadelphia (Pennsylvania), 6/121, 6/146
Philharmonic Ball, 1/107
Philip II (king of Macedonia), 4/82
Philip II (Capetian king), 2/241, 2/247
Philip II (king of Spain), 5/57, 5/210, 5/211, 5/215, 5/215
Philippine Church of Christ, 5/63
Philippine Independent Church, 5/63
Philippine Sea, 5/60
Philippine Trench, 5/60
Philippines, 5/56–71, 5/56–71, **5/59**
    climate, 5/60
    economy, 5/59, 5/68–69, 5/68–69
    government, 5/58
    history, 5/7–59, 5/70–71, 5/70–71
    human migration, 7/58, 7/59, 7/92
    land, 5/60–61, 5/60–61
    people, 5/62–63, 5/62–63
        ethnic communities, 5/64–65, 5/64–65
    population, 5/62
Philipsburg (Sint Maarten), 4/62
Phillip, Arthur, 1/86, 7/106
Philosophy, 3/65, 3/140
Phinda Private Game Reserve, 5/197
Phnom Aural (mountain), 2/12
Phnom Penh (Cambodia), 2/9, 2/11, 2/12, 2/13
Phnom Penh University, 2/13
Phoenicians
    Algeria, 1/24
    Corsica, 2/254
    Gibraltar, 3/46
    Italy, 3/245
    Lebanon, 4/50
    Malta, 4/115
    Monaco, 4/148
    Portugal, 5/94
    Spain, 5/212
    Tunisia, 6/48
Phoenix Islands, 4/16, 4/17
Phoenix Islands Protected Area, 4/17

Phosphate
Morocco, 4/163
Nauru, 4/182, 4/183, *4/183*
Tunisia, 6/53
Western Sahara, 6/192
Photosynthesis, 3/172
Phou Bia mountain, 4/44
Phrygian people, 6/68
Phuan people, 4/46
Phuket Island, *6/36–37*
Phyra Chakri. *See* Rama I
Phytoliths, 7/80
Piast family, 5/76
Picardy (France), 2/238
Picasso, Pablo, 2/253, 4/244
Pico de Orizaba, 4/130
Pico de Teide, 5/203, 5/219, *5/219*
Pico Mogotón, 4/228
Pico Ruivo (Madeira Islands), 1/113
Picts, 6/110
Pidgin English, *5/35*, 5/185, 6/168
Pied Piper of Hamelin, *3/34–35*
Piedmont (Argentina), 1/55
Piedmont (Italy), 3/239
Piedmont (United States), 6/121, 6/124
Piedmontese dialect, 3/233
Pierhead Building (Cardiff), *6/94–95*
Pike, Northern, *2/39*
Pilcomayo River, 5/42
Pilgrims, 6/134
Pillar of Hercules, 3/46
Pillow Book, The (Sei Shonagon), 3/268
Piłsudski, Józef, 5/77
Pinakothek museum (Munich), 3/27
Pinang (Malaysia), 4/94, 4/102
Pinang Island (Malaysia), 4/93
Pinar del Río (Cuba), 2/138
Pinatubo, Mount, 5/58, 5/60
Pindus Mountains, 3/54
Pine Islands, 5/216
Piñera, Sebastián, 2/51
Pines, 4/225
Monterey, 4/224, 4/225
mountain, *1/35*
Scotch, *1/35*
Pinnacle Rock (Bartolomé Island), *2/178–179*
Pinochet Ugarte, Augusto, 2/51
Pinturicchio, 6/171
Pipil Indians, 2/194, 2/198
Pirates
Algeria, 1/24
Bahamas, 1/114
Brunei, 1/178
Cayman Islands festival, 3/49
Copenhagen, 2/158
Galapagos Islands, 2/178
Haiti, 3/96
Jamaica, 3/247
Madagascar, 4/86
Morocco, 4/158
Nicaragua, 4/226
Seychelles, 5/170
Singapore, 5/176

Somalia, 5/187
See also Vikings
Pirin Mountains, 1/183
Pisa (Italy), 3/228
Pissarro, Camille, 2/253
Pitcairn Islands, **2/259,** *2/259*
Piton de la Fournaise, 5/100
Piton des Neiges, 5/100
Pityusae, 5/216
Pius IX (pope), 6/170
Pizarro, Francisco, 5/45, 5/53, 5/214, 6/172, 7/102
Pizarro, Gonzalo, 2/106
Pizza, 3/234
Pizza napoletana verace, 3/234
Place de la Concorde (France), *2/244, 2/247*
Plain of Jars, 4/44
Plains, 1/4
Plains Indians, *2/28–29,* 6/140–141
Plankton, *1/45,* 4/106
Plants
Amazon Region, 1/168
Antarctica, 1/44
Australia, 1/74–75
Himalaya, 4/191
Madagascar, 4/88–89
megafauna extinctions and, 7/28–29
Mississippi River, *6/149*
New Zealand, 4/224–225
Sahara, 4/72–73
See also specific places
Plastic People of the Universe, The, 2/153
Plateau of Mexico, 4/130
Plateau of Tibet, 2/70, 2/88, 2/94
Plateaus, 1/4
Plato, *3/63,* 3/65
Platypuses, 1/74
Plaza Bolívar (Bogotá), *2/105*
Plaza Bolívar (Caracas), 6/178
Plaza Catalunya (Barcelona), *5/204–205*
Plaza de Armas (Peru), *5/47*
Plaza de Armas (Santiago), *2/54*
Plaza de Congreso (Buenos Aires), 1/61
Plaza de España (Madrid), *5/210–211*
Plaza de la Revolución (Havana), *2/132–133*
Plaza de Mayo (Argentina), 1/61, *1/64*
Plaza Navona (Rome), *3/230–231*
Plaza of Three Cultures (Mexico City), 4/142
Pleistocene Epoch, 7/20, 7/33, 7/72
Pliny the Elder, 5/218
PLO. *See* Palestine Liberation Organization
Plovdiv (Bulgaria), 1/184, 1/185
Plymouth (Massachusetts), 6/134
Plywood, 2/230
Plzen (Czech Republic), *2/154–155,* 2/155
Po River, 3/225
Po Valley, 3/225, *3/238–239,* 3/239
Poaching, 6/24, *6/24*
Poás volcano, *2/121,* 2/125
Podgorica (Montenegro), 4/156
Pohamba, Hifikepunye, 4/180

Pohnpei Island, 2/214, 2/215, *2/215*
Point Barrow (Alaska), 6/118, 6/131
Pointe-à-Pitre (Guadeloupe), 3/77
Pointe-à-Noire (Congo), 2/116
Poitiers, Battle of, 5/151
Poitiers, Diane de, 2/242
Pokhara (Nepal), 4/185
Pol Pot, 2/9
Poland, 1/56, 5/72–83, *5/72–83,* **5/75**
economy, 5/82–83, *5/82–83*
history, 5/72–77, *5/76–77,* **5/77**
Communist era, 5/73–75, 5/77
post-Communist era, 5/75
land, 5/78–79, *5/78–79*
power in
Belarus, 1/130
Estonia, 2/204
Latvia, 4/48
Lithuania, 4/76, 4/77
Ukraine, 6/82
Polanie tribe, 5/72
Polders, 4/199, 4/208
Polenta, 3/234
Police Motu language, 5/35
Polisario Front, 4/159, 4/161, *6/192,* 6/193
Polish Committee of National Liberation, 5/77
Polish language, 5/73
Polish people, in United States, 6/138
Pollock, Jackson, 6/158
Pollution
Canada, 2/33
Himalaya, *4/195*
Japan, 3/261
Marshall Islands, 4/117
Mexico City, 4/139, *138–139*
Persian Gulf, 4/37, *4/37*
Russia, 5/119
Spain, 5/210
Thailand, 6/35
United States, 6/131
Venice, 3/243
Polo, *5/21*
Polo, Marco, 2/62, 2/88, 3/168, 3/253
Poltava, Battle of, 5/243
Polygyny, 6/205
Polynesia, 2/170, 4/215, 6/42
Heyerdahl voyage, 5/55
history, 5/12
human settlement, **7/93,** 7/93–95, *7/94–95,* **7/95**
land, 5/10–11
Polynesians
American Samoa, *1/32*
Cook Islands, 2/118
Easter Island, 2/170, 2/171
Federated States of Micronesia, 2/214
Fiji, 2/216
French Polynesia, 2/258, 2/259
Hawaii, 6/129
Nauru, 4/182
New Caledonia, 4/212
origins, 7/93
Pacific Islands, 5/10–11

Papua New Guinea, 5/34
Samoa, 5/138
Tahiti, 6/8
Tonga, 6/44
Tuvalu, 6/77
voyages, **7/93**, 7/93–95, *7/94–95*,
   7/95
Polyps, Coral, 1/92–93, 4/106
Pombal, Marquis of, *5/94*
Pomerania (Poland), 5/76
Pomo Indians, 6/140
Pompeii, 3/224, 3/241
Pompey, 3/226, 6/68
Pompidou Center (France), *2/253*
Ponce de León, Juan, 3/49, 5/97
Pont du Gard (Nîmes), 2/240, *2/241*
Ponte dei Sospiri (Venice), 3/243
Ponte di Rialto (Italy), *3/235*
Pontic Mountains, 6/60
Pontifical Commission for the State of
   Vatican City, 6/171
Poopó, Lake, 5/54
Popalzai clan, 1/13
Popes, 3/227, 3/229, 6/170–171
Popes, Palace of the (France), *2/252*
Popocatépetl, 4/130
Poppies, *3/15*
Popular Action Party (Peru), 5/46
Popular Movement for the Liberation of
   Angola, 1/36–37
Popular Party (Spain), 5/201
Population
  as issue
    Africa, early, 7/98
    Egypt, 2/181, 2/191
    Guatemala, 3/85
    Himalaya, 4/194
    India, 3/139
    Pakistan, 5/20
    Sri Lanka, 5/225
    Turkey, 6/59
  of Earth, 1/2
    prehistoric humans, 7/13, 7/53
  See also specific places
Population Act (South Africa), 5/190
Porcelain, 4/205, *4/205*
Porgera (Papua New Guinea), 5/37
Port-au-Prince (Haiti), 3/97, *3/98–99*
Port Blair (India), 3/156, 3/157
Port Brega (Libya), 4/69
Port Elizabeth (South Africa), 5/196
Port-Gentil (Gabon), *3/10*
Port Harcourt (Nigeria), *4/241*
Port Hedland (Australia), *1/82–83*
Port Louis (Mauritius), 4/125, *4/125*
Port Moresby (Papua New Guinea), 5/35
Port Phillip Bay (Australia), 1/85
Port Rashid (Dubai), 6/87, *6/87*
Port Sudan (Sudan), 5/233, 5/234
Port-Vila (Vanuatu), 6/168, *6/169*
Port wine, 5/92
Porteños, 1/60, *1/61*
Portland (Oregon), 6/128
Porto (Portugal), 5/85, 5/91, 5/94
Porto Santo (Madeira Islands), 1/113

Portobello Road Market (London),
   6/102–103
Portugal, 5/84–95, *5/84–95*, **5/87**
  Australia exploration, 1/86
  China, relations with, 2/61, 2/87
  economy, 5/92–93, *5/92–93*
  history
    early, 5/85, 5/94–95, *5/94–95*,
     5/95
    migration to Americas, 7/102–103
    modern, 5/86–87
  land, **5/88**, 5/88–89, *5/88–89*
  people, 5/90–91, *5/90–91*
  population, 5/90
  power in
    Angola, 5/87
    Azores, 1/112
    Brazil, 1/164–165, 1/170, 1/172,
     1/176
    Cameroon, 2/18
    Cape Verde, 2/40
    Congo region, 2/110, 2/114
    East Timor, 2/168
    Equatorial Guinea, 2/201
    French Guiana, 2/256
    Gabon, 3/8
    Gambia, 3/13
    Ghana, 3/42
    Guinea, 3/86
    Guinea-Bissau, 3/90, 3/91
    India, 3/154–155
    Indonesia, 3/161
    Japan, 3/264
    Kenya, 4/10
    Lesser Antilles, 4/60
    Macau, 2/86, 2/87, 5/87
    Madeiras, 1/112, 1/113
    Malaysia, 4/94, 4/102
    Maldives, 4/104
    Mariana Islands, 3/79
    Morocco, 4/158, 4/167
    Mozambique, 4/170, 5/87
    Nigeria, 4/236
    Portuguese Guinea, 5/87
    Portuguese Timor, 5/87
    Réunion, 5/100
    São Tomé and Príncipe, 5/142–143
    Senegal, 5/153
    Seychelles, 5/170
    Sri Lanka, 5/224
    Tanzania, 6/18
    Togo, 6/41
    Uruguay, 6/160
Portuguese Guinea, 3/90, 5/87
Portuguese language
  Angola, 1/39
  Cape Verde, 2/41
  Macau, *2/87*
  Mozambique, 4/173
  Portugal, 5/94
  São Tomé and Príncipe, 5/143
Portuguese people
  France, 2/239
  Puerto Rico, 5/96
  See also Portugal

Portuguese Timor, 2/168, 5/87
Poseidon, *3/55*
Poseidon, Temple of (Greece), *3/64*
Potala Palace (Tibet), *2/96–97*
Potaro River, 3/94
Potash, 3/211, *6/151*
Potatoes, 1/130, *3/36–37*, 3/203, 3/204,
   5/44, *5/48*, *6/100*
Potomac River, 6/118
Potsdam (Germany), 3/39
Pottery
  Austronesia, 7/92, *7/92*, 7/93
  China, 2/66, *2/66*, 7/56–57
  Great Zimbabwe, *7/100–101*
  Indus Valley, 5/25
  Japan, 7/61, *7/61*
  Korea, 4/31
  Mexico, *4/145*
  Netherlands, 4/205, *4/205*
Poverty
  African Americans, 6/144
  Argentina, 1/61
  Bangladesh, 1/119, 1/120, 1/125
  Belize, 1/145
  Brazil, 1/161, 1/162, 1/170–171
  Chad, 2/46
  Colombia, 2/105
  Comoros, 2/108
  Dominica, 2/165
  Egypt, 2/181
  El Salvador, 2/195, 2/198–199
  Ethiopia, 2/209
  India, 3/126, 3/139
  Libya, 4/67
  Malawi, 4/90
  Mexico, 4/127, 4/139, *4/139*
  Paraguay, 5/38
  Philippines, 5/59
  Senegal, 5/152
  Sri Lanka, 5/224
  Tanzania, 6/20, 6/22
  Thailand, 6/32
  United States, 6/147
  Venezuela, 6/173, 6/175
  See also Slums
Powell, Colin, 6/144, 6/145
Pozzuolana, 2/41
PPN (Niger), 4/233
PPP. *See* Pakistan People's Party; People's
   Power Party (Thailand)
Pra River, 3/40
Prado (Madrid), 5/210–211
Prague (Czech Republic), 2/144, 2/150–
   151, *2/150–151*, 2/155
  art and politics, 2/153, *2/153*
  history, 2/146, 2/148, 2/150–151
  Lesser Town of, 2/150
Prague Castle, *2/148*, 2/150–151
Prague Spring, 2/149, *2/149*, 2/150,
   2/152, 2/153
Prairie Provinces (Canada), 2/26, 2/35,
   2/37
Prajadhipok. *See* Rama VII
Prater (Vienna), 1/109
Prayer flags, *2/97*

Pre-Harappan culture, 7/20
Prehistoric humans
  by modern place
    Afghanistan, 1/10
    Africa, 4/112
    Balearic Islands, 5/216
    Botswana, 1/158
    Congo, 2/110
    Corsica, 2/254
    Cyprus, 2/140
    Guinea, 3/86
    Indonesia, 3/161
    Kenya, 4/10
    Korea, 4/22
    Spain, 5/212
    Tanzania, 6/18
    Ukraine, 6/82
    Zimbabwe, 6/210
  dating living sites, 7/21
  fully modern, 7/48
  hobbits, 7/24–25, 7/25
  skin color changes, 7/45
  species, 7/6, 7/9
  upright gait, 7/8
  volcanic eruption and, 7/13, *7/13*
  See also Human migrations; Tools, Early;
    and specific species
Prenzlauer Berg (Berlin), *3/32*
Presidential Palace (Bolivia), *1/155*
Presley, Elvis, 6/159
Pretoria (South Africa), 5/196
PRI (Mexico), 4/128–129
Pribilof Islands, 6/130
Primary Chronicle, 5/112
Primavera (Botticelli), *3/236*
Prime minister (office), 2/24–25
Primo de Rivera, Miguel, 5/215
Prince, Palace of the (Monaco), *4/149*
Prince, The (Machiavelli), 3/237
Prince Charles Foreland (Norway), 4/256
Prince William Sound (Alaska), 6/131
Princely states, 3/129
Princip, Gavrilo, 1/98, 5/158
Principe. *See* São Tomé and Príncipe
Prinsenhof (Delft), 4/205
Printing, 2/66, 3/38
Prisoner settlements
    Australia, 1/69, 1/84, 1/86, 1/87,
      7/106, 7/106–107, **7/107**
    French Guiana, 2/256, *2/256–257*
    Réunion, 5/100
    Russia, 5/114, 5/122, 5/126, 5/129
    São Tomé and Príncipe, 5/142
Prithwi Narayan Shah, 4/186
Prodi, Romano, 3/223
Producer goods, 6/153
Progressive Democratic Party (Ireland),
    3/202
Progressive Labour Party (Bermuda),
    1/149
Progressive Liberal Party (Bahamas),
    1/114–115
Project Tiger, 3/150–151, *3/150–151*
Proportional representation, 2/224
Prosciutto ham, 3/235

Prose Edda (Snorri Sturluson), 3/122
Prospekts (St. Petersburg), 5/122
Protectorate of Nyasaland, 4/91
Protestant Reformation. *See* Reformation
Protestantism
    Brazil, 1/166
    Cuba, 2/139
    Czech Republic, 2/148, *2/148*
    France, 2/244
    Germany, 3/19
    Haiti, 3/98
    Ireland, 3/204, 3/205
    Korea, 4/30
    Netherlands, 4/202
    Northern Ireland, 6/98
    Pacific Islands, 5/12
    Philippines, 5/63
    United States, 6/156
    Virgin Islands, British, 6/188
Provence (France), *2/248–249*
Provinces (Canada), 2/25, **2/25, 2/33**
Provincia, 2/248
Provodnik attendants, 5/129
Prudhoe Bay (Alaska), 6/131
Prussia, 1/98, 3/21, **3/21**
    power in Poland, 5/72, 5/73, 5/76,
      5/80
Prut River, 4/146
Ptarmigans, *1/35*
Ptolemais River Valley, 3/54
Ptolemy (king of Macedonia), 2/192
Pu Yi, 2/64
Pubs, *3/207*
Puea Thai Party (Thailand), 6/29
Puebla (Mexico), 4/134
Pueblo Indians, *6/126–127,* 6/127,
      6/140
Pueblos jóvenes, 5/48–49, *5/48–49*
Puerta del Sol (Madrid), 5/210, *5/211*
Puerto del Sol (Peruvian carving), 5/44
Puerto Ricans, in United States, 6/142,
      6/143
Puerto Rico, 3/48, 4/58, 7/103
Puerto Vallarta (Mexico), 4/143
Pueyrredón, Prilidiano, 1/58
Puig Mayor, 5/217
Puletasis, 5/138
Pulguk-sa Temple (Korea), *4/31*
Pulses, 3/134, 3/144
Pumas, *6/177*
Puna region (Argentina), 1/55
Puncak Jaya, 3/165
Punic Wars, 3/226
Punjab
    India, 3/126, 3/134, 3/136, 3/141
    Pakistan, 5/15, 5/16, 5/18, 5/20, 5/22
Punjabi language, 3/138, 5/20
Punjabi people, 5/20
Punta del Este (Uruguay), *6/163, 6/165*
Puntarenas province (Costa Rica), *2/123*
Puppet theater, 3/276
Pupusas, 2/199
Puritans, 6/120
Pusan (South Korea), 4/26
Pushers, on trains, 3/271

Pushkin Palace (St. Petersburg), 5/122
Pushtunwali code, 1/14
Pusztas, 3/112, *3/112–113*
Putin, Vladimir, 5/111, *5/115*
Putonghua. *See* Mandarin Chinese
    language
Putrajaya (Malaysia), 4/96
Putumayo River, *2/104–105*
Pygmies, 7/16
    Burundi, 1/188
    Cameroon, 2/16–17
    Central African Republic, *2/43*
    Congo, 2/112, 2/114
    Congo (Kinshasa), 2/116
    Equatorial Guinea, 2/201
    Gabon, 3/8
    Rwanda, 5/130
Pyongyang (North Korea), 4/20, 4/21,
    4/25
Pyramid of the Moon (Mexico), *4/140*
Pyramid of the Sun (Mexico), *7/80*
Pyramiden (Norway), 4/257
Pyramids
    Egypt, *2/180–181,* 2/181, 2/192,
      7/41, *7/41*
    Mexico, *7/80*
    Peru, 5/44, 7/83
Pyramids, Battle of the, 2/192
Pyrenees Mountains, 1/34, *1/35,* 2/237,
    2/238, 5/199, 5/202, 5/205
Pyrite, 4/252
Pythagoras, 3/66
Pyu people, 4/178

Q

Qaboos bin Said, 5/9
Qadan industry, 7/40–41
Qadhafi, Mu'ammar Muhammad al-,
    4/66, 4/67
Qadir, Abd al-, 1/24
Qafzeh Cave, 7/32–34, *7/33*
Qa`ida, Al-, 1/13, 5/17, 6/137
Qajar dynasty, 3/182
Qanats, 1/17, *5/19*
Qarase, Laisenia, 2/221
Qasab reed, 3/196
Qashqai people, 3/178
Qat. *See* Khat
Qatar, 5/98–99, *5/99,* **5/99**
Qin dynasty, 2/62, **2/62,** 2/66, 2/68
Qin Ling Mountains, 2/73
Qin Shi Huang, *2/63*
Qing dynasty, 2/62, 2/62–63
Qinghai Province (China), *2/70–71,* 2/88
Quebec (city, Quebec), *2/30, 2/30,* 2/30,
    2/36, *2/36*
Quebec (province, Canada), 2/23
    early, 2/30–31
    economy, 2/34, 2/35, *2/35*
    land, 2/27, 2/38
    separatism, 2/32, *2/32*
Quebec Act, 2/30
Québécois, 2/32
Quebracho trees, 1/54, 5/42
Quechua Indians, 2/54, *5/52,* 5/55

Quedlinburg (Germany), 3/38
Queen Charlotte Islands (Canada), 2/39
Queensland (Australia), 1/70, 1/80, 1/88, *1/88*
Quemoy Islands, 6/10
Quetico Provincial Park (Ontario), 2/38
Quetta (Pakistan), *5/18–19, 5/23*
Quetzals, 2/125, *2/125*
Quezon City (Philippines), 5/66
Quiapo (Philippines), 5/67
Quiapo Church, 5/67
Quiet Revolution, 2/32
Quirinal hill (Rome), 3/230
Quito (Ecuador), 2/172, 2/174, *2/175*
Qur'ān, 3/185, 4/232

R

Ra II, 5/55
Rabat (Morocco), 4/160, 4/161, *4/161,* 4/166
Rabbani, Burhanuddin, 1/13
Rabbits, 1/79
Rabi, 3/135
Rabin, Yitzhak, 3/213, *3/213*
Rabuka, Sitiveni, 2/220
Raccoons, *2/38, 6/149*
Race relations
 Brazil, 1/166
 South Africa, 5/189–193, *5/194*
 United Kingdom, 6/105
 United States, 6/136, 6/139, 6/144–145
Racine, Jean, 2/252
Radak Chain (Marshall Islands), 4/116
Radama I, 4/86
Radiation contamination, 4/117
Radio Martí, 2/138
Radiocarbon dating, 7/19, 7/21
Raffles, Sir Stamford, 5/174
Raffles Hotel (Singapore), *5/177*
Rahman, Sheik Mujibur, 1/120–121
Raiatea (French Polynesia), 2/259
Railroads
 Canada, 2/31
 Channel Tunnel, *6/113*
 France, **2/251,** *2/251*
 Gabon, *3/10–11,* 3/11
 Hungary, 3/115
 India, *3/148–149*
 Iran, *3/178–179,* 3/183
 Japan, *3/263, 3/270–271, 3/271*
 Netherlands, *4/207*
 Paraguay, *5/42*
 Peru, 5/50–51, *5/50–51,* **5/51**
 Russia, **5/128,** 5/128–129, *5/128–129*
 South Africa, *5/196*
 United States, 6/123, *7/105*
Rain forests
 Barbados, 1/128
 Brazil, 1/161, 1/168, *1/168–169*
 Cameroon, 2/16–17
 Central African Republic, 2/42
 Colombia, 2/98, 2/102
 Congo region, 2/110, *2/112,* 2/114, 2/115

 Costa Rica, 2/125
 Dominica, 2/164
 Ecuador, 2/177
 Equatorial Guinea, 2/200, *2/201*
 Fiji, 2/218, *2/218–219*
 French Guiana, 2/256, 2/257
 Indonesia, 3/161, 3/164, 3/165, 3/172–173, *3/172–173,* **3/173**
 Madagascar, 4/89
 Malaysia, 4/93, 4/96–97, *4/96–97,* 4/100
 Mexico, *4/142–143*
 Paleo-Indian farming, 7/80–81
 Papua New Guinea, 5/36
 Philippines, 5/57, 5/60
 Samoa, 5/138, *5/139*
 Singapore, 5/174
 Sri Lanka, 5/222
 Suriname, 5/238, *5/239*
 Tonga, 6/42
 Trinidad and Tobago, 6/46
 Uganda, 6/80
 Vanuatu, 6/168
 Venezuela, 6/176
 See also Amazon region
Rainier III, 4/148
Rainier III prize for literature, 4/149
Raja Sulayman Park (Philippines), *2/59*
Rajang River, 4/97
Rajasthan (India), *3/136–137, 3/149*
Rajoelina, Andry, 4/87
Rajoy, Mariano, 5/201
Rákóczi, Francis, II, 3/110
Rákosi, Matthias, 3/111
Ralik Chain (Marshall Islands), 4/116
Rama I, 6/36
Rama IV, 6/36
Rama V, 6/36
Rama VII, 6/37
Ramadan, 4/165, *4/242,* 5/21
Ramayana, 3/154, 5/222
Rameau, Jean Philippe, 2/252
Ramkhamhaeng, 6/36
Ramos-Horta, José, 2/168–169
Ramotar, Donald, 3/93
Ramses II, *2/193*
Ramuz, Charles Ferdinand, 5/255
Rana rulers of Nepal, 4/186, 4/189
Ranchos (Paraguay), 5/43
Randstad Holland (Netherlands), 4/200, 4/203
Rangoolam, Navachandra, 4/125
Rangoon (Myanmar). *See* Yangon (Myanmar)
Rano Raraku volcano, 2/171
Rap music, 6/159
Rapa Nui. *See* Easter Island
Rapanui language, 2/171
Raphael, 3/236, 6/171
Rarotonga (Cook Islands), 2/118, 2/119
Ras al Khaymah (United Arab Emirates), 6/84, 6/85
Ras Tanura oil refinery, *5/148–149*
Rastafarians, 3/246, 3/251
Rathaus (Bremen), 3/31

Rats, Pacific, 7/96
Ratsiraka, Didier, 4/87
Rattlesnakes, *6/177*
Raurkela (India), 3/136
Ravalomanana, Marc, 4/87
Ravi River, 5/18, 5/22, 5/24
Rawlings, Jerry, 3/42, 3/45–46
Reagan, Ronald, 2/195, 4/231
Reason, Age of, 2/252
Recession. *See* Economic crisis (2007-)
Recife (Brazil), 1/172
Reconquista, 5/213
Reconstruction (United States), 6/136, 6/144
Recreation, in United States, 6/157
Red Army, 6/74
Red Beard. *See* Frederick I (king of Germany)
Red Brigades (Italy), 3/222
Red Crescent Societies, 5/265
Red Cross, *2/208,* 5/264–265, *5/264–265*
Red Fort (India), 3/148
Red Guard, 2/223
Red Lady of Paviland, 7/50–51
Red River Delta, 6/186, 6/187
Red River settlement, 2/31
Red Sea
 Djibouti on, 2/162
 Eritrea on, 2/202
 Ethiopia on, 2/206
 human migration and, 7/14, *7/15*
 Saudi Arabia on, 5/144
 Sudan on, 5/232, 5/233
 Yemen on, 6/194, 6/200
Red Sea Hills, 5/232
Red Square (Moscow), *5/120–121*
Redonda Island, 1/48
Reds (Russia), 5/114
Reducciones, 5/38
Referendum, 5/256
Reformation, 1/98, 3/20, 3/38, 4/202, 6/111
Refugees
 Afghanistan, 1/12, 1/19
 Andaman and Nicobar Islands, 3/157
 Bulgaria, 1/181
 Germany, 3/22
 Haiti, 2/99
 Iran, 3/179
 Jordan, 3/286
 Lebanon, 4/52
 Mexico, *4/137*
 Vietnam, 6/182
Reggae music, 3/250–251, *3/250–251*
Reggae Sunfest, 3/251
Reggae Sunsplash festival, *3/250, 3/251*
Rehoboth (Namibia), 4/181
Reichstag building, *3/33*
Reindeer, 5/126
Religious freedom, in United States, 6/156
Rembrandt, 4/202, 4/205

Renaissance
Belarus, 1/130
France, 2/242–243, 2/252
Hungary, 3/110
Italy, 3/221, 3/228, 3/236, *3/236–237*
Poland, 1/130
Russia and, 5/112
Venice, 3/242
Renamo, 4/170–171, 4/173
Renoir, Pierre Auguste, 2/253
Republicans (Ireland), 3/205
Reservations, Indian, *6/140,* 6/141
Residence of the Grand Master (Rhodes),
3/68
Residencia de Estudiantes (Madrid),
5/211
Residenz (palace, Austria), 1/101
Resins, 6/198, *6/199*
Retail trade, 6/152
Rethimnon (Crete), 3/68
Réunion, 5/100–101, *5/100–101,* **5/101**
Revolution of 1848, 2/244, 3/21, 3/28
**Revolutionary Armed Forces of
Colombia,** 2/100, 2/101
Revolutionary Command Council, 4/67
Revolutionary Council (Myanmar), 4/176
**Revolutionary Front for an Independent
East Timor,** 2/168
**Revolutionary League for the Independ-
ence of Vietnam.** *See* Vietminh
Revolutionary Party (Tanzania), 6/20
Revolutionary United Front (Sierra
Leone), 5/173
Revolutionary War in America. *See*
American Revolution
Revueltas, Silvestre, 4/145
Reykjavík (Iceland), *3/119,* 3/120
Reza Shah. *See* Pahlavi, Reza Shah
Rhapsody in Blue (Gershwin), 6/159
Rhee, Syngman, 4/24
Rhine River
France, 2/236
Germany, 3/24, 3/34, 3/35, 3/37
Netherlands, 4/200, 4/204, 4/205
Switzerland, 5/257
Rhine Valley (Liechtenstein), 4/74
Rhinoceroses, 2/74
Indian, *4/185*
Rhode Island (United States), 6/118,
6/120, 6/134
Rhodes (Greece), 3/66, 3/67, 3/68–69
Rhodes, Cecil, 5/192, 6/207, 6/208,
6/211
Rhodesia, 4/170, 6/202–203, 6/206,
6/208, 6/211
Rhodope Mountains, 1/182, 1/183
Rhône-Alps region (France), 2/251
Rhône River, 2/236, 5/257
Rhône-Saône Valley, 2/237
Rialto Bridge (Venice), *3/242–243*
Rice
Bangladesh, 1/118
Cambodia, 2/8, 2/12–13
China, 2/73, 2/88, 2/90, *2/90–91,*
7/56

cultivation, *4/45*
East Timor, *2/169*
Guinea, *3/89*
Guyana, 3/94
Haiti, *3/98*
India, *3/134*
Indonesia, *3/167,* 3/170, *3/170–171*
Italy, *3/238*
Japan, 3/260, *3/260–261*
Korea, *4/20–21,* 4/29
Laos, *4/43,* 4/44
Malaysia, 4/101
Myanmar, *4/175*
Philippines, *5/60,* 5/64, 5/68, *5/68*
Thailand, 6/27, 6/30, 6/32, 6/39
Vietnam, 6/186, *6/186–187*
Rice, Condoleezza, *6/144,* 6/145
Richard the Lion-Hearted, 2/143
Richelieu, Cardinal, *2/245*
Richmond (Virginia), 6/124
Riddarholmen Island (Sweden), *5/245*
Riding Mountain National Park
(Manitoba), *2/33,* 2/39
Riebeeck, Jan van, 5/192
Riel, Louis, 2/31
Rif range, 4/162
Rift Valley (Israel), 3/209
See also Great Rift Valley
Riga (Latvia), 4/48, *4/49*
Riga, Gulf of, 2/202
Riga, Treaty of, 1/130, 5/77
Right Bank (Paris), 2/246
Rijksmuseum (Amsterdam), 4/206
Riksdag, 5/243
Rila Monastery, 1/182, *1/183*
Rila Mountains, 1/183
Rímac River, 5/50
Rincón (Puerto Rico), 5/97
Ring of Fire, 1/43, 4/132, 5/36, *5/60*
Ringstrassen (Vienna), *1/108,* 1/109
Rio de Janeiro (Brazil), *1/160–161,*
1/161, **1/170,** 1/170–171,
1/170–171
history, 1/165, 1/166, 7/103
Río de la Plata, 1/60, 1/64, 6/160,
6/164
United Provinces of the, 1/56, 1/65
Rio Grande, 4/127
Río Muni (Equatorial Guinea), 2/200
"Rip Van Winkle" (Irving), 6/158
Risotto alla milanese, 3/234
Rista Falls (Sweden), *5/252–253*
River Plate Stadium, *1/57*
Rivera, Diego, 4/144
Rivera, Julio Adalberto, 2/195
Rivers, 1/4
Rivière Salée strait, 3/76
Riyadh (Saudi Arabia), 5/144, 5/147
Rizal, José, 5/70
Rizal Park (Manila), 5/67
Roatán (Honduras), 5/136
Róbinson Crusoe Island (Chile), 2/53
Rock of Gibraltar, 3/46, *3/47*
Rock-steady, 3/250
Rockefeller Center (New York City), *6/121*

Rocky Mountains
Canada, 2/26, *2/26,* 2/39
United States, 6/127, *6/127,* 6/130,
6/148
Rococo art, 2/252
Rodeos, 1/58, *1/58, 1/78–79*
Rodin, Auguste, 1/61
Rodrigues (Mauritius), 4/124
Rodríguez de Francia, José Gaspar, 5/39
Rodríguez Pedotti, Andrés, 5/40, 5/41
Rodríguez Zapatero, José Luiz, 5/201
Roggeveen, Jacob, 2/171, 5/139
Roh Moo-hyun, 4/25
Roh Tae Woo, 4/25
Roma (people), 2/142, 3/116–117,
3/116–117, 5/107, 5/179
Roman Catholicism
Andorra, 1/35
Angola, 1/39
Argentina, 1/57
Australia, 1/91
Belarus, 1/130
Brazil, 1/166
Burundi, 1/189
Canary Islands, 5/221
Chile, 2/49
Colombia, 2/99
Costa Rica, *2/124*
Croatia, 2/130
Cuba, 2/139
Czech Republic, 2/148
El Salvador, 2/195, 2/199, *2/199*
England, 6/111
France, 2/244
Guatemala, 3/80, *3/81,* 3/84
Haiti, 3/98
Hispanic Americans, 6/143
Honduras, 3/100
Hungary, 3/106, 3/110
India, 3/154
Ireland, 3/201, 3/203
Italy, 3/223, *3/232, 3/233*
Japan, 3/264
Korea, 4/30
Kosovo, 4/32
Luxembourg, 4/80
Macau, 2/86, 2/87
Malta, 4/114
Mexico, 4/136, 4/140, 4/141, 4/142
Mozambique, 4/173
Netherlands, 4/202
Nicaragua, 4/226, 4/229
Northern Ireland, 6/98
Pacific Islands, 5/12
Paraguay, *5/38*
Philippines, *5/62,* 5/63, 5/70
Poland, 5/72–73, 5/75
St. Lucia, 5/135
San Marino, 5/141
São Tomé and Príncipe, 5/143, *5/143*
Seychelles, 5/170
Slovenia, 5/182
Spain, 5/214
Trinidad and Tobago, 6/46
Uganda, 6/79

United States, 6/156
Vatican City, 6/170, 6/171
Venezuela, 6/178
Vietnam, 6/181
Zambia, 6/204
Roman dialect, 3/233
Roman Empire. See Rome, Ancient
Roman Forum, 3/230
Roman mythology, 3/230
Roman Republic, 3/226
Romania, 5/102–107, 5/102–107,
   5/103
history, 5/104–105
Hungarians in, 3/107
land, 5/104–105, 5/104–105
   Danube River, 5/180, 5/181
Moldova, power in, 4/146, 4/147
people, 5/106–107, 5/106–107
prehistoric human migration, 7/44,
   7/44–45
Romanian language, 4/147, 5/106
Romanian Orthodox Church, 5/107
Romanian people, in Serbia, 5/155,
   5/164
Romanov, Michael, 5/112
Romansh language, 5/260
Romanticism, 2/252–253, 3/39
Romany language, 3/117
Rome (Italy), 3/221, 3/230–231,
   3/230–231, 3/232
history, 3/228–231
Vatican City and, 6/170–171
Rome, Ancient
Canary Islands mentioned in, 5/218
Danube River settlements, 5/180
history, 3/226–227, 3/226–227,
   3/227
power in
   Albania, 1/20
   Algeria, 1/24
   Armenia, 1/66
   Austria, 1/98
   Balearic Islands, 5/217
   Bosnia-Herzegovina, 1/156
   Britain, 6/91, 6/102, 6/104, 6/110,
      6/110
   Corsica, 2/255
   Crete, 3/68
   Egypt, 2/192
   France, 2/233, 2/240, 2/241,
      2/248
   Germany, 3/20
   Greece, 3/62–63, 3/226
   Iraq, 3/187
   Israel, 3/216
   Istanbul, 6/63
   Italy, 3/221
   Jordan, 3/282
   Judah, 3/209
   Lebanon, 4/50
   Libya, 4/70
   Macedonia, 4/82
   Madeira Islands, 1/113
   Montenegro, 4/157
   Morocco, 4/158

Netherlands, 4/202
   Portugal, 5/94
   Romania, 5/106
   Spain, 5/212
   Syria, 5/266, 5/274
   Tunisia, 6/48
   Turkey, 6/68, 6/73
Rome (city) in, 3/230–231
Silk Road trade, 1/10
Sweden, trade with, 5/242
Römer town hall (Frankfurt), 3/28–29
Romero, Oscar Arnulfo, 2/195
Romulus and Remus, 3/230
Romulus Augustulus, 3/227
Rongelap (Marshall Islands), 4/117
Ronin warriors, 3/267, 3/275
Roosevelt, Anna, 7/84, 7/84–85
Roosevelt, Franklin D., 4/167, 5/115,
   6/136, 6/137
Roosevelt, Theodore, 5/32, 7/84
Ropotamo River, 1/183
Rosario (Argentina), 1/63
Rosario Sánchez, Francisco del, 2/167
Rose Island (American Samoa), 1/32
Rose Revolution, 3/15
Rosenborg Palace (Copenhagen), 2/161
Roses, 1/185, 1/185
Roses, Wars of the, 6/110
Ross Ice Shelf, 1/43
Rossiya No. 2 Train, 5/128
Rothschild family, 3/28
Rotterdam (Netherlands), 4/200, 4/203,
   4/204
Rouen (France), 2/240
Rousseau, Jean Jacques, 2/252
Rousseff, Dilma, 1/163
Royal Academy of Sciences (Stockholm),
   5/247
Royal Albert Hall (London), 6/109
Royal Ascot Races, 6/108
Royal Avenue (Belfast), 6/99
Royal Dublin Society Horse Show, 3/207
Royal Flying Doctor Service, 1/79
**Royal Guipuzcoana Company of
   Caracas,** 6/173
Royal National Eisteddfod, 6/95
Royal Ontario Museum, 2/37
Royal Palace (Amsterdam), 4/206
Royal Palace (Budapest), 3/114, 3/115
Royal Palace (Madrid), 5/210
Royal Palace (Morocco), 4/161
Royal Palace (Nepal), 4/193–194
Royal Palace (Phnom Penh), 2/13
Royal Palace (Stockholm), 5/251
Royal Palace (The Hague), 4/201
Royalists (Yemen), 6/198
Royle Castle (Poland), 5/81
Rozwi people, 6/210, 7/100
RPF (Rwanda), 5/132
Rua Augusta (Lisbon), 5/90, 5/91
Ruanda-Urundi, 1/188, 5/130
Rub al Khali, 5/8, 5/148, 6/194
Rubber
   Brazil, 1/165, 1/174
   Cambodia, 2/8

Indonesia, 3/170
   Liberia, 4/65
   Malaysia, 4/93
   Thailand, 6/30
Rubens, Peter Paul, 1/133
Rudd, Kevin, 1/77
Rudolf I (Holy Roman emperor), 1/98,
   5/258
Rue de la Régence (Brussels), 1/135
Rufiji River, 6/21
Rugby, 6/108
Rügen Island (Germany), 3/25
Rugs. See Carpets
Ruhr (Germany), 3/36
Ruhuna national park (Sri Lanka), 5/223
Ruiz de Alarcón, Juan, 4/145
Ruíz Massieu, José Francisco, 4/129
Rumania. See Romania
Rumi, 6/71
Rupununi savanna, 3/94
Ruse (Bulgaria), 1/185
Russia, 1/2, 5/108–129, 5/109–129,
   5/111
history
   early, 5/109, 5/112–113, 5/112–
      113, **5/113**
   modern, 5/109–111, 5/114–115,
      5/114–115, **5/115**
people and way of life, 5/116–117,
   5/116–117
power in
   Afghanistan, 1/11, 1/18
   Alaska, 6/131
   Armenia, 1/66–67
   Azerbaijan, 1/111
   Belarus, 1/130, 1/131
   Estonia, 2/204
   Finland, 2/222–223, 2/228
   Georgia, 3/14, 3/15
   Kazakhstan, 4/8
   Kyrgyzstan, 4/38
   Latvia, 4/48, 4/49
   Lithuania, 4/76
   Moldova, 4/146
   Mongolia, 4/151
   Poland, 5/72, 5/76–77, 5/80
   Tajikistan, 6/16
   Turkmenistan, 6/74
   Ukraine, 6/82
   Uzbekistan, 6/166
prehistoric human settlement, 7/42,
   7/44–47, 7/44–47, **7/62,**
   7/62–65, 7/63–65, **7/65**
relations with
   Greece, 3/63
   Kazakhstan, 4/9
   Serbia, 5/156
   Ukraine, 6/83
Trans-Siberian Railroad, 5/128–129,
   5/128–129
Volga River, 5/118–119, 5/118–119,
   5/119
See also Soviet Union
Russian language, 5/109
Russian Orthodox Church, 5/117

Russian people
Armenia, 1/67
Azerbaijan, 1/111
Estonia, 2/204
Kazakhstan, 4/9
Kyrgyzstan, 4/39
Moldova, 4/147
Mongolia, 4/154
Siberia, 5/126
Turkmenistan, 6/74
Ukraine, 6/83
See also Russia
Russian Republic, 5/114
Russian Revolution, 2/223, 5/109,
5/112, 5/114, *5/114,* 5/122–123
Russian Social Democratic Labor Party,
5/114
Russian State Library (Moscow), 5/121
Russification programs, 5/109
Russo-Japanese War, 3/264
Russo-Turkish War, 5/158
Ruthenia (Czech Republic), 2/148
Ruwenzori Range, 6/80
Rwanda, 5/130–131, *5/130–131,* **5/131**
Congo conflict, 2/111, *2/113*
Rwandan Patriotic Front, 5/132
Ryoanji Temple (Kyoto), 3/281
Ryukyu island chain (Japan), 3/258–259,
7/61

S

Saadian dynasty, 4/158
Saakashvili, Mikheil, 3/15
SAARC, 4/104
Saaremaa Island (Estonia), 2/204
Saarinen, Eliel, 2/229
Sab clan, 5/186
Saba (Netherlands Antilles), 4/58, 4/63,
4/63, 4/200, 4/210
Sabaean people, 5/144
**Sabah, Sheik Sabah al-Ahmad al-Jabir
al-,** 4/35
Sabah (Malaysia), 4/93, 4/94, 4/96, 4/97
Sabah I, 4/34
Sabi River, 6/206
Sables, 5/126
Saborna Crkva (Belgrade), *5/168*
Sachsenhausen district (Frankfurt), 3/29
Sacré Coeur, Basilique du (Paris), 2/233
Sacred Lake of Apollo, *3/66–67*
Sacsahuaman fortress, *5/52–53*
Sadat, Anwar el-, 2/193, *2/193,* 3/213
Sadza, 6/207
Saemund the Learned, 3/122
Safavid kings, 1/11, 3/183, 6/71
Safed Koh mountains, *1/19*
Saffron, 3/158
Saga of Egill Skallagrimsson, The
(Snorri Sturluson), 3/122
Sagas, 3/122–123, *3/122–123*
Sagres (Portugal), 5/95
Sahara, 4/72, 4/72–73, *4/72–73*
Algeria, 1/26, 1/28, *1/29,* 1/31
caravan routes, **4/110**
Chad, 2/46

Egypt, 2/181
human migration, 7/8, **7/9,** 7/10, 7/33,
7/40, 7/98–99, **7/99**
Libya, 4/67–69, *4/68,* 4/72
Mali, 4/110, 4/111
Mauritania, 4/120, 4/121
Morocco, 4/162, *4/164,* 4/169
Niger, *4/233, 4/234, 4/235*
Sudan, 5/232
Tunisia, 6/53
Western Sahara, 6/192
Saharawis people, *6/192*
Sahel, 2/44, 4/110, 4/111, 4/235, 5/232
Saibou, Ali, 4/233
Said (ruler of Egypt), 2/193
Said bin Taimur, 5/9
Saigon (Vietnam). *See* Ho Chi Minh City
(Vietnam)
Saimaa (lake), 2/226
St. Anthony monastery (Brazil), 1/172
St. Augustine (Florida), 7/102, *7/103*
St.-Barthélemy (Guadeloupe), 3/76, 4/58,
4/63, *4/63*
St. Cajetan church, *3/155*
St. Catherine, Mount, 3/74
St. Croix (U.S. Virgin Islands), 6/190,
6/191
St. David's Island (Bermuda), 1/148
St.-Denis (Réunion), 5/100, *5/101*
St. Domingue colony, 3/96
St.-Étienne (France), 2/251
St. Eustatius (Netherlands Antilles), 4/58,
4/63, 4/200, 4/210
St. Gallen (Switzerland), 1/100
St. George's (Grenada), 3/75
St. George's Island (Bermuda), 1/148,
1/149
St. Helier (Jersey), *6/101*
St. Isaac of Dalmatia Cathedral
(St. Petersburg), *5/122–123*
St. John (U.S. Virgin Islands), 6/190
St. John, Cathedral of (Warsaw), 5/81
St. John the Baptist Christian Church
(Damascus), 5/275
St. John's (Antigua and Barbuda), 1/49,
1/49
St. John's Cathedral (Cyprus), 2/143
St. Kitts and Nevis, 4/58, 5/132, *5/133,*
5/133
St.-Laurent du Maroni (French Guiana),
2/256
St. Lawrence, Gulf of, 2/30
St. Lawrence Lowlands (Canada), 2/27,
2/34, 2/35
St. Lawrence River, 2/23, 2/30, 2/36
St. Lawrence River Valley, 2/27
St. Lawrence Seaway, 2/31
St. Louis (Missouri), 6/123, *6/149*
St. Lucia, 3/75, 4/58, *4/58, 5/134,*
5/134–135, **5/135**
St. Maria Wörth Church (Austria), *1/101*
St. Mark's Square (Venice), *3/243*
St. Martin (Leeward Islands), 3/76, 4/58,
4/62, 4/200, 4/210, 4/211
St. Martin Cathedral (Utrecht), 4/205

St. Martin Church (Bremen), *3/30–31*
St. Michael, Cathedral of (Brussels),
1/141
St. Moritz (Switzerland), *5/262–263*
St. Paul River (Liberia), 2/65
St. Paul's Cathedral (London), 6/102,
6/103
St. Peter Cathedral (Bremen), 3/31
St. Peter's Church (Rome), 3/237, 6/170
St. Peter's Square (Vatican City), *6/171*
St. Petersburg (Russia), 2/228, 5/116,
5/120, 5/122–123, *5/122–123*
history, 5/114, *5/114*
St. Pölten (Austria), 1/102
St. Sergius Trinity Monastery (Russia),
5/113
St. Stephen's Cathedral (Vienna), 1/108,
1/108–109
St. Thomas (U.S. Virgin Islands), 6/190,
6/191, *6/191*
St. Vincent and the Grenadines, 3/75,
4/58, *5/136,* 5/136–137, **5/137**
St. Vitus Cathedral (Prague), *2/150–151,*
2/151
Ste.-Anne (Martinique), 4/118
Ste.-Pierre (Martinique), *4/118*
Saipan (Northern Marianas), 3/79, *3/79*
Saisyat people, *7/58*
Sajudis, 4/77
Sakiehs, 2/185
Sakkara (Egypt), 2/191
Saladin, 2/192, 3/282, 5/267, 5/274
Salamanders, *6/177*
Alpine, *1/35*
giant, 2/74–75
Salamasina, 5/138
Salamis (ancient city), 2/143
Salamis Island (Greece), *3/64*
Salazar, António de Oliveira, 1/36, 5/95
Salih, Ali Abdullah, 6/195, 6/201
Sall, Macky, 5/153
Salmon, 2/35, 2/38, 3/261, 6/97
Salonika Plain, 3/54
Salsa, *2/136, 6/179*
Salsa music, 3/250
Salt
Australia, *1/82–83*
Austria, 1/101, 1/102
Djibouti, *2/163*
Thailand, *6/33*
Salta, 6/200
Salvador (Brazil), *1/172,* 7/103
Salvagens (Madeira Islands), 1/113
Salween River, 2/94, 4/178
Salzach River, 1/101
Salzburg (city, Austria), 1/97, 1/100,
1/100–101, 1/101, 1/107
Salzburg (province, Austria), 1/101
Salzburg Festival, 1/107
Salzkammergut (Austria), 1/102
Samaal clan, 5/186
Samar Island (Philippines), 5/69
Samarqand (Uzbekistan), 6/166
Samarra (Iraq), 3/187, 3/190
Samba, 1/171

Sambre River, 1/142
Sambungs, 6/197
Sami, 2/222, 4/246, 5/245
Samnorsk Norwegian language, 4/248
Samo people, 1/186, 1/187
Samoa, 5/11, 5/12, 5/138–139, 5/139,
    5/139, 6/44, 7/94
    See also American Samoa
Samoan language, 1/32, 5/138
Samoans, in Kiribati, 4/16
Samos (Greece), 3/66
Samothraki (Greece), 3/66
Samovar stoves, 5/128, 5/129
Samsun (Turkey), 6/67
Samurai, 3/264, 3/266–267, 3/266–267,
    3/275
Samurai films, 3/276
San Agustín (Colombia), 2/98
San Agustin Church (Philippines), 5/66
San Antonio (Texas), 6/142
San Blas Mountains, 5/28
San Diego (California), 6/128, 6/129
San Fermin, Festival of, 5/206–207,
    5/207
San Fernando (Trinidad and Tobago), 6/47
San Francisco (California), 6/128, 6/128–
    129, 6/129
San Gimignano (Italy), 3/228
San Giorgio Maggiore church of (Venice),
    3/243
San Jose (California), 6/128
San José (Costa Rica), 2/124, 2/125
San Juan (Puerto Rico), 5/96
San Juan Teotihuacán (Mexico), 4/142
San Marino, 3/229, 5/140–141,
    5/140–141, 5/141
San Marino (San Marino), 5/140, 5/141
San Martín, José de, 1/65, 2/49, 5/45
San Pedro (Côte d'Ivoire), 2/129
San Pedro Sula (Honduras), 3/105
San people, 1/158, 4/181, 5/189
San River, 5/76
San Salvador (El Salvador), 2/195, 2/196,
    2/198
San Salvador Island (Bahamas), 1/114
Sanaa (Yemen), 6/195, 6/196, 6/197,
    6/201
Sanaga River, 2/20
Sanandaj (Iran), 3/198
Sanchez de Lozada, Gonzalo, 1/155
Sancti Spiritus fort, 1/64
Sand, George, 2/252, 5/216
Sandinistas, 4/227, 4/230, 4/230–231
Sandino, Augusto César, 4/226, 4/227
Sandjar, 3/198
Sandoy (Faroe Islands), 2/212
Sangha people, 2/117
Sango language, 2/42
Sanguinetti, Julio María, 6/163
Sankoh, Foday, 5/173
Sans Souci, Palace of, 3/39, 3/39
Sanskrit language, 3/128, 3/138, 4/49,
    4/189, 5/176
Sanskrit literature, 3/141
Santa Ana (El Salvador), 2/196

Santa Anna, Antonio López de, 4/140
Santa Catarina (Brazil), 1/169
Santa Clara Cathedral (Bogotá), 2/105
Santa Cruz de Tenerife (Canary Islands),
    5/218, 5/220
Santa Maria del Fiore Cathedral (Italy),
    3/228
Santa Marta (Colombia), 2/102
Santa Prisca, Church of (Mexico), 4/144
Santals, 3/138
Santería, 2/139
Santiago (Chile), 2/49, 2/54, 2/54, 2/55,
    2/56
Santiago de Compostela, cathedral of
    (Spain), 5/205
Santiago de Cuba (Cuba), 2/134, 2/138
Santiago del Estero (Argentina), 1/64
Santo (Vanuatu), 6/168
Santo Domingo (Dominican Republic),
    2/166, 2/167
Santo Tomás, University of, 5/67
Santos, Juan Manuel, 2/101
São Francisco River, 1/176
São Miguel (Azores), 1/112
São Paulo (Brazil), 1/170, 1/172, 1/173,
    1/173, 1/174, 1/176
São Paulo Cathedral (Macau), 2/86,
    2/86
São Tomé and Príncipe, 5/87, 5/142–
    143, 5/142–143, 5/143
São Vicente Island, 2/41
Sapporo (Japan), 3/272
Saqqaq culture, 7/75
Sara language, 2/47
Sara people, 2/45–47
Sarajevo (Bosnia-Herzegovina), 1/157
Sarawak (Malaysia), 4/93, 4/94, 4/96,
    4/97, 4/103
Sardinia (Italy), 3/225, 3/244–245,
    3/245–246
Sark (Channel Islands), 6/100, 6/101
Sarkozy, Nicolas, 2/245
Sarmatians, 4/146
Sarnath (India), 3/147
Sarney, José, 1/163
Sarongs, 1/179, 3/166, 4/98
Sarqaq culture, 3/72
Sartre, Jean-Paul, 2/253
Saskatchewan (Canada), 2/26, 2/33, 2/39
Sassanians, 1/11, 3/182, 3/184
Sassou-Nguessou, Denis, 2/117
Sata, Michael, 6/202
Sati, 3/146
Satori, 3/279
Satsuma Rebellion, 3/267
Saud (king of Saudi Arabia), 5/145
Saud, Abd al-Aziz ibn, 5/144/145
Saud, Muhammad ibn, 5/144
Saud dynasty (Saudi Arabia), 5/144
Saudi Arabia, 1/13, 5/144–151, 5/144–
    151, 5/147
    government, 5/146
    history, 5/144–145
        growth of Islam, 5/150–151,
        5/150–151, 5/151

    land and economy, 5/148–149,
        5/148–149
    people, 5/146–147
    relations with
        Bahrain, 1/116, 1/117
        Yemen, 6/198
Saudi Arabian Oil Company, 5/149
Saudi Mosque (Mauritania), 4/121
Sauer, Carl, 7/80
Sauna, 2/227
Sausage, 3/235
Sava River, 2/130, 5/154, 5/163, 5/168
Savai'i (Samoa), 5/138, 5/139
SAVAK police force, 3/183
Savane des Petrifactions (Martinique),
    4/118
Savannas
    Chad, 2/44
    Ghana, 3/40
    Guinea, 3/86
    Guinea-Bissau, 3/90
    Guyana, 3/94
    Malawi, 4/90
    Niger, 4/234, 4/235
    Serengeti, 6/25
    Somalia, 5/186
    Uganda, 6/80
Saw Maung, 4/176
Sawau people, 2/220–221
Saxons, 3/20, 3/38, 6/91, 6/104,
    6/110
Sayun (Yemen), 6/200
Sayyids, 6/201
Scafell Pike, 6/92
Scaligeri Castle, 3/224–225
Scandinavia. See Denmark; Norway;
    Sweden; Vikings
Scandinavian Peninsula, 4/246, 5/242,
    5/248, 7/53
Scandinavian people, in United
    Kingdom, 6/104
Scenery, Mount, 4/62–63
Schaumainkai (Frankfurt), 3/29
Schelde River, 1/138, 1/142–143
Scheveningen resort (Netherlands),
    4/199, 4/204–205
Schiller, Johann Christoph Friedrich von,
    3/39
Schloss Hochosterwitz (Austria), 1/101
Schmidt, Klaus, 7/36, 7/37, 7/37
Schmitt, Pál, 3/109
Schoenberg, Arnold, 1/106
Schönbrunn Palace (Vienna), 1/94–95,
    1/95, 1/109
Schönbrunn Zoo (Vienna), 1/109
Schools of the air, 1/78
Schouten, Willem Cornelis, 6/44
Schröder, Gerhard, 3/18
Schubert, Franz, 1/106
Schweitzer, Albert, 3/8, 5/246
Schwejk (literary character), 2/152
Schwyz canton (Switzerland), 5/258
Schwyzerdütsch dialect, 5/260
Scilly Islands (England), 6/92
Scotch whisky, 6/97

Scotland, 6/96, 6/96–97, 6/96–97, 6/109
history, 6/110, 6/112
Scottish people, in United States, 6/138
Sculpture
Africa, 4/113
Belgium, 1/133, 1/140–141
Brazil, 1/172
China, 2/66, 2/66–67
Colombia, 2/98, 2/106–107, 2/106–107
Denmark, 2/160, 2/161
Indus River Valley, 5/25, 5/25
Nigeria, 4/244–245, 4/244–245
Northern Ireland, 6/99
Peru, ancient, 5/44
Vatican City, 6/171
See also Woodcarving
Scythians, 3/65, 3/128, 4/146, 5/14
Sea Dayaks, 4/98, 4/99
Sea lilies, 4/106–107
Sea Lion Island (Falkland Islands), 2/211
Sea of Galilee, 3/218, 3/218–219, 3/286
Sea of Japan, 4/22
Sea slugs, 1/93
Seaga, Edward, 3/249
Seal hunting, 3/72
Seals, 1/44, 1/45, 6/130
Antarctic fur, 1/44
southern elephant, 1/44
Weddell, 1/44
Seashells, 4/125
Seattle (Washington), 6/128
Sechin Alto site, 7/82–83
Second Balkan War, 3/63, 4/82
Secoton village, 7/77
Sedges, 5/124
Seeiso, 4/57
Sei Shonagon, 3/268
Seine River, 2/246
Sejm (Poland), 5/75
Sejong, 4/31
Selassie, Haile. *See* Haile Selassie I
Sele River, 3/225
Selebi-Pikwe (Botswana), 1/159
Seleucid dynasty, 3/182
Seleucus, 3/182
Seljuk Empire
Armenia, 1/66
Iran, 3/182
Kurdistan, 3/198
Saudi Arabia, 5/151
Turkey, 6/68, 6/71, 6/73
Uzbekistan, 6/166
Selkirk, Alexander, 2/53
Selva, 1/168, 5/19
Selvagen Islands, 1/113
Semana Santa, 5/206, 5/206
Semipalatinsk (Kazakhstan), 4/9
Semitic people
Jordan, 3/282
Saudi Arabia, 5/144
Sumer, 3/197
Syria, 5/266, 5/272
Yemen, 6/195
See also Arabs; Jews

Sena people, 4/90
Senate
Canada, 2/24
United States, 6/116
Senate Square (Finland), 2/228, 2/228–229
Sendero Luminoso. *See* Shining Path (Peru)
Senegal, 3/12, 4/109, 5/152–153, **5/153,** 5/153–153
arts, 4/112, 4/113
Sénégal River, 4/110, 4/120, 4/123
Senegalese Democratic Party, 5/152, 5/153
Senegambia, 5/152
Senghor, Léopold Sédar, 5/153
Sennar Dam, 5/232
Sentebale, 4/57
Senufo people, 1/186, 1/187
Seoul (South Korea), 4/22, 4/25, 4/27
Separatism, in Canada, 2/32, 2/32
Sephardim, 3/214, 3/215
Sepoys, 3/129
September 11 terrorist attacks, 1/11, 1/13, 5/17, 6/137, 6/137
Sequoia National Park (California), 6/129
Serahuli people, 3/12–13
Serbia, 1/99, 5/154–169, 5/154–169, **5/157**
economy, 5/162–163, 5/162–163
government, 5/156–157
history
early, 5/155, 5/158–159, 5/158–159, **5/159**
recent, 5/156–157, 5/160–161, 5/160–161
land, 5/166–167, 5/166–167
people, 5/155, 5/164–165, 5/164–165
power in
Albania, 1/20
Kosovo, 4/32–33
Macedonia, 4/82
relations with, 4/157
Bosnia-Herzegovina, 1/156, 1/157
Croatia, 2/130, 2/131
Slovenia, 5/182
Serbia and Montenegro, 4/156, 4/157, 5/155, 5/157, 5/161
Serbian language, 1/156, 4/32, 4/156, 5/164
Serbian people
Bosnia-Herzegovina, 1/156–157
Kosovo, 4/32
See also Serbia
Serbian Radical Party, 5/156
Serbs, Croats, and Slovenes, Kingdom of the
Bosnia-Herzegovina in, 1/156
Croatia in, 2/130
Hungary in, 3/107
Kosovo in, 4/33, 5/157
Macedonia in, 4/82
Montenegro in, 4/157
Serbia in, 5/155, 5/156, 5/158
Slovenia in, 5/182

Serengeti National Park, 6/24–25, 6/24–25, **6/25**
Serengeti Plain, 6/18, 6/20, 6/24
Serenissima, La, 3/242
Serfs, 2/240, 5/112, 5/113
Sergels Torg (Stockholm), 5/250
Sergiyev Posad (Russia), 5/113
Serra da Estrela range, 5/88, 5/89
Serra Pelada (Brazil), 1/176–177
Serrano Elías, Jorge, 3/81, 3/83
Sertão (Brazil), 1/169
Service industries, 6/152
See also specific places
Seven Pagodas (India), 3/148
Seven Wonders of the Ancient World, 3/69, 3/187, 6/70, 6/73
Seven Years' War, 6/112
See also French and Indian War
Seventh-day Adventists, 5/164
Severn River, 6/94
Seville (Spain), 5/204, 5/206, 5/206, 5/207
Sèvres, Treaty of, 3/199
Seward, William, 6/131
Seychelles, 5/170–171, 5/170–171, **5/171**
Seychelles Peoples United Party, 5/170
Sfax (Tunisia), 6/52
Sforza, Lodovico, 3/236
Shaanxi (China), 2/64
Shadoofs, 2/185
Shadow of a Gunman, The (O'Casey), 3/206
Shadow-puppet dramas, 3/169
Shafii sect, 6/195, 6/201
Shah-Nama, 3/184–185
Shakespeare, William, 2/143, 6/111, 6/111
Shammas, 2/209
Shan people, 4/174, 4/178
Shandong Peninsula, 2/72
Shang dynasty, 2/62, 2/64, 7/57
Shanghai (China), 2/72, 2/81, 2/81, 2/89
Shanghai World Financial Center, 2/81
Shankar, Ravi, 3/147
Shankaracharya, 3/147
Shannon, River, 3/201
Shanshui art, 2/66
Sharchop people, 1/151
Shari`ah, 2/189, 4/239, 5/146, 6/51
Sharif, Nawaz, 5/17
Sharifs, 4/158, 4/166
Sharon, Ariel, 3/213
Sharon plain (Israel), 3/209
Sharp, Granville, 5/172
Sharpeville (South Africa), 5/193
Shatt al Arab, 3/186, 3/190–191, 3/191
Shatts, 6/53
Shaw, George Bernard, 3/201
Shaykh Lutfullah Mosque, 3/185
Sheba, Queen of, 6/198
Sheberghan area (Afghanistan), 1/17
Sheep and wool
Afghanistan, 1/15, 1/16, 1/17
Argentina, 1/55, 1/62

Australia, 1/78, 1/80, *1/80*, 1/87
Benin, *1/147*
Chile, *2/55*
Ecuador, *2/176*
Falkland Islands, 2/211
Italy, *3/238–239*
Morocco, *4/168–169*
New Zealand, 4/218, *4/218*, 4/219
Norway, 4/250, 4/253
South Africa, 5/196
Spain, *5/209*
Tibet, 2/96
Turkmenistan, 6/75
Uruguay, *6/160, 6/165*
Sheffield (England), 6/93
Sheiks, 5/95
Shenandoah National Park (United States), 6/133
Sherpas, 4/188, 4/192, *4/197*
Shetland Islands, 6/108
Shevardnadze, Eduard A., 3/15
Shi Huangdi, 2/66, 2/69
Shibam (Yemen), 6/195
Shi`ite Muslims
Afghanistan, *1/14–15*, 1/15
Azerbaijan, 1/110
Egypt, 2/188
Iran, 3/175, 3/179
Iraq, 3/192
Lebanon, 4/52, 4/54, *4/55*
Pakistan, 5/21
Saudi Arabia, 5/146, 5/151
Syria, 5/272
Tajikistan, 6/17
Yemen, 6/195
Shikaras, *3/158–159*
Shikoku Island (Japan), 3/253, 3/256–257
Shilluk people, 5/229, 5/236
Shinawatra, Thaksin, 6/28
Shinawatra, Yingluck, 6/29
Shining Path (Peru), 5/47
Shinkansen bullet train, *3/263, 3/270–271*
Shinto, 3/268–269, *3/278–279, 3/278–279*
Shipbuilding, *3/122–123,* 4/28, 4/204, 5/77
Shipley, Jenny, 4/217
Shipping
Danube River, 5/181
Djibouti, 2/162
Mississippi River, 6/149
Norway, 4/246, 4/247
Singapore, 5/175
Yemen, 6/196
Ships, Refrigerator, 1/59, 1/62
Shipyards, *2/230, 2/231*
Shirley Heights (Antigua and Barbuda), *4/59*
Shirvan Shahs, Palace of (Azerbaijan), 1/110
Shiva, 3/143, 3/146, 4/193, 5/227
Shoals, 4/107
Shoes, Early, 7/54
Shogun, 3/264, 3/266

Shogunate, 3/264, 3/266
Shona people, 6/207, 6/209, 6/210, 7/100
Shonekan, Ernest, 4/239
Shrews, *2/75*
Shrikes, Grey, *4/185*
Shrove Tuesday festivals, 3/35, *3/35*
Shudras, 3/140
Shugakuin Palace (Kyoto), *3/281*
Shwe Dagon pagoda (Myanmar), 4/174, *4/177, 4/178*
Siad Barre, Mohamed, 5/186
Siam. *See* Thailand
Siam Paragon sculpture, *6/31*
Siamangs, *3/173*
Sibelius, Jean, 2/222
Siberia, 5/112, **5/124, 5/128,** 5/128–129, *5/128–129*
human migration, 7/17, 7/60, **7/62,** 7/62–63, *7/62–63,* 7/75
labor camps, 5/114, 5/116
people, 5/126–127, *5/126–127*
Trans-Siberian Railroad, **5/128,** 5/128–129, *5/128–129*
See also Bering Land Bridge
Sichuan Basin, 2/72, *2/72,* 2/73, 2/74, 2/89
Sicilian dialect, 3/233
Sicily (Italy), 3/224, 3/225, 3/228, 3/232, 3/244–245, *3/245–246*
Malta, control of, 4/115
Siddhartha Gautama. *See* Buddha
Sidra, Gulf of, 4/68
Siecha, Lake, *2/106*
Siegfried (count of Ardennes), 4/78
Sienkiewicz, Henryk, 5/73
Sierra (Ecuador), 2/176, 2/177
Sierra de Escambray range, 2/134
Sierra de Levante range, 5/217
Sierra de los Órganos range, 2/134
Sierra del Rosario range, 2/134
Sierra Leone, 5/95, 5/172–173, *5/172–173,* **5/173**
Sierra Madre del Sur range, 4/130
Sierra Madre Occidental range, 4/130
Sierra Madre Oriental range, 4/130, *4/131*
Sierra Madre range, 2/197
Sierra Maestra range, 2/134
Sierra Nevada range, *5/202*
Siesta, 5/204
Sifakas, *4/88–89*
Siguiri (Guinea), 3/88
Sikhism, 3/126, 3/141, *6/105*
Sikkim (India), **3/152,** 3/152–153, *3/152–153*
Sikkimese language, 3/153
Silesia, 2/144
Siljan, Lake, 5/248–249, *5/249*
Silk
China, 2/62, 2/66, *2/71*
Iran, 3/184
Japan, 3/260
Thailand, *6/33*
Turkey, *6/65*

Silk Road, 1/10, *2/71,* 2/80, 6/70, 6/166
Silla, 4/22–23, 4/30
Silver
Bolivia, 1/153, 1/154
Mexico, 4/130, 4/135, *4/135,* 4/140
Navajo jewelry, *6/141*
Silvia (Colombia), *2/104*
Simi (Greece), 3/66
Simon, Winston "Spree," 4/61
Sinai Peninsula, 2/181, 3/212, 7/14
Sindh (Pakistan), 5/15, 5/16, 5/18, *5/19*
Sindhi language, 3/138, 5/20
Sindhi people, 5/20
Sing-sings, 5/34
Sing-songs, 6/77
Singapore, 4/93, 4/94, 4/102, 5/174–177, *5/174–177,* **5/175**
economy, 5/175
history, 5/174–175
people, 5/176–177, *5/176–177*
Singapore (city, Singapore), 5/175, 5/176
Singapore River, *5/175*
Sinhala language, 5/226
Sinhalese, 4/105, 5/224, 5/225, 5/226
Sino-Tibetan languages, 3/139
Sint Maarten (Netherlands Antilles). *See* St. Martin
Sioux Indians, 6/141, 7/77
Siqueiros, David, 4/144
Sir Richard Squires Memorial Park (Newfoundland), 2/38
Sirocco, 1/28
Sisal, *2/175*
Sistan Basin, 1/16
Sistine Chapel, 6/171, *6/171*
Sitars, 3/147
Sitrah (Bahrain), 1/117
Sittang River, 4/178
Six-Day War, 3/283, 4/36, 5/268
Sixth Street (Austin), *6/127*
Sjælland Island (Denmark), 2/158, 2/160
Ska, 3/250
Skåne (Sweden), 5/249, 5/253
Skansen Park (Sweden), 5/251
Skerries, 4/251
Skhul Cave, 7/33, 7/34
Skiing, 4/249, 5/263
Skikda (Algeria), 1/30
Skopje (Macedonia), 4/82
Skotkonung, Olof, 5/242
Skoura (Morocco), *4/169*
Skulls, Human. *See* Human migrations; Prehistoric humans
Skunks, *6/177*
Skvorecky, Josef, 2/152
Skyscrapers, 6/153, 6/158
Slash-and-burn agriculture, 4/172–173, 7/81
Slavery
Angola, 1/36
Antigua and Barbuda, 1/48
Bahamas, 1/114
Barbados, 1/129
Benin, 1/146
Bermuda, 1/149

Brazil, 1/165, 1/167, 7/102
Cameroon, **2/17**, 2/18
Cape Verde, 2/40
Central African Republic, 2/42
Chad, 2/44
Colombia, 2/99, 2/104
Congo, 2/110
Côte d'Ivoire, 2/126
Cuba, 2/132
Dominica, 2/165
Dominican Republic, 2/167
Easter Island, 2/171
French Guiana, 2/257
Gabon, 3/8
Ghana, 3/44, *3/44–45*
Guinea, 3/86
Guyana, 3/92, 3/95
Haiti, 3/96
Honduras, 3/100
Jamaica, 3/246, 3/247
Lesser Antilles, 4/59, 4/60, 4/61
Liberia and, 4/64
Madagascar, 4/86
Malawi, 4/91
Martinique, 4/118
Mauritius, 4/125
Mozambique, 4/170
Nigeria, 4/236–237
Panama, 5/30
Roma, 3/116
St. Kitts and Nevis, 5/132
St. Vincent and the Grenadines, 5/137
São Tomé and Príncipe, 5/142–143
Senegal, 5/153
Sierra Leone, 5/172–173
Sudan, 5/228, 5/229
Suriname, 5/238
Tanzania, 6/18–19
Thailand, 6/36
Togo, 6/41
Trinidad and Tobago, 6/47
Tuvalu, 6/77
United States, 6/124, 6/125, 6/135,
    6/136, 6/138, 6/144, 7/104,
    7/105
Slavic people
    Austria-Hungary, 1/98
    Belarus, 1/130
    Bosnia-Herzegovina, 1/156
    Bulgaria, 1/181
    Croatia, 2/130
    Czech Republic, 2/148
    Italy, 3/232
    Kosovo, 4/32
    Poland, 5/72
    Romania, 5/106
    Russia, 5/109, 5/112, 5/118
    Serbia, 5/155, 5/158
    Slovenia, 5/178, 5/182
    Ukraine, 6/83
Sleeping Bear Dunes, 6/133
Sleeping sickness, 2/42
Sloths, *6/177*
Slovak Evangelical Christian Church,
    5/164

Slovak language, 2/145, 5/179
Slovak people, 5/178, 5/179
    Czechoslovakia, 2/144, 2/145
    Serbia, 5/155, 5/164
Slovakia, 2/144, 2/148, 5/178–179,
    5/179, *5/179*
    Danube River, 5/180–181, *5/180–181,*
        5/181
    history, 5/178–179
Slovene people, 3/232, 5/182
Slovenia, 2/131, 5/182–183, *5/183,*
    5/183
    Serbia, relations with, 5/155, 5/159
Slovenian language, 5/182
Slums
    Botswana, 1/159
    Colombia, 2/105
    India, *3/145*
    Jamaica, *3/247*
    Manila, *5/66–67*
    Mexico City, *4/139*
    Morocco, 4/166
    Nicaragua, *4/230*
    Nigeria, 4/242
    Peru, 5/48
    Thailand, 6/35
    Venezuela, 6/179
Smartphones, *6/152*
Smetana, Bedřich, *2/149*
Smith, Bruce D., 7/76
Smith, Ian, 6/208
Smuts, Jan Christiaan, *5/192,* 5/193
Snake Alley (Taipei), *6/14–15*
Snakeheads, Pyrenean, *1/35*
Snakes, 2/75, 4/224
    bushmaster, *6/177*
Sniardwy, Lake, 5/785
Śnieżka mountain, 5/79
Snorri Sturluson, 3/122
Snow Festival (Japan), *3/272*
Snow leopards, 2/74, *2/74,* 4/184
Snowdon, 6/94
Snowdonia National Park, 6/94
Snowy Mountains, 1/72
Soares, Mário, *5/94*
Sobhuza II, 5/241
Soccer, *1/57,* 1/61, 1/171, 2/199, 6/108,
    6/161
Social Christian Party (Venezuela), 6/173
Social Democratic Party (Germany), 3/18
Social Democratic Party (Japan), 3/254
Social Democratic Party (Mongolia),
    4/153
Social Democratic Party (Portugal), 5/86
Social Democrats (Nigeria), 4/239
Socialism
    Algeria, 1/26, 1/30
    Bolivia, 1/155
    Chile, 2/50, 2/51
    Guinea, 3/86–87
    Laos, 4/42
    Tanzania, 6/19, 6/23
    United Kingdom, 6/107
    See also Communism
Socialist Party (Albania), 1/21

Socialist Party (Chile), 2/51
Socialist Party (France), 2/234, 2/250
Socialist Party (Italy), 3/222
Socialist Party (Portugal), 5/86
Socialist Party (Uruguay), 6/163
Socialist Party of Senegal, 5/152, 5/153
Socialist Party of Serbia, 5/156
Socialist Workers' Party (Spain), 5/201
Society Islands, 2/258, 2/259, 6/8, 7/94
Socotra Island (Yemen), 6/197
Socrates, 3/58, 3/65
Sócrates, José, 5/66
Södermalm Island, 5/250
Sofala, 7/101
Sofia (Bulgaria), 1/180, 1/183, 1/185
Sogdiana Province, 6/166
Soglo, Nicephore, 1/146
Sogne Fiord (Norway), 4/251
Soka City (Japan), *3/263*
Sokoto Plains, 4/240
Sokoto River, 4/240
Soleib (Sudan), *5/234–235*
Solidarity (Poland), 5/74–75
Solomon (Biblical king), 3/216, 3/282
Solomon Islands (country), 5/10,
    5/184–185, *5/184–185,* **5/185**
Solomon Islands (island chain), 5/12,
    5/37, 5/184
Solutrean culture, 7/71
Sólyom, Lászlo, 3/109
Somali language, 2/163, 5/186
Somali people, 2/162, 2/163, 2/209
Somalia, 5/186–187, *5/187,* **5/187**
    relations with Ethiopia, 2/206–207,
        5/186–187
Somaliland Republic, 5/187
Somerset Island (Bermuda), 1/148
Somoza Debayle, Anastasio, 4/227,
    4/231
Somoza Garcia, Anastasio, 4/226–227
Song dynasty, 2/62
Songhai Empire, 3/86, 4/108, 4/120,
    4/232, 4/236, 5/152
Songhai people, 4/110
Soninké people, 4/110, 4/122
Sons Brancion, Chapel of (France), 2/236
Sophia (queen of Spain), *5/215*
Sorak-san National Park (South Korea),
    4/26–27
Sorb people, 3/19
Soroche, 5/51
Sotho people, 5/194
Sotomayor, Sonia, *6/143*
Soufrière, Mount, 3/77, 5/136
Soufrière Hills (Montserrat), 4/63
Souphanouvong, 4/41, 4/42
Sousse (Tunisia), 6/53
South Africa, 5/188–197, *5/188–197,*
    5/191
    economy, 5/196–197, *5/196–197*
    history, 5/189–193, *5/192–193,* 7/99
    people, 5/189, 5/194–195, *5/194–*
        195
    power in
        Namibia, 4/180

Swaziland, 5/240
relations with
    Angola, 1/37
    Botswana, 1/158, 1/159
    Lesotho, 4/56
    Mozambique, 4/170–171
South America, 1/4
    See also Human migrations; *and specific*
        countries
South Asian Association for Regional
    Cooperation, 4/104
South Australia, 1/70, 1/81, 1/89
South Carolina (United States), 6/124
South China Sea, 1/178, 4/93, 4/97,
    5/60, 6/10, 6/180, 6/182, 6/186
South Dakota (United States), 6/122
South Georgia (Falkland Islands), 2/210
South Island (New Zealand), 4/215,
    4/219
South Korea. *See* Korea, South
South magnetic pole, 1/41, 1/43
South Orkney Islands (British Antarctic
    Territory), 2/210
South Ossetia (Georgia), 3/15, 5/111
South Pacific, University of the, 2/216,
    6/45
South Pole, 1/41, 1/43
South Sandwich Islands (Falkland
    Islands), 2/210
South Shetland Islands, 2/210
South Sudan. *See* Sudan, South
South Vietnam. *See* Vietnam
South West Africa, 4/180
South West African People's
    Organization, 4/180
South Yemen. *See* Yemen (Aden)
South Yorkshire (England), 6/93
Southern Alps, *4/225*
Southern Rhodesia, 6/208
Souvanna Phouma, 4/41
Soviet of Workers' and Soldiers'
    Deputies, 5/114
Soviet Union
    former republics
        Armenia, 1/66, 1/67
        Azerbaijan, 1/110, 1/111
        Belarus, 1/130–131
        Estonia, 2/202–203
        Georgia, 3/14
        Kazakhstan, 4/8, 4/9
        Kyrgyzstan, 4/38, 4/39
        Latvia, 4/48
        Lithuania, 4/76–77
        Moldova, 4/146–147
        Russia, 5/109, 5/110, 5/114–115
        Tajikistan, 6/16
        Turkmenistan, 6/74–75
        Ukraine, 6/82–83
        Uzbekistan, 6/166–167
    history, 5/109–115, *5/114–115,*
        5/115
    power in
        Afghanistan, 1/11, 1/12, 1/14
        Finland, 2/223, 2/230
        Germany, 3/22, 3/36

        Iran, 3/182
        Korea, 4/23
        Latvia, 4/48–49
        Lithuania, 4/76–76
        Poland, 5/72, 5/74
    relations with
        Angola, 1/36–37
        Bulgaria, 1/184
        China, 2/64, 2/92, 4/151
        Cuba, 2/133, 2/136
        Czech Republic, 2/149, *2/149,*
            2/150, 2/153
        Hungary, 3/111
        Korea, North, 4/18
        Nicaragua, 4/231
        Poland, 1/130, 5/74, 5/77,
            5/80–81
        Romania, 5/102
        United States, 6/136
        Vietnam, 6/182, 6/185
        Yugoslavia, 5/159, 5/162
    Siberia policy, 5/126
    Wallenberg and, 5/246
    way of life, 5/116–117
    See also Russia
Soviets, 5/114
Soweto (South Africa), 5/193, *5/195*
Space exploration, *6/136*
Spain, *5/198–199,* 5/198–221,
    5/198–221, **5/201**
    Amazon region exploration, 7/84
    Australia exploration, 1/86
    economy, 5/208–209, *5/208–209,*
        5/209
    history
        early, 5/212–213, *5/212–213*
        migration to Americas, 7/102–103,
            7/102–103, **7/103**
        modern, 5/200–201, **5/214,**
            5/214–215, 5/214–215
    land, 5/202–203, *5/202–203,* **5/203**
    Line of Demarcation and, 1/164
    people, 5/204–205, *5/204–205*
        fiestas, 5/206–207, *5/206–207*
    power in
        Algeria, 1/24
        American Southwest, 6/126
        Andorra, 1/34
        Antigua and Barbuda, 1/48
        Argentina, 1/56, 1/64–65
        Bahamas, 1/114
        Bolivia, 1/152, 1/154
        Chile, 2/49
        Colombia, 2/98, 2/99, 2/104
        Costa Rica, 2/120, 2/123
        Cuba, 2/132, 2/136
        Dominican Republic, 2/167
        Ecuador, 2/172
        El Salvador, 2/194, 2/198
        Equatorial Guinea, 2/201
        Federated States of Micronesia,
            2/215
        Gibraltar, 3/46, 3/47
        Greater Antilles, 3/48
        Guatemala, 3/80, 3/84

        Haiti, 3/96
        Honduras, 3/100
        Italy, 3/228
        Jamaica, 3/246
        Kiribati, 4/16
        Lesser Antilles, 4/59, 4/60
        Luxembourg, 4/79
        Mariana Islands, 3/79
        Marshall Islands, 4/116
        Mexico, 4/127, 4/136, 4/138,
            4/140–141
        Morocco, 4/158–159, 4/167
        Netherlands Antilles, 4/210
        New Guinea, 1/86
        Nicaragua, 4/226
        North America, 6/142
        Pacific Islands, 5/12
        Paraguay, 5/38–39
        Peru, 5/45
        Philippines, 5/57, 5/64, 5/66, 5/70
        Portugal, 5/95
        Puerto Rico, 5/97
        Sardinia, 3/245
        Trinidad and Tobago, 6/46
        Uruguay, 6/160, 6/161
        Vanuatu, 6/169
        Venezuela, 6/172, 6/173, 6/178
        Virgin Islands, U.S., 6/190
        Western Sahara, 6/193
    relations with
        England, 6/111
        France, 7/103
        Roma of, 3/117
    See also Balearic Islands; Canary Islands
Spanish-American War, 2/132, 3/79,
    5/12, 5/57, 5/66, 5/97
Spanish Armada, 6/111
Spanish Civil War, 5/210, 5/211, 5/215
Spanish Inquisition, 5/214
Spanish language
    Argentina, 1/56
    Belize, 1/145
    Bolivia, 1/152
    Colombia, 2/104
    Costa Rica, 2/120
    Cuba, 2/136
    Easter Island, 2/171
    El Salvador, 2/198
    Equatorial Guinea, 2/200
    Guatemala, 3/85
    Hispanic Americans, *6/142,* 6/143
    Honduras, 3/103
    Lesser Antilles, 4/60
    Mexico, 4/136
    Morocco, 4/164
    Panama, 5/30
    Spain, 5/204–205, 5/212
    Trinidad and Tobago, 6/47
    United States, 6/143
    Uruguay, 6/160
    Venezuela, 6/178
Spanish people
    Argentina, 1/51
    Azores Islands, 1/112
    Canary Islands, 5/221

France, 2/239
Peru, 5/46
Puerto Rico, 5/96
Uruguay, 6/160, 6/165
See also Spain
Spanish Riding School, 1/108, *1/109*
Spanish Sahara, 4/159, 6/193
Spanish Steps (Rome), *3/230–231,* 3/231
Spanish Succession, War of, 5/215
Sparta, 3/62
SPDC (Myanmar), 4/177
Spears, 1/76, *1/76*
early, 7/9, 7/46, 7/68, *7/68–69*
See also Clovis points
Special administrative regions, 2/85–87
Specialized households, 2/91
Spice Islands. *See* Moluccas (Indonesia)
Spiders, Fishing, *6/149*
Spiegel, Der (newspaper), 3/30
Spinach, *3/234–235*
Spitak (Armenia), 1/66
Spitsbergen Island (Norway), 4/256,
4/256, 4/256–257
Sponges, 4/107
Sporades Islands (Greece), 3/66
Spree River, 3/33
Sputnik 1 (satellite), 4/9
Spyri, Johanna, 5/255
Squash, 5/270
Squid, 1/44
Squills, Pyrenean, *1/35*
Squirrels, Gray, *2/38*
Sranan Tongo language, 5/239
Sredna Gora mountains, 1/183
Sri Lanka, 5/225
history, 5/224–225
land, 5/222–223
people, 5/226–227, *5/226–227*
population, 5/225
prehistoric humans, 7/14, 7/19
wildlife, 5/222, *5/222–223,* 5/223
Sri Lankan people, in Malaysia, 4/99
Sri Panda (Sri Lanka), 5/227
Srinagar (India), 3/159
Stadsholmen Island (Sweden), 5/250
Stalactites, *6/133*
Stalagmites, *6/133*
Stalin, Joseph, 2/204, 5/109, 5/114–
115, *5/115,* 5/116, 5/123, 5/126
Stalingrad (Russia). *See* Volgograd (Russia)
Standard Oil of California, 5/149
Stanley (Falkland Islands), 2/210,
2/210–211, 2/211
Stanley, Henry M., 2/110, 2/114
Stanley Falls, 2/114
Stanley Pool, 2/116
Stanovoy range, 5/125
Star of David, 5/265
Staré Město (Prague), 2/150
Starfish, *4/107*
crown-of-thorns, 1/92
Stari Grad (Belgrade), 5/168
State farms, 2/136
State Historical Museum (Moscow),
5/121

**State Law and Order Restoration
Council,** 4/176–177
State Mosque (Malaysia), *4/102–103*
State Opera House (Berlin), 3/32
State Peace and Development Council
(Myanmar), 4/177
State Shinto, 3/279
Statia. *See* St. Eustatius
Stations, in Australia, 1/78, *1/80,* 1/91
Stavanger (Norway), 4/254
Stavanger Fiord, 4/254
Steel. *See* Iron and steel
Steel bands, *4/60, 4/61*
Stendhal, 2/252
Step Pyramid (Egypt), *7/41*
Stephen I (king of Hungary), 3/106,
3/110, *3/110,* 3/111
Steppes, 2/70, 4/154, 5/124, 6/52–53,
6/82, 7/62
Stevenson, Robert Louis, 6/9, 6/189
Steveston (British Columbia), 2/35
Stigfoss waterfall (Norway), *4/251*
Stilt houses, *2/259, 6/179*
Stingray City (Caiman Islands), *3/48*
Stirling, Sir James, 1/86
Stitchie, Lieutenant, *3/251*
Stock Exchange (Copenhagen), 2/161
Stockholm (Sweden), 5/250–251,
5/250–251
Stoker, Bram, *5/104*
Stone Age. *See* Mesolithic Period;
Paleolithic Period
Stonehenge, 7/50–51, *7/51*
Storting, of Norway, 4/248, *4/249,*
5/247
Stoss, Veit, 3/39
Strahov Library, *2/148*
Strait of Gibraltar, 3/46, 4/160, 5/199
Strait of Malacca, 4/96, 4/102
Strait of Messina, 3/244
Strandvägen (Stockholm), *5/251*
Strasser, Valentine, 5/173
Strategic Arms Limitation Talks, 1/95
Strauss, Johann, 1/95, 1/106
Strauss, Johann, Jr., 1/95
Streymoy (Faroe Islands), 2/212
Stroessner, Alfredo, 5/40
Stromboli, 3/224
Strong, John, 2/210
Sturgeon, 5/105, 5/119
Sturgeon eggs, *3/181*
Styria (Austria), 1/103
Subic Bay Naval Station (Philippines),
5/58–59
Subsistence farmers, *3/134*
Suburbanization, in United States,
6/147, 6/147
Suchetgarh (Kashmir), 3/159
Sucre, Antonio José de, 1/153, 2/172,
5/45
Sudan, 5/228–235, *5/228–235,* **5/231**
economy, 5/232–233
history, 5/228–231, 5/237
land, 5/232
Liberia and, 4/64

people, 5/234–235, *5/234–235,*
5/235
Sudan, South, 5/233, **5/235,** 5/236–
237, *5/237,* **5/237**
history, 5/228–229, 5/231, 5/237
people, 5/236–237
Sudan People's Liberation Party, 5/230
Sudd (Sudan), 5/232
Sudeten Mountains, 2/154, 2/155, 5/78,
5/79
Sudetenland, 2/148, 3/26
Sudhuroy (Faroe Islands), 2/212
Suez, Isthmus of, 2/193
Suez Canal, 2/162, 2/186, 2/193, 5/246,
6/199
Sufis, 6/71
Sugar beets, 3/239
Sugar cane
Antigua and Barbuda, 1/48
Barbados, 1/128, 1/129, *1/129*
Brazil, 1/164, 1/167, 1/174, *1/177*
Colombia, 2/98
Cuba, 2/136, 2/137, *2/137*
Ecuador, *2/177*
Fiji, 2/216–217, *2/217,* 2/221
Guadeloupe, 3/76
Guyana, 3/94, *3/95*
Haiti, 3/99
Jamaica, 3/247, 3/249
Japan, 3/258
Lesser Antilles, 4/59
Mauritius, 4/124
Réunion, *5/100*
São Tomé and Príncipe, 5/142
Trinidad and Tobago, 6/46
Sugar Loaf Mountain (Brazil), *1/161,*
1/170, *1/170*
Suharto, 3/162, 3/171
Sui dynasty, 2/62, 2/69, 2/80
Sukarno, 3/162, 3/166
Sukothai, 6/36
Sukothai Candle Festival, 6/36
Sukuma people, 6/22
Sulawesi (Indonesia), 3/164, 7/24
Sulawesi people, *3/166*
Suleyman I, the Magnificent, 6/69, *6/69*
Süleymaniye Mosque (Turkey), *6/63,*
6/70
Sultan Ahmet Mosque (Turkey), 6/70
Sultans, 1/178, 4/158, 4/166, 6/58,
6/69
Sulu Archipelago, 5/61, 5/64
Sulu Sea, 5/60, *5/64–65*
Sumadija (Serbia), 5/163
Sumatra (Indonesia), 3/163, 3/164,
3/164, 3/170, 7/13
Sumela, Monastery of (Turkey), 6/70
Sumer, 3/186, 3/196–197, 7/40, *7/41*
Summer Palace (Beijing), 2/82
Summit Ridge (Gibraltar), 3/47
Sumo wrestling, 3/276, *3/276–277*
Sun Belt region (United States), 6/125
Sun Valley (Idaho), 6/127
Sun Yat-sen, 2/63, 2/64, *2/80*
Sundanese people, 3/166

173

Sunflowers, *1/184–185*
Sunni Muslims
  Afghanistan, 1/14, 1/15
  Cyprus, 2/143
  Iran, 3/179
  Iraq, 3/192, 3/198
  Jordan, 3/286
  Lebanon, 4/52, 4/54
  Libya, 4/70
  Maldives, 4/105
  Pakistan, 5/21
  Saudi Arabia, 5/146, 5/151
  Sudan, 5/234
  Syria, 5/272
  Tajikistan, 6/17
  Turkmenistan, 6/74
  Yemen, 6/195
Supreme Council for the Restoration of Democracy (Niger), 4/233
Supreme Court (Canada), 2/25
Supreme Court (United States), 6/116–117, *6/143*, 6/144
Suqs, 3/192, 4/164, 4/168
Surf lifesaving, *1/90–91*
Suriname, 4/206, 5/238–239, *5/238–239*, **5/239**
Suriname people, in Netherlands, 4/202–203
Surtsey Island, 3/120
Susah (Tunisia), *6/54*
Susak (Croatia), *2/131*
Susu people, 3/88
Sutlej River, 4/191, 5/18
Suva (Fiji), 2/218, *2/218*
Suwalki (Poland), 5/83
Suzhou (China), 2/81
Svalbard (Norway), 4/250, 4/256–257, *4/256–257*, **4/257**
Svalbard Treaty, 4/257
Svartsengi geothermal power station, *3/121*
Swahili culture, 6/18
Swahili language, 2/112, 4/14, 6/22, 7/99
Swains Island (American Samoa), 1/32
Swan River (Australia), *1/89*
Swans, Black, 1/74
Swansea (Wales), 6/95
SWAPO, 4/180
Swat (Pakistan), 5/18
Swayambhunath Temple, *4/188–189*
Swazi people, 5/240, *5/241*
Swaziland, 5/240–241, *5/240–241*, **5/241**
Sweden, 5/242–249, *5/242–253*, **5/245**
  climate, 5/248
  Denmark, relations with, 2/156, 2/160
  economy, 5/242, 5/244, **5/252,** *5/252–253*, *5/252–253*
  history
    early, 5/242–243
    modern, 5/244
    neutrality and peace efforts, 5/246–247, *5/246–247*
  land, 5/248–249, *5/248–249*, **5/249**

people, 5/244–245
power in
  Estonia, 2/204
  Finland, 2/222, 2/223, 2/228
  Latvia, 4/48
  Leeward Islands, 4/63
  Norway, 4/247
  Poland, 5/80
Swedish language, 2/225
Swedish people
  Finland, 2/225
  United States, 6/123
  See also Sweden
Swedish People's Party, 2/224
Swift, Jonathan, 3/205
Swimming, *1/90–91*
Swiss Confederation, 5/255, 5/258
Swiss Family Robinson, The (Wyss family), 5/255
Swiss Plateau, 5/257
Swiss Re Building (London), *6/102–103*
Switzerland, 5/254–265, *5/254–265*, **5/257**
  cantons, 5/258–259, *5/258–259*, **5/259**
  climate, 5/257
  economy, 5/262–263, *5/262–263*
  government, 5/256
  history, 5/255, 5/258–259
  international organization headquarters, 5/264–265, *5/264–265*
  land, 5/256–257
  Liechtenstein, relations with, 4/75
  people, **5/260,** 5/260–261, *5/260–261*
  population, 5/261
  Voralberg tie to, 1/100
Swords, 3/266, *3/266*
Sydney (Australia), 1/69, 1/83, 1/84, *1/84–85*, **1/85,** 1/86, 1/88, 7/106
Sydney Harbour, 1/84, *1/84–85*
Sydney Opera House, 1/84
Sydneysiders, 1/84
Sylvester II (pope), 3/106
Synge, John, 3/206
Synyk Kala minaret (Azerbaijan), 1/110
Syr Darya River, 6/16, 6/166, 6/167
Syria, 1/67, 5/266–275, *5/266–275*, **5/269**
  history, 5/266–269
    crusades and, 2/192
    Islam, growth of, 5/144
  land and economy, 5/270–271, *5/270–271*
  Lebanon, relations with, 4/53, 5/268
  people, 5/272, *5/272–273*
  population, 5/272
  prehistoric human migration, 7/38, 7/38–39
Syrian Desert, 3/191, 3/286, 5/148
Syrian people, in Jamaica, 3/246

T

Taal Volcano, *5/61*
Tabasará Mountains, 5/28

Table Mountain, 5/189
Tadzhikistan. *See* Tajikistan
Taebaek Mountains, 4/22
Tafari, Ras. *See* Haile Selassie I
Tagus River, *5/87*, 5/88, 5/90, 5/92
Tahiti, 2/259, 5/137, 6/8–9, *6/8–9*, **6/9,** *7/93*, 7/94
T'ai chi ch'uan, *2/76–77*, 2/77
Tai people, 4/40, 4/46
T'ai people, 2/79
Tai Ping Shan, 2/84
Taiga, 5/124, 5/126
Tain (Scotland), 6/97
Taipei (Taiwan), 6/13, *6/14–15*
Taipei 101 skyscraper, *6/11*
Taiping Rebellion, 2/63
Taiwan, 2/64, 3/264, 6/10–15, *6/10–15*, **6/13**
  economy, 6/11
  government, 6/12–13
  history, 6/14–15, *6/14–15*, **6/15**
  human settlement, *7/58*, 7/58–59, **7/59**
  land, 6/10, *6/10*
  migration to Pacific Islands, 7/91, 7/92
Taj Mahal (India), 3/125, 3/148
Tajik language, 6/17
Tajik people, 1/15, 6/17
Tajik Soviet Socialist Republic, 6/16
Tajikistan, 6/16–17, *6/16–17*, **6/17**
Taki-Taki language, 5/239
Takins, 2/74
Takiyya as-Süleimaniyya, 5/275
Taklimakan Desert, 2/70, 2/71, *2/71*
Taktsang Monastery, *1/151*
Talabani, Jalal, 3/193, 3/199
Talayots, 5/216
Tale of Genji, The (Murasaki), 3/268, 3/280
Taliban, 1/11, 1/13
Tallinn (Estonia), *2/205*
Tamanduas, *6/177*
Tamati Waka Neni, *7/96*
Tamazight dialect, 1/25
Tambomachay (Peru), *5/52–53*
Tambons, 6/28
Tambouras, 5/165
Tamerlane, 3/14, 6/166
Tamil language, 3/139, 5/176, 5/226
Tamil Nadu (India), 3/140, *3/148*, 3/149
Tamil Tigers. *See* Liberation Tigers of Tamil Eelam
Tamils, 5/224, 5/225, 5/226
Tana, Lake, *2/207*, 2/208
Tandja, Mamadou, 4/233
Tang dynasty, 2/62, *2/66–67*, 2/67, 2/80
Tanganyika, 6/19
Tanganyika, Lake, 1/188, 6/18
Tanganyika African National Union, 6/19
Tanggula Mountains, 2/88
Tangier (Morocco), *4/162*, 4/167
Tango, 1/61
Tanjung Priok (Indonesia), 3/171
Tannin and tanning, 1/54, 5/42, *6/164*

Tano River, 3/40
TANU (Tanganyika), 6/19
Tanzania, 6/18–25, *6/18–25,* **6/20**
  fossil, *7/8*
  government, 6/20
  history, 6/18–20
  land, 6/20–21, *6/21*
  people and economy, 6/22–23,
    6/22–23
  Serengeti National Park, 6/24–25,
    6/24–25, **6/25**
Taoiseach, 3/202
Taoism
  China, 2/59, 2/62, 2/66, 2/81
  Korea, 4/30
  Malaysia, 4/98
  Singapore, 5/176
  Taiwan, *6/15*
Taormina (Sicily), *3/244–245*
Tarai plain (Nepal), 4/192
Tarantulas, *2/125*
Tarascan language, 4/137
Taraua mountain range, *4/219*
Tarawa (Kiribati), 4/17
Taro, 7/92
Taroko Gorge (Taiwan), *6/10–11*
Tarranto (Italy), 3/233
Tarsiers, 5/61
Tartooshes, 6/200, *6/201*
Tashkent (Uzbekistan), 6/166
Tasman, Abel Janszoon, 1/86, 2/220,
    7/106
Tasmania, 1/70, 1/86, *1/88,* 1/89, *7/106*
Tassili-n-Ajjer (Algeria), 1/28
Tatami, 3/272, 3/273
Tatars
  Georgia, 3/14
  Russia, 5/112, 5/116, 5/120, 5/126
  Turkmenistan, 6/74
  Ukraine, 6/82
Tatra Mountains, *5/78–79,* 5/79
Tattooing, 7/96, 7/97
Tau Island (American Samoa), 1/32
Taurus Mountains, 3/191, 6/60, 6/73
Tausug people, 5/64
Tavernas, *3/59*
Taxation, 4/149, 4/263
Taxco (Mexico), 4/134, 4/144
Tay people, 6/187
Tay Son brothers, 6/181
Taya, Maawiya Ould Sid Ahmed, 4/121
Taylor, Charles, 4/65
Te Kanawa, Kiri, *4/222*
Tea
  Argentina, 1/55
  India, *3/134–135*
  Japan, 3/260, 3/279, 3/281
  Kenya, 4/15, *4/15*
  Malawi, 4/90
  Morocco, 4/165
  Paraguay, *5/38–39*
  Sri Lanka, 5/222, 5/224
  Turkey, *6/62,* 6/65
  Uruguay, *6/160*
  Virgin Islands, British, 6/188

Tea gardens, 3/280, 3/281
Teak, 6/32, 6/33
Teakwood, 6/27
Tebassah Mountains, 6/52
Tectonic plates, 3/120, 3/224, 4/132,
    4/132, **4/133**
Tedzhen River, 6/74
Tegucigalpa (Honduras), 3/102, *3/103,*
    3/105
Tehran (Iran), 3/176
Tehran University, 3/175
Tel Aviv (Israel), 3/209, 3/215, *3/215*
Tela (Honduras), 3/102
Telephones, *6/152*
Tell (Algeria), 1/28, 1/31
Tell, William, 5/258
Telugu language, 3/139
Tema (Ghana), 3/41
Temne people, 5/172
Temple (Jerusalem), 3/216
Temple of Abu Simbel, *2/193*
Temple of Apollo (Cyprus), 2/143
Temple of Artemis (Turkey), 6/70
Temple of Athena Nike, *3/64*
Temple of Hadrian (Turkey), 6/67
Temple of Heaven (Beijing), 2/82, *2/82*
Temple of Hera (Sicily), 3/244
Temple of Poseidon (Greece), *3/64*
Temple of the Emerald Buddha
    (Bangkok), 6/35
Temple of the Giant Jaguar (Guatemala),
    3/80–81
Temple of the Tooth, *5/226–227,*
    5/227
Temple of the Wooden Lintel (Belize),
    1/145
Temple of Zeus (Cyprus), 2/143
Temppeliaukio Church (Finland), *2/229*
Temüjin. *See* Genghis Khan
Ten Years' War, 2/132
Tenayuca (Mexico), 4/142
Tenerife (Canary Islands), 5/203, 5/218,
    5/219, *5/219,* 5/220
Tennessee (United States), 6/124, 6/125
Tenochtitlán, 4/130, 4/138, 4/140,
    7/102
Tenryu-ji temple, *3/268*
Tenzing Norgay, 4/188
Teosinte, 7/78
Teotihuacán, *4/140,* 7/80
Tepes, Vlad, 5/105
Tepoztlan (Mexico), 4/142
Terai, 4/184, 4/185
Teresa, Mother, *5/247*
Termessos (Turkey), 6/73
Terns, Whiskered, *4/185*
Terra preta, 7/84, 7/85
Terracing
  Armenia, 1/66
  China, 2/90, *2/90–91*
  early agriculture, 7/81
  Indonesia, *3/164*
  Nepal, *4/184–185*
  Philippines, 5/64
Terre-de-Haut (Guadeloupe), *3/76*

Terrorism
  Afghanistan, 1/13
  Argentina, 1/52
  Armenia, 1/67
  Colombia, 2/100, 2/101
  Cyprus, 2/140
  India, 3/159
  Ireland, 3/205
  Israel, 3/209, 3/213
  Italy, 3/222
  Kenya, 4/11
  Mexico, 4/129
  Pakistan, 3/159
  Peru, 5/47
  Russia, 5/111
  Spain, 5/200, 5/201
  United Kingdom campaign, 6/113
  United States, 1/13, 6/137, *6/137*
  Uruguay, 6/162
  Yemen, 6/199, 6/201
Tesla, Nicola, *5/159*
Teutonic Knights, 4/48
Teutons, 3/19, 3/34
Tewhatewas, *7/96*
Texas (United States), 4/140, 6/126,
    6/135, 6/153
Texcoco, Lake, 4/138, 4/140
Textiles
  Bangladesh, 1/119, 1/121
  Belgium, 1/138, 1/140–143
  Cambodia, *2/12*
  China, *2/92*
  France, 2/251
  Inca, *2/172*
  India, 3/136, 3/137, *3/137*
  Turkey, *6/67*
  See also Cotton; Sheep and wool
TGV high-speed train, *2/251,* *2/251*
Thai language, 6/36
Thai people, 6/27, 6/31
  See also Thailand
Thai Rak Thai Party (Thailand), 6/28
Thailand, 6/26–39, *6/26–39,* **6/29**
  Chinese culture, 2/79
  economy, 6/32–33, *6/32–33*
  government, 6/28
  history, 6/28–29, 6/36–37, *6/36–37*
  human settlement, *7/22–23,* 7/23,
    7/92
  land, 6/30–31
  people, 6/30–31, *6/30–31*
    Buddhism, 6/38–39, *6/38–39*
  population, 6/31
  relations with
    Cambodia, 2/15
    Laos, 4/45
Thailand, Gulf of, 6/30, 6/33
Thakins, 4/179
Thames, River, 6/92, *6/102–103,*
    6/108–109
Thammasat University, 6/28
Thanh Hoa (Vietnam), 6/186
Thanksgiving Day, *6/157*
Thar Desert, 5/18, 5/19
Tharp, Twyla, 6/158

Thatcher, Margaret, 6/107, *6/107*
Theater
  ancient Greece, 3/65
  China, 2/67
  Indonesia, 3/169
  Ireland, 3/206
  Japan, 3/276
  Korea, 4/31
  Turkey, 6/73
  United States, 6/158
Théâtre de la Monnaie (Brussels), 1/141
Theravada school, 4/47, 4/178, 6/38–39, *6/38–39*
Theseus, 3/68
Thessaloniki (Greece), 3/60, 3/61, 3/71
Thessaly (Greece), 3/54
Thimphu (Bhutan), *4/193–194*
Thingvellir (Iceland), *2/119*
Thinker, The (Rodin), 1/61
Thira (Greece), 3/66, 3/68
Third Reich, 3/21
Thirteen Colonies (United States), 6/120, **6/134**
Thirty Years' War, 1/98, 2/148, *2/148*, 2/150, 2/151, 3/20, 5/246
Tholos (Greece), *3/63*
Thorgilsson, Ari, 3/122
Thrace (Turkey), 6/57, 6/60, 6/61, 6/68
Thracian people, 1/180
Three Gorges Dam, 2/89
Three Kingdoms, 4/22, 4/30
Throni, Hill of, *1/143*
Thule culture, 7/75
Thuringia (Germany), 3/34
Thuringian Fair, 3/34
Thuringian Forest (Germany), 3/24
Thutmose III, 2/192
Tiahuanaco Indians, 1/152
Tian Shan Mountains, 2/71, 4/38, *4/39*, 6/166
Tiananmen Square (Beijing), 2/60–61, 2/65, *2/65*, 2/82, 2/83, *2/83*
Tianyuan Cave, 7/54
Tiber River, 3/226, 3/230, 6/170
Tibesti Mountains, 2/46
Tibet (China), 2/88, 2/94–95, *2/94–95*, **2/95**, 4/193
  Bangladesh history, 1/124
  Bhutan, power in, 1/150, 1/151
  Chinese minorities, 2/78–79
  land, 2/70–71, *2/70–71*, 2/88, 2/94
Tibet, Plateau of, 2/70, 2/88, 2/94
Tibetan Highlands (China), 2/70–71, *2/70–71*, **2/71**, 2/73
Tibetan language, 2/79, 2/96
Tibetan people, 2/78, 2/79
Tibeto-Burman languages, 1/124, 4/46
Tichka Pass, 4/168
Ticlio (Peru), 5/51
Tidal waves
  Bangladesh, 1/120, 1/126–127
  Chile, 2/52
  Japan, 3/256
  Portugal, 5/90
Tiergarten (Berlin), 3/32

Tierra del Fuego, 1/55–57, *1/56, 2/52–53, 2/53*
  early settlement, **7/88,** 7/88–89, *7/88–89*
Tigers, 2/74, *2/74,* 3/150–151, *3/150–151, 4/185*
Tigran II, 1/66
Tigray (Ethiopia), 2/206, 2/209
Tigrinya language, 2/209
Tigris-Euphrates Valley, 3/186, 7/38
Tigris River, 3/190–191, 3/193, 3/196, 3/197, 6/61, 7/38
Tîgu Mures (Romania), 5/107
Tihamah region, 5/148, 6/194
Tijuana (Mexico), 4/134, 4/143
Tikal (Guatemala), *3/80*
Timber. See Forests and forest products
Timbuktu (Mali), 4/108, *4/111,* 4/169
Timor (Indonesia), 3/164
Timor-Leste. See East Timor
Timur, 1/18
Timurids, 1/11
Tin
  Bolivia, 1/153, 1/154
  Congo, *2/112*
  Malaysia, 4/93, 4/100
  Thailand, 6/32, 6/33
Tipperas tribe, 1/124
Tiranë (Albania), 1/23, *1/23*
Tisissat Falls, 2/208
Tista River, 1/122, 3/153, 4/191
Tisza River, 3/112
Titano, Mount, 5/140
Titian, 6/171
Titicaca, Lake, 1/152, 5/44, **5/54,** 5/54–55, *5/54–55*
Tito, Josip Broz, 5/158, 5/159, *5/159,* 5/162
Tiv people, 4/243
Tivoli Gardens (Copenhagen), 2/161, *2/161*
Tiwanaku Indians, 5/44, 5/54
Tlingit Indians, 2/28–29, 2/29
Toba, Lake, 3/164
Toba, Mount, 7/12, 7/13, *7/13*
Tobacco, 1/34, 2/136, 6/124, 6/134
Tobago. See Trinidad and Tobago
Toberua Island, *2/221*
Tocantins River, 1/176
Togo, 6/40–41, *6/40–41,* **6/41**
Togo Mountains, 6/40
Tokelau, 5/11
Tokugawa Ieyasu, 3/264, *3/265,* 3/270
Tokugawa shogunate, 3/264, *3/264*
Tokyo (Japan), 3/253, 3/270–271, *3/270–271,* **3/271,** *3/272–273*
  history, 3/264, 3/269
  land, 3/256
Tokyo Bay, 3/270
Tokyo Stock Exchange, *3/263*
Tolbert, William R., Jr., 4/65
Tolimán volcano, *3/83*
Toluca (Mexico), 4/134, *4/134*
Tomatoes, *1/81,* 3/235, *5/92*

Tomb of the Unknown Soldier (Greece), *3/53*
Tonga, 5/11, 6/42–45, *6/42–45,* **6/43,** 7/94
  economy, *6/45*
  government, 6/42–43
  land, 6/42
  people, 6/44–45, *6/44–45*
Tonga people, in Malawi, 4/90
Tongatapu (Tonga), 6/42, 6/44, 6/45
Tonkin, 6/181
Tonkin, Gulf of, 6/185
Tonle Sap (lake), 2/8, 2/12
Tonle Sap River, 2/8, 2/12
Tontons Macoutes, 3/97
Töölönlahti Bay, *2/229*
Tools, Early
  Amazon region, 7/84
  Arctic region, 7/64, 7/65, 7/74–75
  Australia, 7/29
  Bantu people, 7/98
  Clovis, 7/68–69, *7/68–69*
  Eastern Europe, 7/44, 7/46
  India, *7/12*
  Japan and Korea, 7/60–61
  Mexico, 7/78
  Middle East, 7/10, 7/34, 7/37, 7/39–41
  migration and, 7/8–9, 7/11, 7/14
  Paleolithic, 7/19
  Southeast Asia, 7/23, 7/24
  Western Europe, 7/49, 7/50
Topa Inca Yupanqui, 5/52
Topkapi Palace (Istanbul), 6/63, 6/70
Topolánek, Mirek, 2/147
Toraja people, *3/166*
Tordesillas, Treaty of, 1/164, 7/102
Torii gates, *3/278–279*
Tornadoes, 1/120
Toronto (Ontario), 2/23, 2/27, 2/34, 2/36–37, *2/37*
Torrijos Herrera, Omar, 5/29
Tórshavn (Faroe Islands), 2/213
Tortillas, 4/134
Tortoises, 2/178, *2/179*
Tortola Island, 6/188, 6/189
Toshogu Shrine (Tokyo), *3/279*
Tosk people, 1/22
Totonac Indians, 4/144
Totora boats, 5/54, 5/54–55
Toubou people, 2/46, *2/46–47*
Toucouleur people, 4/122
Touggourt oasis (Algeria), *1/31*
Toungoo (Myanmar), 4/178
Touré, Amadou Toumani, 4/109
Touré, Sékou, 3/45, 3/86–87
Tourism
  Aegean Islands, 3/67
  Andorra, 1/35, *1/35*
  Antigua and Barbuda, 1/48
  Austria, 1/100–104
  Bahamas, 1/114, 1/115
  Belize, 1/144
  Bermuda, 1/148, *1/149*
  Bulgaria, *1/184*
  Canary Islands, 5/218, 5/221, *5/221*

176

Cayman Islands, 3/49
Channel Islands, 6/101
China, 2/69
Cook Islands, 2/118
Costa Rica, **2/124,** 2/124–125,
    2/124–125
Cuba, 2/137
Cyprus, 2/142–143, *2/142–143*
Fiji, 2/216
French Polynesia, 2/259
Gibraltar, 3/47
Guadeloupe, 3/77
Himalaya, 4/195
India, 3/148–149
Indonesia, *3/170–171*
Italy, 3/241, 3/245
Jamaica, 3/248
Kenya, 4/12, 4/15
Lesser Antilles, 4/60
Luxembourg, 4/81
Madagascar, 4/85
Madeira, 1/113
Mexico, 4/138, 4/142–143,
    4/142–143
Monaco, 4/149
Montenegro, 4/157
Morocco, 4/163, 4/168–169,
    4/168–169
Pacific Islands, 5/11
Poland, 5/83
Puerto Rico, 5/96, *5/96*
Réunion, 5/101
Seychelles, 5/171
Spain, 5/199, 5/208
Switzerland, 5/263
Thailand, 6/33, 6/35
Trinidad and Tobago, 6/47
Tunisia, 6/53
Turkey, 6/67, *6/67,* 6/72–73
United States, 6/127
Virgin Islands, British, 6/188, 6/189
Virgin Islands, U.S., 6/191
Tours, Battle of, 5/151
Tower of London, *6/102–103,* 6/103
Town Hall Square (Copenhagen),
    2/160–161
Trabzon (Turkey), 6/70
Trade
    Belgium, 1/143
    Djibouti, 2/162
    Hong Kong, 2/85
    Netherlands, 4/202
    Taiwan, 6/11
    United States, 6/152
    Yemen, 6/198
    See also specific countries
Trafalgar Square (Barbados), 1/128
Tragopans, Temminck's, *2/75*
Trail (British Columbia), *2/34*
Trains. *See* Railroads
Trajan, 3/230–231, 5/106, 5/212
Tran family, 6/180
Transantarctic Mountains, 1/41–43
Trans-Canada Highway, 2/31
Transcaucasian Federation, 1/110, 3/14

Transdanubia, 3/112
Trans-Gabon Railway, *3/10–11,* 3/11
Trans-Iranian Railway, 3/182
Transitional Mountains, 1/182
Transjordan, 3/282, 5/267
Transjordan Plateau, 3/286, *3/286–287*
Trans-Siberian Railroad, **5/128,** 5/128–
    129, *5/128–129*
Transvaal (South Africa), 5/192
Transylvania (Romania), 3/110,
    5/104–105
    people, 5/106, 5/107
Transylvanian Alps, 5/104, *5/104*
Transylvanian Plateau, 5/104
Traoré, Dioncounda, 4/109
Traoré, Moussa, 4/109
Trapping, 5/126
Treasure Island (Stevenson), 6/189
Treaty of ... *See under key word, for
    example* Windsor, Treaty of
Tree of Life, 5/216
Trekking, in Himalaya, 4/196–197,
    4/196–197
Trench Town (Kingston), *3/250*
Trentino-Alto Adige (Italy), 3/232
Tresnjica Canyon, *5/167*
Trianon, Treaty of, 3/107
Tribhuwan Shah, 4/186
Trieste (Italy), 3/232
Trinh family, 6/181
Trinidad and Tobago, 1/129, 4/58,
    6/46–47, *6/46–47,* **6/47**
    music, 3/250, 4/61
Trinidad English language, 6/47
Trinità dei Monti, Church of (Rome),
    3/231
Trinity College (Dublin), *3/203*
Trinity Monastery of St. Sergius (Russia),
    5/112
Trinkhaus, Erik, 7/54
Triple Alliance, War of the, 5/39
Tripoli (Libya), 4/67, 4/68, 4/69, *4/70–71*
Tripolitania, 4/67
Trobriand Islands (Papua New Guinea),
    5/36
Tromsø (Norway), 4/254, *4/255*
Trondheim (Norway), 4/254, 4/255,
    4/255
Trondheim Fiord (Norway), 4/255
Trondheim Lowlands (Norway), 4/250
Troodos Massif, 2/142, *2/142,* 2/143
Trooping of the Colour ceremony,
    6/102, 6/108
Trotsky, Leon, 5/114
Trout, Rainbow, *2/39*
Trucial States, 6/84–85
Trudeau, Pierre, 2/32
Trujillo Molina, Rafael Leonidas, 2/167
Truk Islands. *See* Chuuk Islands
Truong Son mountains, 6/186
Truth and Reconciliation Commission
    (Solomon Islands), 5/185
Tsbulyovka (Moldova), *4/147*
Tsetse flies, 2/42
Tshiluba language, 2/112

Tshisekedi, Etienne, 2/111
Tsiranana, Philibert, 4/87
Tsunamis
    Indian Ocean (2004), 3/127, 3/157,
        3/163, 4/104, 5/225, 6/29
    Japan, 3/255, 3/256, *3/257*
Tsvangirai, Morgan, 6/209
Tswana people, 1/158*1/158,* 4/181,
    5/194
Tuamotu Islands, 2/258, 2/259, 5/55
Tuareg people, 1/187, *4/68,* 4/109,
    4/111, 4/232, 4/233, 4/234, 4/235
Tuataras, 4/224
Tubman, William V., 4/65
Tucumán (Argentina), 1/65
Tugurinos, 2/105
Tu'i Tonga, 6/44
Tuileries Gardens (France), 2/247
Tuk-tuk vehicles, 6/34
Tula (Mexico), 4/142
Tulcea (Romania), *3/116*
Tulips, *5/121*
Tumbuka people, 4/90
Tuna, 1/32, 3/261, *3/261*
Tundra
    Alaska, 6/130
    Canada, 2/26, 2/39
    Finland, 2/227
    Greenland, *3/72–73*
    Norway, 4/257
    Siberia, 5/124, 5/126
Tunis (Tunisia), 6/48, 6/50, *6/51,* 6/52,
    6/53, 6/55
Tunis, University of, 6/51
Tunisia, 1/24, 6/48–55, *6/48–55,* **6/51**
    government, 6/50–51
    history, 6/48–49
    land and economy, 6/52–53, *6/52–53*
    people, 6/54–55, *6/54–55*
Tunisian people, in France, 2/239
Tunjos, 2/98
Tupac Amaru II, 1/152
Tupac Amaru Revolutionary Movement,
    5/47
Tupamaros, 6/162
Tupou I, George, 6/44
Tupou IV, 6/43
Turbans, *1/14*
Turin (Italy), 3/235, 3/245
Turkana, Lake, 4/15
Turkey, 6/56–73, *6/56–73,* **6/59**
    Bulgarian refugees, 1/181
    climate, 6/61
    economy
        agriculture, 6/64–65, *6/64–65,*
            6/67
        industry, 6/66–67, *6/66–67*
    historic sites, 6/70–71, *6/70–71,*
        6/71
    history, 6/58–59
    land, 6/60–61, *6/60–61,* **6/61**
    people, 6/64–65, *6/64–65*
    population, 6/59
    prehistoric human migration, **7/36,**
        7/36–37, 7/36–37, 7/42

relations with
    Cyprus, 2/141
    Greece, 3/52
    Turkish Riviera, **6/72**, 6/72–73, *6/72–73*
    See also Ottoman Empire
Turkic language, 1/15
Turkish baths, 6/63
Turkish language, 1/181, 3/193, 6/58
Turkish people
    Azerbaijan, 1/111
    Bulgaria, 1/181
    Cyprus, 2/141, 2/142
    France, 2/239
    Germany, 3/37
    Kyrgyzstan, 4/38
    Uzbekistan, 6/166
    See also Turkey
Turkish Riviera, 6/72, 6/72–73, *6/72–73*
Turkmen, 6/74
Turkmen Soviet Socialist Republic, 6/74
Turkmenistan, 6/74–75, *6/74–75,* **6/75**
Turkoman people, 1/15, 3/178, 3/192
Turks and Caicos Islands, 3/48–49, **3/49**
Turku (Finland), *2/230*
Turner's Hall Woods (Barbados), 1/128
Turpan Depression (China), 2/71
Turquoise Coast, 6/72, *6/72–73*
Turtles, 2/219
    Atlantic green, *2/125*
    slider, *6/149*
    See also Tortoises
Tuscan dialect, 3/233
Tuscany (Italy), 3/221, *3/224,* 3/225, *3/228, 3/232, 3/238–239,* 3/239
Tutsi people, 1/188–189, *1/189,* 5/130, 5/131
Tutu, Osei, 3/44
Tutuila Island (American Samoa), 1/32
Tuvalu, 4/182, 5/11, 6/76–77, *6/76–77,* **6/77**
Tuvaluan language, 6/77
Tver (Russia), *5/118*
Twa people, 1/188, 5/130
Twelve-tone technique, 1/106
Tyne and Wear (England), 6/93
Tyne Bridge (England), *6/92–93*
Tynwald Ceremony, *6/101*
Tynwald Court (Isle of Man), 6/101
Typhoons
    Japan, 3/257
    Korea, South, 4/27
    Philippines, 5/57, 5/60–61, *5/61*
    Taiwan, 6/11
Tyrol (Austria), 1/100
Tyrolean Summer Festival, 1/107
Tyrrhenian Sea, 3/224, 3/230
Tyska Kyrka, 5/251

U

U Nu, 4/179
UAE. *See* United Arab Emirates
U.A.R. *See* United Arab Republic
Ubangi River, *2/43,* 2/114, 2/116
Ubangi-Shari, 2/42–43

Ucagizli Cave, 7/36–37
Udaipur (India), *3/136–137*
Udmart people, 5/116
Ugali porridge, 6/22
Uganda, 2/111, 6/78–81, *6/78–81,* **6/79**
    government, 6/78
    history, 6/78–79
    land and people, 6/80–81, *6/80–81*
    Tanzania, relations with, 6/19
Uigur people. *See* Uyghur people
Ujamaa system, 6/19, 6/23
U.K. *See* United Kingdom
Ukraine, 1/131, 6/82–83, *6/82–83,* **6/83**
Ukrainian language, 6/83
Ukrainian people
    Canada, 2/23
    Moldova, 4/147
    Russia, 5/116
    Turkmenistan, 6/74
    See also Ukraine
Ulaanbaatar (Mongolia), 4/152, *4/153,* 4/154
Ulama, 2/188, 2/189
Ulbricht, Walter, 3/23
Ulemas, 5/146
Ulster. *See* Northern Ireland
Ulu Cami Mosque (Turkey), 6/70
Ulua-Chamelecón River Basin, 3/102
Uluru, *1/73, 1/76–77*
Um El Ma lake, *4/67*
Umanak (Greenland), *3/72–73*
Umayyad dynasty, 5/267, 5/274, 5/275
Umayyad Mosque (Syria), *5/274–275,* 5/275
Umbria (Italy), 3/225, 3/239
Umbrian dialect, 3/233
Umiaks, *7/74*
Umm al Qaywayn (United Arab Emirates), 6/84
Unemployment
    African Americans, 6/145
    Botswana, 1/159
    Italy, 3/241
    Korea, South, 4/28
    Tajikistan, 6/17
    United Kingdom, 6/93
    United States, 6/136
UNESCO, 5/64, 7/101
Union, Treaty of, 6/96
Union of Democratic Forces (Bulgaria), 1/180
Union of Kalmar, 2/156, 4/247
Union of South Africa, 5/193
Union of Soviet Socialist Republics.
    See Soviet Union
Union Solidarity and Development Party (Myanmar), 4/177
UNIP (Zambia), 6/202
UNITA (Angola), 1/36–37
United Arab Emirates, 1/116, *6/84–87,* **6/85,** 6/85–87
    economy, 6/86–87, *6/86–87*
    land, 6/86
United Arab Republic, 5/268
United Democratic Front (Malawi), 4/91

United Fruit Company, 3/80
United Kingdom, 6/88–113, *6/88–113,* **6/91**
    economy, 6/106–107, *6/106–107,* **6/107**
    festivals, 6/108–109, *6/108–109*
    government, 6/90
    history
        Australia and New Zealand settlement, 7/106–107, *7/106–107,* **7/107**
        early, 6/110–111, *6/110–111*
        modern, 6/112–113, *6/112–113*
        North America settlement, 7/104, *7/105*
    people, 6/90–91, 6/104–105, *6/104–105,* **6/105**
    population, **6/91**
    power in
        Afghanistan, 1/11, 1/18
        American colonies, 6/115, 6/134, 7/104
        Antigua and Barbuda, 1/48
        Australia, 1/69, 1/86–87
        Bahamas, 1/114–115
        Bahrain, 1/116
        Bangladesh, 1/118, 1/119
        Barbados, 1/128, 1/129
        Belize, 1/144
        Bermuda, 1/148, 1/149
        Bhutan, 1/150
        Botswana, 1/158
        Brunei, 1/178
        Cameroon, 2/16, 2/18–19
        Canada, 2/23, 2/25, 2/29–32
        Channel Islands, 6/100–101
        China, 2/61, 2/63, 2/85
        Cook Islands, 2/119
        Corsica, 2/255
        Cyprus, 2/140
        Dominica, 2/164–165
        Egypt, 2/192–193
        Eritrea, 2/202
        Falkland Islands, 2/210–211
        Fiji, 2/217, 2/220
        French Guiana, 2/256
        Gambia, 3/13
        Germany, 3/22
        Ghana, 3/42, 3/44
        Gibraltar, 3/46–47
        Greater Antilles, 3/48
        Grenada, 3/74
        Guyana, 3/92
        Hong Kong, 2/85
        India, 3/126, 3/129, 3/148–149, 3/152
        Indonesia, 3/161
        Iran, 3/182, 3/183
        Iraq, 3/187
        Ireland, 3/201, 3/204–205
        Isle of Man, 6/100–101
        Jamaica, 3/247, 3/249
        Jordan, 3/282
        Kenya, 4/10–11
        Kiribati, 4/17

Lebanon, 4/51
Lesser Antilles, 4/59, 4/60
Madagascar, 4/87
Malawi, 4/91
Malaysia, 4/94, 4/102–103
Maldives, 4/104
Malta, 4/114, 4/115
Marshall Islands, 4/116
Martinique, 4/118
Mauritania, 4/120
Mauritius, 4/124, 4/125
Morocco, 4/167
Myanmar, 4/176, 4/178–179
Nepal, 4/186
New Zealand, 4/215
Nicaragua, 4/226
Nigeria, 4/236–237
Pacific Islands, 5/12
Pakistan, 5/15
Palestine, 3/212
Papal States, 6/170
Papua New Guinea, 5/35
Rhodesia, 6/208
Russia, 5/114
St. Kitts and Nevis, 5/132–133
St. Lucia, 5/134, 5/135
St. Vincent and the Grenadines, 5/136
Samoa, 1/32, 5/139
Saudi Arabia, 5/144
Senegal, 5/153
Seychelles, 5/170
Sierra Leone, 5/172–173
Singapore, 5/174
Solomon Islands, 5/185
Somalia, 5/186
South Africa, 5/192–193, 7/99
Sri Lanka, 5/224
Sudan, 5/229
Suriname, 5/238
Swaziland, 5/240
Syria, 5/274
Tahiti, 6/8
Tanzania, 6/19
Togo, 6/41
Tonga, 6/42, 6/43, 6/44
Trinidad and Tobago, 6/46
Turkmenistan, 6/74
Tuvalu, 6/77
Uganda, 6/78
United Arab Emirates, 6/84–85
Vanuatu, 6/169
Virgin Islands, British, 6/188
Yemen, 6/195, 6/199
Zambia, 6/202
Zimbabwe, 6/211
relations with
Argentina, 1/52, 1/65
Azores, 1/112
Denmark, 2/160
France, 2/238, 2/241, 2/245
Germany, 3/26
Greece, 3/63
Iraq, 3/189
New Zealand, 4/216
Oman, 5/9

Portugal, 5/94, 5/95
See also British people; England; Great
Britain (island); Northern Ireland;
Scotland
United National Independence Party
(Zambia), 6/202
United Nations
actions in
Afghanistan, 1/13
Albania, 1/21
Angola, 1/37
Bangladesh, 1/126
Bosnia-Herzegovina, 1/157
Cambodia, 2/11
Cape Verde, 2/40
Chad, 2/45
Congo, Democratic Republic of the,
2/113
East Timor, 2/168, 2/169
El Salvador, 2/197
Ethiopia, 2/209
Haiti, 3/97
India, 3/126
Indonesia, 3/162, 3/163
Iran, 3/177
Iraq, 3/188, 3/189, 3/195, 4/35,
6/136
Israel, 3/212, 3/213, 4/53
Jordan, 3/286
Korea, South, 4/23
Kosovo, 4/32, 4/33, 5/161
Liberia, 4/65
Libya, 4/67
Macedonia, 4/83
Marshall Islands, 4/117
Morocco, 4/159
Namibia, 4/180
Nauru, 4/183
Pacific Islands, 5/12, 5/27
Palestine, 5/268
Rwanda, 5/131
Samoa, 5/139
Solomon Islands, 5/185
Somalia, 5/186, 5/187
Sudan, 5/231
Vietnam, 6/183
Western Sahara, 6/193
China's replacement of Taiwan, 6/12
Disarmament Conference, 5/247
role of
Sweden, 5/246, 5/247
Switzerland, 5/255, 5/264, 5/265
Vienna center, 1/108
United Nations Development
Programme, 7/5
United Provinces of Central America,
2/123, 2/194, 3/80, 3/100, 4/226
United Provinces of the Río de la Plata,
1/56, 1/65
United Self-Defense Forces of
Colombia, 2/101
United States, 6/114–159, *6/114–159,*
6/118–119
arts and entertainment, 6/158–159,
6/158–159

economy
agriculture, **6/154,** 6/154–155,
6/154–155
energy and resources, 6/150–151,
6/150–151
industry, 6/152–153, *6/152–153*
government, 6/116–117, 6/152
history
beginning to 1865, 6/112, **6/134,**
6/134–135, *6/134–135*
immigration, 7/104–105, *7/105*
since 1865, 6/136–137, *6/136–*
137, 6/184, 6/185
military bases
Azores, 1/112
New Caledonia, 4/212
Okinawa, 3/259
Pacific Islands, 5/11
Philippines, 5/58–59
national parks, **6/132,** 6/132–133,
6/132–133
people
African Americans, 6/144–145,
6/144–145
American way of life, 6/156–157,
6/156–157
Hispanic Americans, 6/142–143,
6/142–143
migration patterns, 7/5
multiethnicity, **6/138,** 6/138–139,
6/138–139
Native Americans, 6/140–141,
6/140–141, **6/141**
urban life, 6/146–147, *6/146–147*
power in
American Samoa, 1/32–33
Cuba, 2/132
Dominican Republic, 2/167
Federated States of Micronesia,
2/215
Germany, 3/22
Guam, 3/78–79
Guatemala, 3/80
Haiti, 3/96–97
Japan, 3/258
Kiribati, 4/17
Korea, 4/23
Mariana Islands, 3/78
Marshall Islands, 4/117
Nicaragua, 4/226–227
Pacific Islands, 5/12–13
Palau, 5/27
Panama, 5/28–29, 5/32
Philippines, 5/57–59, 5/66,
5/70–71
Puerto Rico, 5/97
Russia, 5/114
Samoa, 5/139
Virgin Islands, U.S., 6/190
relations with
Afghanistan, 1/11, 1/13
Bolivia, 1/154
Cambodia, 2/9
Canada, 2/23, 2/32, 2/34
China, 2/65

Colombia, 2/100–101, 2/101
Cuba, 2/133, 2/136
El Salvador, 2/195
Federated States of Micronesia,
   2/215
Georgia, 3/15
Grenada, 3/75
Honduras, 3/100–101
Indonesia, 2/168, 3/162
Iran, 3/176–177
Iraq, 3/188–189, 3/188–189
Japan, 3/264, 3/276
Korea, North, 4/19
Kuwait, 4/36, 4/37
Libya, 4/67
Mexico, 4/140, 6/135
Nicaragua, 4/230–231
Saudi Arabia, 5/145, 5/149
Taiwan, 6/11
Thailand, 6/32
Ukraine, 6/83
Vietnam, 6/182, 6/183, 6/184–
   185, 6/184–185
states of, 6/118–119, **6/118–119**
   East Coast, **6/120**, 6/120–121,
      6/120–121
   governments, 6/117
   Midwestern, **6/122**, 6/122–123,
      6/122–123
   Pacific Coast, **6/128**, 6/128–129,
      6/128–129
   Southern, **6/124**, 6/124–125,
      6/124–125
   Western interior, **6/126**, 6/126–
      127, 6/126–127
United States Virgin Islands. See Virgin
   Islands, U.S.
Unity Day (Germany), 3/18
University of Adelaide, 7/28
University of Belgrade, 5/168
University of Brunei Darussalam,
   1/179
University of Cocody, 2/129
University of Glasgow, 7/12
University of Helsinki, 2/228
University of London, 6/93
University of Santo Tomás, 5/67
University of Tehran, 3/175
University of the South Pacific, 2/216,
   6/45
University of Tunis, 6/51
University of Utrecht, 4/205
University of Wittenberg, 3/38
University of Wollongong, 7/8–9
University of Zambia, 6/205
Unter den Linden (Berlin), 3/32
Unterwalden canton (Switzerland), 5/258
Untouchables, 3/140–141
Up-Helly-A' festival, 6/108
Upernavik (Greenland), 3/72
Upland District (Finland), 2/226
Upolu (Samoa), 5/138
Upper Austria, 1/102, 1/102
Upper Volta, 1/86, 1/87
Ur (Sumer), 3/196–197

Ural Mountains, 5/78, 5/109, 5/112,
   5/118, 5/124, 5/125, 5/128
Uralic-Altaic language, 3/106
Uranium, 1/71, 2/34, 4/233
Urdu language, 1/124, 3/95, 3/138,
   5/20
Urgel (Spain), 1/34
Urhobo people, 4/243
Uri canton (Switzerland), 5/258
Uribe Vélez, Álvaro, 2/101
Ursula (saint), 6/190
Uru Indians, 5/55
Uruguay, 6/160–165, 6/160–165, **6/163**
   government, 6/162
   history, 6/160–163
   land and economy, 6/164–165,
      6/164–165
   Paraguay, relations with, 5/39
   people, 6/160, 6/160
Uruguay River, 6/160, 6/164
Uruk (Sumer), 7/40, 7/41
Urundi, 1/188
Usa River, 7/64
USDP (Myanmar), 4/177
Useful Chad, 2/44
Ushuaia (Tierra del Fuego), 1/55–56
Uspensky Cathedral (Finland), 2/229
Utah (United States), 6/126
Utrecht (Netherlands), 4/205
Utrecht, Treaty of, 3/47, 3/245
Utrecht, University of, 4/205
Uvs Lake, 4/154
Uyghur people, 2/78, 2/79, 2/79
Uyghur script, 4/154–155
Uzbek Empire, 6/166
Uzbek people, 1/15, 4/39, 6/74, 6/167
Uzbek Soviet Socialist Republic, 6/166
Uzbekistan, 6/166–167, 6/166–167,
   6/167

V

Vaduz (Liechtenstein), 4/74, 4/75
Vaez de Torres, Luis, 7/106
Vágar (Faroe Islands), 2/212
Vail (Colorado), 6/127
Vaisakhi festival, 6/105
Vaishyas, 3/140
Vakhsh River, 6/17
Valai Hills, 5/118
Valdez (Alaska), 6/131
Valdivia, Pedro de, 2/49
Valdivia (Chile), 2/49
Vale of Kashmir, 3/158, 3/159, 4/192
Valencia (Spain), 5/206, 5/208–209,
   5/209
Valencia (Venezuela), 6/173
Valencia, Lake, 6/177
Valle del General (Costa Rica), 2/122
Valletta (Malta), 4/115
Valley of Mexico, 4/130–131
Valleys, 1/4
Valparaíso (Chile), 2/49, 2/52, 2/56
Van (Turkey), 3/198
Van, Lake, 6/60–61, 6/61
Vancouver (British Columbia), 2/37, 2/37

Vancouver Island (Canada), 2/39
Vandals
   Algeria, 1/24
   Balearic Islands, 5/217
   Libya, 4/70
   Morocco, 4/158
   Tunisia, 6/48
Van Diemen's Land, 1/86, 7/106
Vänern, Lake, 5/249
Van Eyck, Jan, 1/133
Van Gogh Museum, 4/206
Vanilla, 4/84, 5/100, 5/101
Vanua Levu (Fiji), 2/218
Vanuatu, 5/10, 6/168–169, **6/169,**
   6/169, 7/92
Varanasi (India), 3/146–147, 3/146–147
Varangian Russes, 5/112
Vargas, Getúlio, 1/165
Varna (Bulgaria), 1/183, 1/184, 1/185
Varnas, 3/140
Vasa **museum** (Sweden), 5/251
Vatican City, 1/2, 3/229, 3/236,
   6/170–171, 6/170–171, **6/171**
Vatican Museums, 6/171
Vatican Palace, 6/170, 6/171
Vatnajökull glacier, 3/120
Vatoa Island, 2/220
Vättern, Lake, 5/249
Vatulele Island, 2/219
Vava'u (Tonga), 6/42
Vázquez Rosas, Tabaré Ramón, 6/163
Vedas, 3/141
Veddah people, 5/226
Vega, Lope de, 5/215
Vega Real (Dominican Republic), 2/166
Velázquez, Diego, 5/210, 5/210–211,
   5/215
Veld (Zimbabwe), 6/206, 6/207
Velvet revolution, 2/146, 2/152–153,
   2/153
Venetia (Italy), 3/229, 3/232
Venetian dialect, 3/233
Venezuela, 6/172–179, 6/172–179,
   6/175
   Guyana, relations with, 6/176
   history, 2/99, 2/172, 6/172–175
   land and economy, 6/176–177,
      6/176–177
   people, 6/178–179, 6/178–179
Venice (Italy), 1/139, 2/140, 3/221,
   3/234–235, 3/240, 3/242–243,
   3/242–243, **3/243**
   as city-state, 3/228, 3/236, 3/242
   Crete, rule in, 3/68
   tourism, 3/241
Venus de Milo (sculpture), 2/247
Venus figures, 7/46, 7/47
Veracruz (Mexico), 4/140, 4/141
Verbunkos, 3/117
Verde River, 5/42
Verkhoyansk range, 5/125
Vermeer, Jan, 4/202
Vermont (United States), 6/120, 6/121
Versailles, Palace of, 2/244, 2/252,
   2/252–253, 3/20

Versailles, Treaty of, 3/21, 5/76
Vesterålen Islands, 4/251
Vestmannaeyjar (Iceland), *3/120*
Vesuvius, 3/224
Via Dolorosa (Israel), 3/217
Viceroyalty of La Plata, 1/64–65, 5/39, 6/160
Viceroyalty of Peru, 2/49
Vichy (France), 1/25
Victor Emmanuel II, 3/222, 3/229
Victor Emmanuel III, 3/222
Victoria (Australia), 1/70, 1/83, 1/89
Victoria (Cameroon), 2/18
Victoria (Hong Kong), 2/84
Victoria (queen of United Kingdom), 1/134, 2/36, 6/80, 6/112, *6/113,* 6/206
Victoria, Lake, 4/11, 6/21, 6/25, 6/80
Victoria Falls, *6/203,* 6/206
Victoria Harbour (Hong Kong), 2/84
Victoria Nile River, 6/80
Video games, *3/263*
Vieira, João Bernardo, 3/91
Vienna (city, Austria), 1/95, 1/96, *1/97,* 1/102, 1/108–109, *1/108–109*
    history, 1/98, *1/98–99*
    music, 1/106–107, *1/106–107*
Vienna (province, Austria), 1/102
Vienna Basin (Austria), 1/96, 1/103, 1/105
Vienna Festival, 1/106
Vienna State Opera House, 1/106, *1/106–107,* 1/108
Vienna Woods, 1/109
Vientiane (Laos), 4/40, 4/46
Viet Cong, 6/184, 6/185
Vietminh, 6/181, 6/184
Vietnam, 6/180–187, *6/180–187,* **6/183**
    Chinese culture, 2/79
    history, 6/180–183
        wars, 6/184–185, *6/184–185,* **6/185**
    human settlement, 7/23, 7/92
    relations with
        Cambodia, 2/9, 2/10, 2/13
        China, 6/185
        Laos, 4/41
        Soviet Union, 6/185
        United States, 6/184–185
Vietnam War, 4/41, 6/181, 6/182, 6/184–185, *6/184–185,* 6/187
    battles, **6/185**
    Cambodia's role, 2/8
    in U.S. history, 6/136
    Thailand's role, 6/32
Vietnamese people, 4/212, 6/31, 6/186–187
    See also Vietnam
Vijaya (India), 5/226
Vikings
    Britain, 6/110
    Canada, 2/30
    Denmark, 2/157, 2/161
    Faroe Islands, 2/213
    Greenland, 3/72

Iceland, 3/118, *3/122–123,* **3/123**
Ireland, 3/204
Latvia, 4/48
Norman heritage, 2/238
Norway, 4/246, 4/256
Russia, 5/118
Sweden, 5/242, 5/246
Vila Velha Park (Brazil), *1/168*
Vilcanota Valley (Peru), *5/50–51*
Villa, Francisco (Pancho), 4/141, *4/141*
Villa of Dionysus (Cyprus), 2/143
Villarrica (Chile), 2/49
Villeda Morales, Ramón, 3/100
Vilnius (Lithuania), 4/76, *4/77*
Viminal Hill (Rome), 3/230
Vinci, Leonardo da. *See* Leonardo da Vinci
Vindhya range, 3/131
Vinland, 2/30
Vinson Massif (Antarctica), 1/43
Violencia, La (Colombia), 2/99, 2/100
Viracocha, 5/52
Virgin Gorda Island, 6/188, 6/189
Virgin Islands, 4/58, 6/188
Virgin Islands, British, 6/188–189, *6/188–189,* **6/189**
Virgin Islands, U.S., 6/188, 6/190–191, *6/190–191,* **6/191**
Virgin Islands National Park, 6/191
Virgin Lands project (Kazakhstan), 4/9
Virginia (United States), 6/124, 6/132, 6/134, 6/137
    Clovis artifacts, 7/70, *7/71*
Virginia Falls (Canada), 2/39
Visayas Islands (Philippines), 5/61
Vishnu, 4/189
Visigoths, 2/240, 5/213, 5/217
Vistula River, 5/76, 5/80
Vitamin D, 7/45
Viti Levu (Fiji), 2/218
Vivekananda, 3/147
Vjosa Valley, *1/22*
Vladimir I, 5/109, 6/82
Vladivostok (Siberia), 5/127, 5/128
Vltava River, 2/150, 2/155
Vojvodina (Serbia), 5/154, 5/155, 5/159, 5/163, 5/164, 5/166
Volador dance, *4/144*
Volcán Tajumulco, 3/82
Volcanic Axis (Mexico), 4/130
Volcanic plugs, 5/134
Volcano Islands (Japan), 3/258
Volcanoes
    Antarctica, 1/43
    Antigua and Barbuda, 1/49
    Azores, 1/112
    Canary Islands, 5/218, 5/219
    Chile, 2/52, *2/52*
    Colombia, 2/98
    Costa Rica, *2/121,* 2/125
    Crete, 3/68
    Dominica, 2/164, *2/165*
    Easter Island, 2/170, 2/171
    El Salvador, 2/196, 2/197
    Fiji, 2/218, 2/219

Guadeloupe, 3/77
Guatemala, 3/82, *3/83*
Iceland, 3/118, 3/120, *3/120,* 3/121
Indonesia, 3/161, 3/164, *3/164, 3/165,* 7/13, *7/13*
Italy, 3/224
Japan, 3/256
Madeiras, 1/113, *1/113*
Martinique, 4/118
Mauritius, 4/124
Mexico, 4/130, 4/132–133, **4/133**
Montserrat, 4/63
New Zealand, 4/224
Nicaragua, 4/228
Palau, 5/26
Papua New Guinea, 5/36
Philippines, 5/57, 5/60, *5/61*
Réunion, 5/100, *5/100*
St. Lucia, 5/134
St. Vincent and the Grenadines, 5/136
Samoa, 5/138
Siberia, 5/125
Sicily, 3/244
Tonga, 6/42
Turkey, 6/61
United States, 6/130
Virgin Islands, U.S., 6/190, *6/190–191*
Volga-Don Canal, 5/118
Volga River, 1/131, 5/112, 5/118–119, *5/118–119,* **5/119**
Volgelherd Cave, 7/47
Volgograd (Russia), 5/118, *5/118*
Volksgarten (Vienna), 1/109
Volta, Lake, 3/40, 3/41
Volta Redonda (Brazil), 1/176
Volta River, 3/40
Voltaic people, 1/186, 2/128, 4/110
Voltaire, 2/252
Volubilis (Morocco), 4/166
Voodoo, 3/98
Vorarlberg (Austria), 1/100
Vosges mountains, 2/236
Vostok Station (Antarctica), 1/41
Voyage of the Beagle, The (Darwin), 7/88
Vultures, Griffon, *1/35*
Vyšehrad, Palace of, 2/150

W

Wa people, 4/174
Wachau Valley (Austria), *1/102–103*
Wad Madani (Sudan), 5/234
Wade, Abdoulaye, 5/152, 5/153
Wadis, 3/286, 6/194
Wailers (pop group), 3/250
Wailing Wall. *See* Western Wall (Jerusalem)
Wairakei power plant, *4/218–219*
Wairarapa region (New Zealand), *4/225*
Waitangi, Treaty of, *4/221,* 7/107
Waitangi Day, *4/221*
Wajda, Andrzej, 5/73
Wakizashis, 3/266
Walachia (Romania), 5/104, 5/105, *5/105*
Wales, 6/94, **6/94,** 6/95, 6/106, 6/109
    history, 6/110, 6/111
    prehistoric humans, 7/50

Wałęsa, Lech, 5/74, 5/75
Walid, Khalid ibn al-, 5/274, 5/275
Walker, Alice, 6/145
Walker, William, 2/124, 4/226
Wallabies, 1/74, 1/79, 4/224
Walled City (Manila), 5/66
Wallenberg, Raoul, 5/246
Wallis and Fortuna, 5/11
Wallonia (Belgium), 1/133, 1/134, 1/136, 1/137, 1/142
Walloon people, 1/133, 1/134, 1/136–137, *1/136–137*
Walpole, Sir Robert, 6/112
Walt Disney World (Orlando), *6/159*
"Waltzing Matilda" (song), 1/91
Wang Kon, 4/23
Wang Wei, 2/67
Wangchuk, Ugyen, 1/150
War lords, in China, 2/64
War of the Pacific (1879-1883), 1/153
War of the Triple Alliance, 5/39
Waratahs, 1/75
Warblers, 6/149
Warhol, Andy, 6/158
Warrau Indians, 3/92
Wars. *See specific wars and places*
Wars of the Roses, 6/110
Warsaw (Poland), 5/79, 5/80–81, *5/80–81*
Wartburg Castle, *3/38–39*
Washington (state, United States), 6/128, *6/128*
Washington, D.C. (United States), 6/118, *6/120–121*, 6/121, 6/124, *6/138*, *6/147*
Washington, George, *6/134*, 6/135, *6/135*
Washington, Harold, 6/145
Wat Phra Singh monastery (Thailand), *6/37*
Watches, 5/255, 5/262
Water
　　Antarctic ice sheets, 1/42
　　Australia, 1/79, *1/79*, 1/81
　　Bermuda, 1/148
　　Burundi, *1/189*
　　desalination, *6/86*
　　Earth's surface, 1/4
　　India, *3/144*, 3/145
　　Kuwait, 4/34
　　Netherlands battle with sea, 4/208–209, *4/208–209*
　　United States, 6/151
　　See also Drought; Floods; Irrigation
Water buffalo, *4/100*, *5/62–63*
Water gardens, 3/280
Water hyacinths, 6/149
Water power. *See* Hydroelectric power
Water wheels, 2/185, *2/185*, 3/190, *3/191*
Waterfall climbing, *3/246–247*
Waterford (Ireland), 3/201
Waterloo, Battle of, 2/244, 6/112
Waterlooplein (Amsterdam), 4/207
Wats, *6/38*

Wattle houses, *2/198*
Wattles, 1/75
Watusi people. *See* Tutsi people
Wau kites, *4/103*
Waza National Park, 2/20
Weapons of mass destruction, 3/189
Weaving, *5/53*
Webern, Anton, 1/106
Weimar (Germany), 3/39
Weimar Republic, 3/21, 3/30, 3/33
Welchman Hall Gully (Barbados), 1/128
Welfare system
　　Denmark, 2/156
　　Finland, 2/224
　　Norway, 4/248
　　Sweden, 5/242, 5/244
　　United Kingdom, 6/107
　　United States, 6/117
Wellington (New Zealand), *4/217*, 4/219
Wellington, Duke of, 6/112, *6/113*
Wells, Spencer, 7/11
Welsh language, 6/94
Wenceslas Square (Prague), 2/150
Weno (Federated States of Micronesia), 2/214
Weser River, 3/24, *3/30–31*, 3/31
Wessex (United Kingdom), 6/110
West Bank, 3/209, 3/212, 3/213, 3/283, 3/286
West Bengal (India), 1/119, 1/151, 3/137
West Cameroon, 2/19
West Falkland Island, 2/210
West Germany, 3/17, 3/22–23, 3/32, 3/36
West Indies, 1/149, 3/48–49, 4/58
　　United Kingdom immigrants, 6/104
　　U.S. immigrants, 6/139
　　See also specific islands
West Indies Associated States, 1/48
West Indies Federation, 1/128–129
West Midlands (England), 6/93
West Pakistan, 1/119, 1/120, 1/124, 5/15, 5/16
West Quoddy Head (Maine), 6/118
West Roman Empire, 3/226–227
West Siberian Plain, 5/125
West Virginia (United States), 6/124
West Yorkshire (England), 6/93
Western Australia, 1/70, *1/72*, 1/79, 1/80, 1/86, 1/89
Western Cape (South Africa), 5/195
Western Carpathian Mountains, 5/78, *5/78–79*, 5/79, 5/178
Western civilization
　　Greece and, 3/51, 3/64
　　Rome and, 3/230
Western Dvina River, 1/130
Western Ghats, 3/154
Western Ghats range, 3/131
Western Sahara, 4/67, 4/121, 4/159–161, **6/192**, 6/192–193, *6/192–193*
Western Samoa, 5/12, 5/139
Western Wall (Jerusalem), 3/216–217, *3/216–217*

Westminster, City of, 6/102, 6/103
Westminster Abbey (London), 6/103
Westphalia, Peace of, 3/20
Whales, 1/44, *2/211*
　　blue, 1/44
　　fin, *1/45*
　　humpback, 1/44
　　killer, 1/44, *1/45*
　　minke, 4/253
Whaling, 3/261, 4/253, 4/257, *7/74*, *7/75*
Wheat
　　Afghanistan, 1/17
　　Argentina, 1/62, *1/63*
　　Australia, 1/73, 1/78, 1/80, 1/81, *1/81*
　　Bulgaria, 1/185
　　Canada, 2/31, 2/35
　　China, 2/72, 2/73, 2/97
　　France, 2/248, *2/248–249*
　　India, *3/134–135*
　　Italy, 3/238
　　Serbia, *5/163*
　　United States, 6/122, 6/154, *6/155*
Whirling dervishes, 6/71
White, Patrick, *1/87*
White Australia policy, 7/107
White Guard, 2/223
White House (Washington, D.C.), *6/117*
White Moors, 4/122
White Mountains, 6/120
White Nile River, 2/208, 5/232, 5/234, 5/236, 6/80
White Sea, 5/118
White Volta River, 3/40, *3/41*
Whites (Russia), 5/114
Whites, in South Africa, 5/194–195
　　See also Apartheid
Whitsuntide, *6/108*, 6/109
Wholesale trade, 6/152
Wiener Neustadt (Austria), 1/102
Wiener schnitzel, 1/97
Wild West, American, 6/126
Wilde, Oscar, 3/201
Wilder, L. Douglas, 6/145
Wildflowers, 1/75
Wildlife. *See* Animals; Conservation; Plants
Wildlife reserves, for tigers, 3/150–151
Wilhelm I, *3/20*, 3/21
Wilhering Abbey (Austria), *1/102*
Wilkes Land (Antarctica), 1/41
Willemstad (Curaçao), *4/59*, *4/210*, 4/211
William I, Prince of Orange, 4/205, 6/111, 6/112
William I, the Conqueror, 2/238, 6/110
Wills, William, 1/86
Wilmington (Delaware), 6/124
Wilpattu national park (Sri Lanka), 5/223
Wilson, Woodrow, *5/247*, 6/136
Windhoek (Namibia), 4/180, 4/181
Windmills, 2/159, *4/198–199*, 4/208, *4/209*
Windsor, Treaty of, 5/94
Windward Islands, 3/74, 4/58
Wine. *See* Grapes and wine
Winfrey, Oprah, 6/145

Winter Palace (St. Petersburg), *5/113, 5/114*, 5/122, *5/122–123*
Winter War (Finland), 2/223
Wisconsin (United States), 6/122, *6/122–123*, 6/123
Wittelsbach, House of, 3/26
Wittenberg (Germany), 3/38
Wittenberg, University of, 3/38
Witwatersrand gold field (South Africa), *5/196*
Wolds, 6/92
Wollongong (Australia), 1/88
Wollongong, University of, 7/8–9
Wolof people, 3/12, 4/122
Wombats, 1/74
Women's roles
  Algeria, *1/28–1/29*, 1/29
  Brunei, 1/179, *1/179*
  China, 2/76
  Egypt, 2/187
  Ghana, *3/45*
  Japan, 3/273
  Kurds, 3/199
  Libya, 4/71
  Liechtenstein, 4/75
  Malawi, 4/90
  Morocco, 4/165, *4/165*
  prehistoric, 7/47
  Saudi Arabia, 5/146–147
  Syria, 5/273
  Tunisia, 6/51, 6/55
  Turkey, 6/58
  United States, 6/156
Wood, Grant, 6/158
Wood Buffalo National Park (Alberta), 2/39
Woodcarving, *4/223, 5/27*
Wooden Lintel, Temple of the (Belize), *1/145*
Woodlark Island (Papua New Guinea), 5/36
Woodpeckers, Hairy, *2/38*
Woods, Tiger, 6/145
Wool. *See* Sheep and wool
World. *See* Earth
World Bank, 2/47
World Council of Churches, 5/265
World Health Organization, 5/265
World Trade Center, 1/13, 6/137, *6/137*
World Trade Organization, 2/61, 2/93, 4/43, 5/149
World War I
  role of
    Albania, 1/20
    Algeria, 1/24
    Armenia, 1/67
    Belarus, 1/130
    Cameroon, 2/18
    Canada, 2/31
    Czechoslovakia, 2/148, 2/152
    Denmark, 2/157
    Federated States of Micronesia, 2/215
    Iraq, 3/187
    Jordan, 3/282

    Lithuania, 4/76
    Luxembourg, 4/79
    Macedonia, 4/83
    Marshall Islands, 4/116
    Mesopotamia, 3/197
    Namibia, 4/180
    Norway, 4/247
    Palestine, 3/212
    Papua New Guinea, 5/34
    Sweden, 5/246
    Switzerland, 5/255
    Tunisia, 6/48
  warring nations
    Austria-Hungary, 1/98, 1/104, 3/107, 3/111
    Belgium, 1/133, 1/140
    France, 2/233, 2/245, 2/247
    Germany, 3/21, 3/33
    Japan, 3/264
    Montenegro, 4/157
    Ottoman Empire, 6/69
    Russia, 5/113, 5/122
    Serbia, 5/158, 5/169
    United Kingdom, 6/112
    United States, 6/136
World War II
  role of
    Albania, 1/20–21
    Algeria, 1/24–25
    Austria, 1/99, 1/104, 1/108
    Azores, 1/111
    Belarus, 1/131
    Casablanca meeting, 4/167
    Federated States of Micronesia, 2/215
    Finland, 2/223, 2/224, 2/230
    Gabon, 3/9
    Gibraltar, 3/46–47
    Jewish people, 3/21, 4/206
    Jews, 3/212
    Kiribati, 4/17
    Korea, 4/23
    Laos, 4/40
    Macedonia, 4/83
    Malaysia, 4/95
    Malta, 4/114
    Marshall Islands, 4/117
    Myanmar, 4/179
    New Hebrides, 6/169
    New Zealand Maori, 4/223
    Pacific Islands, 5/12
    Palau, 5/27
    Papua New Guinea, 5/34–35
    Philippines, 5/57, 5/66, 5/71, *5/71*
    Qatar, 5/98
    Serbia, 5/158–159, 5/169
    Singapore, 5/174
    Solomon Islands, 5/185
    Sweden, 5/246
    Switzerland, 5/255
    Thailand, 6/37
    Vietnam, 6/181
    Yalta Conference, 5/77, *5/115*
  warring nations
    Belgium, 1/133, 1/140

    Canada, 2/31
    China, 2/64, 7/105
    Czechoslovakia, 2/148, 2/150, 2/152–153, 3/26
    Denmark, 2/157
    France, 2/233, *2/244*, 2/245, 2/250, 2/255
    Germany, 3/21, 3/22, 3/26, 3/29, 3/31, 3/32, 3/33, 3/36
    Greece, 3/51
    Iran, 3/182
    Italy, 3/229
    Japan, 3/253, 3/254, 3/258–259, 3/262, 3/264–265, 3/267, 3/270
    Luxembourg, 4/79
    Netherlands, 4/207
    Norway, 4/247
    Poland, 5/72, 5/77, 5/78, 5/80
    Soviet Union, 5/114, *5/115*, 5/123, 5/126
    United Kingdom, 6/89, 6/107, 6/112, *6/113*
    United States, 6/136
World Wildlife Fund, 6/24
Wörther, Lake, 1/101, *1/101*
Wouri River, 2/18
Wren, Sir Christopher, 6/102
Wright, Frank Lloyd, *6/158–159*
Writing
  cuneiform, 3/186, 3/196
  hieroglyphics, 2/181
  Indus River Valley, 5/24
  Japan, 3/268
  Korea, 4/31
  Maya, 3/100
  Minoan culture, 3/62
  See also Alphabet
Wrocław (Poland), 5/79
WTO. *See* World Trade Organization
Wu dialect, 2/76
Wuhan (China), 2/88, 2/89
Wye River, 6/94
Wyeth, Andrew, 6/158
Wyoming (United States), 6/126, 6/132
Wyspiański, Stanisław, 5/73
Wyss family, 5/255

X

Xavier, St. Francis, 3/154
Xhosa people, 5/194
Xi Jiang River, 2/73
Xi Xia (China), 4/150
Xia culture, 2/62, 7/57
Xi'an (China), 2/66, 2/80
Xinjiang (China), 2/71, 2/79
Xinjiang-Mongolian Uplands (China), 2/70, 2/70–71
Xochimilco, Lake, *4/138–139*

Y

Y-chromosome DNA, 7/11–13
Yablonovyy range, 5/125
Yahya Khan, Agha Mohammad, 5/16
Yaks, 2/70, *2/75, 2/95, 2/95*, 2/97

Yaksa tribe, 5/226
Yakut people, 5/126
Yakutsk (Russia), 5/124
Yakuza, 3/275, 3/276
Yalta Conference, 5/77, *5/115*
Yalu River, 4/20
Yamato clan, 3/264
Yamoussoukro (Côte d'Ivoire), 2/126, 2/128
Yams, *6/45*
Yana River, 7/65
Yan'an (China), *2/72–73*
Yanbu (Saudi Arabia), *5/149*
Yang di-pertuan agong, 4/95
Yangon (Myanmar), 4/174, *4/177, 4/178*
Yangshao culture, 2/62
Yangtze River, 2/70, 2/73, 2/81, **2/88**, 2/88–89, *2/88–89*, 2/94
Yangtze River Delta, 2/89
Yangtze River Valley, 2/72–74, 7/56, 7/57
Yangzhou (China), *2/81*
Yantra River, 1/182
Yanukovych, Viktor, 6/83
Yao people, 4/90–91, 6/18, *6/30–31*
Yaoundé (Cameroon), 2/19, 2/21
Yap Islands, 2/215, *2/215*
Yar'Adua, Umaru, 4/239
Yaravi, 5/54
Yasaka Shrine, *3/268*
Yathrib, 5/150
Yaxchilan, *7/81*
Yazidis people, 3/192, 3/193
Yeats, William Butler, 3/201, *3/205*, 3/206
Yellow fever, 5/32
Yellow Hat sect, 2/96–97
Yellow River. *See* Huang He
Yellow Sea, 4/20, 4/22, 4/26
Yellowstone National Park (United States), 6/132, *6/133*
Yeltsin, Boris N., 4/8, 5/110–111, *5/114*, 5/115, 6/75
Yemen, *5/150*, 6/196–201, *6/196–201*, **6/197**
    economy, 6/196–197
    government, 6/96, 6/201
    history, 6/195, 6/198–199
    land, 6/194–195
    people, 6/200–201, *6/200–201*
Yemen (Aden), 6/195, 6/196, 6/199
Yemen (Sanaa), 6/195, 6/196, 6/199
Yemeni Socialist Party, 6/199
Yenisey River, 5/125
Yerba maté, *6/160*
Yerebatan Sarayi, 6/63
Yeti, 4/190
Yi Songgye, 4/23
Yiddish language, 3/214
Yilmaz, Mesut, 6/59
Yin and yang, 4/30

Yiris, 1/186
Yoido Plaza (Seoul), *4/27*
Yokohama (Japan), 3/270, 3/272
Yom Kippur War, 3/213
Yoruba people, 1/147, 4/236, 4/243, *4/244*, 4/244–245
Young, Andrew, 6/145
Young Turks, 6/68–69
Younger Dryas period, 7/72–73
Ypacaraí, Lake, *5/42–43*
Yü Shan Mountain, 6/10
Yuan dynasty, 2/62, 4/150, 4/151
Yuan Shikai, 2/64, *2/64*
Yucatán Peninsula, 1/144, 1/145, 4/130–131, 4/133, 4/137, 4/142
    human settlement, 7/78, **7/80**, 7/80–81, *7/80–81*
Yuchanyan Cave, 7/56, *7/56–57*
Yue dialect, 2/76
Yugoslavia
    Bosnia-Herzegovina in, 1/156–157
    Croatia in, 2/130–131
    Kosovo in, 4/33
    Macedonia in, 4/82–83
    Montenegro in, 4/156, 4/157
    Serbia in, 5/155, 5/158–161, **5/159**
    Slovenia in, 5/182–183
    See also by republic
Yuit, 5/126
Yukon (Canada), 2/26, 2/39
Yunggang Grotto (China), *2/67*
Yunnan province (China), 2/75, 2/79, 2/89
Yurts, 2/78, 4/150
Yushchenko, Viktor, 6/83

Z

Zafy, Albert, 4/87
Zagora (Morocco), 4/169
Zagreb (Croatia), 2/130
Zagros Mountains, 3/178, 3/180, *3/180–181, 3/190–191, 3/191*, 3/198
    prehistoric settlement, 7/38–39
Zagwé dynasty, 2/206
Zahir, Muhammad, 1/12
Zaibatsu, 3/262
Zaire. *See* Congo, Democratic Republic of the
Zambezi River, *4/172*, 6/202, 6/204, 6/206, 6/207
Zambia, *6/202–205*, **6/203**
    economy, 6/203
    government, 6/202
    history, 6/202–203
    land and people, 6/204–205, *6/204–205*
Zambia, University of, 6/205
Zamkowy Square (Warsaw), 5/80
Zandvoort resort (Netherlands), 4/199
Zanj people, 3/197

ZANU-PF (Zimbabwe), 6/208
Zanussi, Krzysztof, 5/73
Zanzibar (Tanzania), 6/18, 6/19, 6/21, *6/23*
Zapata, Emiliano, 4/141
Zapotec Indians, 4/140
Zapotec language, 4/136
Zardari, Asif Ali, 5/17
Zauditu, 2/206
Zaydi sect, 6/195, 6/201
Zedillo Ponce de León, Ernesto, 4/129
Zeebrugge (Belgium), 1/139, 1/142
Zeit, Die (newspaper), 3/30
Zelaya, José Santos, 4/226
Zen Buddhism, 3/279, *3/279*
Zenobia, 5/266
Zero kilometer stone, 1/61
Zeroual, Liamine, 1/26
Zeta River, 4/156
Zeus, Temple of (Cyprus), 2/143
Zhao Tuo, 6/180
Zhelev, Zhelyu, 1/180
Zhivkov, Todor, 1/180
Zhou Enlai, *2/64*
Zhou people, 2/62
Zhu Jiang, 2/84, 2/86
Zia-ul-Haq, Mohammed, 5/16–17
Zielona Góra (Poland), 5/82
Ziggurats, *3/196–197*
Ziggy Marley & the Melody Makers, 3/251
Zihuatanejo (Mexico), 4/143
Zijin Mountain, *2/80*
Zimbabwe, 2/111, 6/206–211, *6/206–211*, **6/207, 6/209**
    history, 6/208–209, 7/99
    land, 6/206, *6/206–207*
    Mozambique, relations with, 4/170
    people and economy, 6/206, *6/206–207, 6/209*
    See also Great Zimbabwe
Zimbabwe African National Union-Patriotic Front, 6/208
Zinc, *2/34,* 4/175
Zionist movement, 3/212
Ziyad, Tariq ibn, 3/46
Zócalo (Mexico City), 4/138
Zog I, 1/20
Zola, Émile, 2/253
Zonguldak (Turkey), 6/67
Zoroastrianism, 3/179
Zoser, *7/41*
Zuazo, Siles, 1/155
Zugspitze mountain, 3/25
Zuid-Holland (Netherlands), 4/203
Zuider Zee, 4/208, 4/209
Zulu people, 5/192, 5/193, 5/194, *5/194*
    empire of, 7/99
Zunil (Guatemala), *3/84*
Zurich (Switzerland), 2/140, 5/255, 5/261, 5/263, *5/264–265*

THE WORLD ON THE WEB

Websites for independent countries included in *The World Book Encyclopedia of People and Places* are listed below. Please note that World Book, Inc., is not responsible for the content of websites other than its own.

Afghanistan, *Afghanistan's Web Site,* http://www.afghanistans.com

Albania, *Ministry of Foreign Affairs,* http://www.mfa.gov.al/

Algeria, *Embassy of Algeria,* http://www.algeria-us.org

Andorra, *Andorra, Tourisme SAU,* http://www.andorra.ad

Angola, *Embassy of the Republic of Angola,* http://www.angola.org

Antigua and Barbuda, *Antigua and Barbuda Department of Tourism,* http://www.antigua-barbuda.org

Argentina, *Ministry of Tourism,* http://www.turismo.gov.ar

Armenia, *Ministry of Foreign Affairs,* http://www.armeniaforeignministry.com

Australia, *Australian Government,* http://australia.gov.au

Austria, *Embassy of Austria,* http://www.austria.org

Azerbaijan, *Eurasianet, Azerbaijan,* http://www.eurasianet.org/resource/azerbaijan

The Bahamas, *The Commonwealth of The Bahamas,* http://www.bahamas.gov.bs

Bahrain, *Embassy of the Kingdom of Bahrain,* http://www.bahrainembassy.org

Bangladesh, *Bangladesh Government,* http://www.bangladesh.gov.bd

Barbados, *Barbados Government Information Service,* http://www.gisbarbados.gov.bb

Belarus, *President of the Republic of Belarus,* http://www.president.gov.by

Belgium, *Belgium Federal Government, Official Information and Services,* http://www.belgium.be

Belize, *Belize Government Official Portal,* http://www.belize.gov.bz

Benin, *University of Pennsylvania African Studies Center, Benin Page,* http://www.africa.upenn.edu/Country_Specific/Benin.html

Bhutan, *Bhutan Tourism Corporation Limited,* http://www.kingdomofbhutan.com

Bolivia, *Embassy of Bolivia,* http://bolivia-usa.org

Bosnia-Herzegovina, *Embassy of Bosnia and Herzegovina,* http://sarajevo.usembassy.gov/

Botswana, *Republic of Botswana,* http://www.gov.bw

Brazil, *Embassy of Brazil,* http://washington.itamaraty.gov.br/en-us/

Brunei, *The Prime Minister's Office of Brunei Darussalam,* http://www.pmo.gov.bn/

Bulgaria, *Embassy of the Republic of Bulgaria,* http://www.webhousing.biz/~bulgaria

Burkina Faso, *University of Pennsylvania African Studies Center, Burkina Faso Page,* http://www.africa.upenn.edu/Country_Specific/Burkina.html

Burundi, *University of Pennsylvania African Studies Center, Burundi Page,* http://www.africa.upenn.edu/Country_Specific/Burundi.html

Cambodia, *Kingdom of Cambodia,* http://www.cambodia.gov.kh

Cameroon, *Consulate of the Republic of Cameroon,* http://www.cameroonconsul.com

Canada, *Government of Canada,* http://canada.gc.ca

Cape Verde, *University of Pennsylvania African Studies Center, Cape Verde Page,* http://www.africa.upenn.edu/Country_Specific/C_Verde.html

Central African Republic, *University of Pennsylvania African Studies Center, Central African Republic Page,* http://www.africa.upenn.edu/Country_Specific/CAR.html

Chad, *Chad Embassy,* http://www.chadembassy.org

Chile, *Embassy of Chile,* http://www.chile-usa.org

China, *Embassy of the People's Republic of China in the United States of America,* http://www.china-embassy.org/eng

Colombia, *Embassy of Colombia,* http://www.colombiaemb.org

Comoros, *University of Pennsylvania African Studies Center, Comoros Page,* http://www.africa.upenn.edu/Country_Specific/Comoros.html

Congo, Democratic Republic of the, also known as Congo (Kinshasa), *University of Pennsylvania African Studies Center, Democratic Republic of Congo Page,* http://www.africa.upenn.edu/Country_Specific/Zaire.html

Congo, Republic of the, also known as Congo (Brazzaville), *University of Pennsylvania African Studies Center, Congo Page,* http://www.africa.upenn.edu/Country_Specific/Congo.html

Costa Rica, *Costa Rican Institute of Tourism,* http://www.visitcostarica.com

Côte d'Ivoire, *University of Pennsylvania African Studies Center, Côte d'Ivoire Page,* http://www.africa.upenn.edu/Country_Specific/Cote.html

Croatia, *Embassy of the Republic of Croatia,* http://www.croatiaemb.org

Cuba, *Cuba Ministry of Tourism,* http://www.cubatravel.cu/otroe

Cyprus, *Cyprus Tourism Organisation,* http://www.visitcyprus.com

Czech Republic, *Ministry of Foreign Affairs,* http://www.mzv.cz/jnp

Denmark, *Ministry of Foreign Affairs,* http://www.denmark.dk

Djibouti, *University of Pennsylvania African Studies Center, Djibouti Page,* http://www.africa.upenn.edu/Country_Specific/Djibouti.html

Dominica, *Discover Dominica Authority,* http://www.dominica.dm

Dominican Republic, *State Secretariat of Tourism, Dominican Republic,* http://www.godominicanrepublic.com

East Timor, also known as Timor-Leste, *Government of Timore-Leste,* http://www.gov.east-timor.org/

Ecuador, *Embassy of Ecuador,* http://www.ecuador.org

Egypt, *Embassy of the Arab Republic of Egypt,* http://www.egyptembassy.net

El Salvador, *Embassy of El Salvador,* http://sansalvador.usembassy.gov/

Equatorial Guinea, *University of Pennsylvania African Studies Center, Equatorial Guinea Page,* http://www.africa.upenn.edu/Country_Specific/Eq_Guinea.html

Eritrea, *World Factbook, Eritrea,* https://www.cia.gov/library/publications/the-world-factbook/geos/er.html

Estonia, *President of the Republic of Estonia,* http://www.president.ee/en

Ethiopia, *Embassy of Ethiopia,* http://www.ethiopianembassy.org/

Federated States of Micronesia, *Government of the Federated States of Micronesia,* http://www.fsmgov.org

Fiji, *Government of Fiji,* http://www.fiji.gov.fj

Finland, *Embassy of Finland,* http://www.finland.org/Public/Default.aspx

France, *Ministry of Foreign Affairs,* http://www.diplomatie.gouv.fr

Gabon, *University of Pennsylvania African Studies Center, Gabon Page,* http://www.africa.upenn.edu/Country_Specific/Gabon.html

Gambia, *University of Pennsylvania African Studies Center, Gambia Page,* http://www.africa.upenn.edu/Country_Specific/Gambia.html

Georgia, *Ministry of Foreign Affairs,* http://www.mfa.gov.ge

Germany, *German Missions in the United States,* http://www.germany.info

Ghana, *Ghana Embassy,* http://www.ghanaembassy.org

Greece, *Ministry of Foreign Affairs,* http://www.mfa.gr/en

Grenada, *Board of Tourism,* http://grenadagrenadines.com

Guatemala, *Department of Tourism,* http://www.visitguatemala.com

Guinea, *University of Pennsylvania African Studies Center, Guinea Page,* http://www.africa.upenn.edu/Country_Specific/Guinea.html

Guinea-Bissau, *University of Pennsylvania African Studies Center, Guinea-Bissau Page,* http://www.africa.upenn.edu/Country_Specific/G_Bissau.html

Guyana, *Guyana News and Information,* http://www.guyana.org/

Haiti, *Embassy of Haiti,* http://www.haiti.org

Honduras, *U.S. Department of State, Honduras Desk,* http://travel.state.gov/travel/cis_pa_tw/cis/cis_1135.html

Hungary, *Hungarian National Tourist Office,* http://gotohungary.com

Iceland, *Embassy of Iceland,* http://www.iceland.is/iceland-abroad/us

India, *Ministry of External Affairs,* http://www.meaindia.nic.in

Indonesia, *The Embassy of the Republic of Indonesia,* http://www.embassyofindonesia.org

Iran, *Ministry of Foreign Affairs,* http://www.mfa.gov.ir

Iraq, *Iraq State Company for Internet Services,* http://www.uruklink.net

Ireland, *Department of Foreign Affairs and Trade,* http://foreignaffairs.gov.ie

Israel, *Ministry of Foreign Affairs,* http://www.mfa.gov.il/mfa

Italy, *Embassy of Italy,* http://www.ambwashingtondc.esteri.it/ambasciata_washington

Jamaica, *Embassy of Jamaica,* http://www.embassyofjamaica.org

Japan, *Embassy of Japan in the United States of America,* http://www.us.emb-japan.go.jp

Jordan, *Embassy of the Hashemite Kingdom of Jordan,* http://www.jordanembassyus.org

Kazakhstan, *President of the Republic of Kazakhstan,* http://www.akorda.kz/en/

Kenya, *Government of Kenya,* http://www.information.go.ke/

Kiribati, *National Tourism Organisation,* http://www.kiribatitourism.gov.ki/

Korea, North, *Officeal webpage of the Democratic People's Republic of Korea,* http://www.korea-dpr.com/

Korea, South, *Office of the President,* http://english.president.go.kr

Kosovo, *Embassy of the Republic of Kosovo,* http://ambasada-ks.net/us/

Kuwait, *Embassy of Kuwait,* http://www.kuwaitembassy.us/

Kyrgyzstan, *Embassy of the Kyrgyz Republic,* http://www.kgembassy.org

Laos, *Embassy of the Lao People's Democratic Republic,* http://www.laoembassy.com/

Latvia, *Embassy of Latvia,* http://www.latvia-usa.org

Lebanon, *Embassy of Lebanon,* http://www.lebanonembassyus.org

Lesotho, *University of Pennsylvania African Studies Center, Lesotho Page,* http://www.africa.upenn.edu/Country_Specific/Lesotho.html

Liberia, *Embassy of the Republic of Liberia,* http://www.embassy.org/embassies/lr.html

Libya, *World Factbook, Libya,* https://www.cia.gov/library/publications/the-world-factbook/geos/ly.html

Liechtenstein, *Principality of Liechtenstein,* http://www.welcome.li

Lithuania, *Ministry of Foreign Affairs,* http://www.urm.lt

Luxembourg, *National Tourist Office,* http://www.ont.lu

Macedonia, *Government of the Republic of Macedonia,* http://www.vlada.mk

Madagascar, *Embassy of Madagascar,* http://www.madagascar-embassy.org

Malawi, *University of Pennsylvania African Studies Center, Malawi Page,* http://www.africa.upenn.edu/Country_Specific/Malawi.html

Malaysia, *Ministry of Foreign Affairs,* http://www.kln.gov.my

Maldives, *The President's Office,* http://www.presidencymaldives.gov.mv

Mali, *Embassy of Mali, Washington, DC,* http://www.maliembassy.us

Malta, *Government of Malta,* http://www.gov.mt

Marshall Islands, *Embassy of the United States: Majuro-Marshall Islands,* http://majuro.usembassy.gov

Mauritania, *University of Pennsylvania African Studies Center, Mauritania Page,* http://www.africa.upenn.edu/Country_Specific/Mauritania.html

Mauritius, *Portal of the Republic of Mauritius,* http://www.gov.mu

Mexico, *Ministry of Foreign Affairs,* http://www.sre.gob.mx/en

Moldova, *Republic of Moldova Official Page,* http://romania-on-line.net/general/moldova.htm

Monaco, *Official Government Tourist Office,* http://www.visitmonaco.com/us

Mongolia, *Embassy of Mongolia,* http://www.mongolianembassy.us

Montenegro, *Government of Montenegro,* http://www.gov.me

Morocco, *Government of the Kingdom of Morocco,* http://www.maroc.ma

Mozambique, *University of Pennsylvania African Studies Center, Mozambique Page,* http://www.africa.upenn.edu/Country_Specific/Mozambique.html

Myanmar, *Ministry of Foreign Affairs,* http://www.mofa.gov.mm

Namibia, *Embassy of the Republic of Namibia,* http://www.namibianembassyusa.org/

Nauru, *Government of the Republic of Nauru,* http://www.naurugov.nr/

Nepal, *Tourism Board,* http://www.welcomenepal.com

The Netherlands, *Board of Tourism & Conventions,* http://us.holland.com

New Zealand, *New Zealand Government,* http://www.govt.nz

Nicaragua, *World Factbook, Nicaragua,* https://www.cia.gov/library/publications/the-world-factbook/geos/nu.html

Niger, *University of Pennsylvania African Studies Center, Niger Page,* http://www.africa.upenn.edu/Country_Specific/Niger.html

Nigeria, *Embassy of the Federal Republic of Nigeria,* http://www.nigeriaembassyusa.org

Norway, *Information from the Government and Ministries,* http://www.regjeringen.no

Oman, *Ministry of Information,* http://www.omanet.om

Pakistan, *Government of Pakistan,* http://202.83.164.25/wps/portal

Palau, *Palau Visitors Authority,* http://www.visit-palau.com

Panama, *World Factbook, Panama,* https://www.cia.gov/library/publications/the-world-factbook/geos/pm.html

Papua New Guinea, *Embassy of Papua New Guinea to the Americas,* http://www.pngembassy.org

Paraguay, *Embassy of Paraguay in Canada,* http://www.embassyofparaguay.ca

Peru, *Peruvian Embassy, United Kingdom,* http://www.peruembassy-uk.com

Philippines, *Government of the Republic of the Philippines,* http://www.gov.ph

Poland, *Polska, Official Promotional Website of the Republic of Poland,* http://en.poland.gov.pl/

Portugal, *Presidency of the Portuguese Republic,* http://www.presidencia.pt

Qatar, *Embassy of the State of Qatar,* http://www.qatarembassy.net

Romania, *Government of Romania,* http://www.guv.ro

Russia, *Embassy of the Russian Federation,* http://www.russianembassy.org

Rwanda, *Office of the President,* http://www.paulkagame.com

Saint Kitts and Nevis, *Office of the Prime Minister,* http://www.cuopm.com

Saint Lucia, *Government of Saint Lucia,* http://www.stlucia.gov.lc

Saint Vincent and the Grenadines, *Government of Saint Vincent and the Grenadines,* http://www.gov.vc/

Samoa, *Samoa Tourism Authority,* http://www.samoa.travel/

San Marino, *World Factbook, San Marino,* https://www.cia.gov/library/publications/the-world-factbook/geos/sm.html

São Tomé and Príncipe, *University of Pennsylvania African Studies Center, São Tomé and Príncipe Page,* http://www.africa.upenn.edu/Country_Specific/Sao_Tome.html

Saudi Arabia, *Royal Embassy of Saudi Arabia,* http://www.saudiembassy.net

Senegal, *Senegal Tourist Office USA,* http://www.senegal-tourism.com

Serbia, *Official Web Site of the Serbian Government,* http://www.srbija.gov.rs/?change_lang=en

Seychelles, *University of Pennsylvania African Studies Center, Seychelles Page,* http://www.africa.upenn.edu/Country_Specific/Seychelles.html

Sierra Leone, *Government of Sierra Leone,* http://www.sierra-leone.org

Singapore, *Government of Singapore,* http://www.gov.sg

Slovakia, *The Embassy of the Slovak Republic in Washington,* http://www.mzv.sk/washington

Slovenia, *Ministry of Foreign Affairs,* http://www.mzz.gov.si/en

Solomon Islands, *Visitors Bureau,* http://www.visitsolomons.com.sb

Somalia, *University of Pennsylvania African Studies Center, Somalia Page,* http://www.africa.upenn.edu/Country_Specific/Somalia.html

South Africa, *South Africa Government Online,* http://www.gov.za

Spain, *Sí, Spain,* http://www.sispain.org

Sri Lanka, *Government of Sri Lanka,* http://www.priu.gov.lk

Sudan, *Embassy of the Republic of the Sudan,* http://www.sudanembassy.org/

Sudan, South, *Government of the Republic of South Sudan,* http://www.goss.org/

Suriname, *Embassy of the Republic of Suriname,* http://www.surinameembassy.org/

Swaziland, *Government of the Kingdom of Swaziland,* http://www.gov.sz

Sweden, *Official Gateway to Sweden,* http://www.sweden.se

Switzerland, *The Federal Authorities of the Swiss Confederation,* http://www.admin.ch

Syria, *World Factbook, Syria,* https://www.cia.gov/library/publications/the-world-factbook/geos/sy.html

Taiwan, *Government Information Office,* http://www.gio.gov.tw

Tajikistan, *Embassy of Tajikistan,* http://www.tjus.org/

Tanzania, *Embassy of the United Republic of Tanzania,* http://www.tanzaniaembassy-us.org

Thailand, *Ministry of Foreign Affairs,* http://www.mfa.go.th

Togo, *Embassy of Togo,* http://www.togoleseembassy.com/

Tonga, *Tonga Visitors Bureau,* http://www.thekingdomoftonga.com/

Trinidad and Tobago, *Government of the Republic of Trinidad and Tobago,* http://www.gov.tt

Tunisia, *Tunisia Online,* http://www.tunisiaonline.com

Turkey, *Ministry of Foreign Affairs,* http://www.mfa.gov.tr

Turkmenistan, *Embassy of Turkmenistan,* http://www.turkmenistanembassy.org

Tuvalu, *Talofa!* http://www.tuvaluislands.com

Uganda, *Embassy of the Republic of Uganda,* http://ugandaemb.org/

Ukraine, *Government Portal,* http://www.kmu.gov.ua/control/en

United Arab Emirates, *Embassy of the United Arab Emirates,* http://www.uae-embassy.org

United Kingdom, *Prime Minister's Office,* http://www.number10.gov.uk

United States of America, *Government of the United States,* http://www.usa.gov

Uruguay, *World Factbook, Uruguay,* https://www.cia.gov/library/publications/the-world-factbook/geos/uy.html

Uzbekistan, *Embassy of Uzbekistan,* http://www.uzbekistan.org

Vanuatu, *Vanuatu, Tourism Office,* http://www.vanuatu.travel

Vatican City, *The Holy See,* http://www.vatican.va

Venezuela, *Embassy of the Bolivarian Republic of Venezuela,* http://venezuela-us.org

Vietnam, *Embassy of the Socialist Republic of Vietnam,* http://www.vietnamembassy-usa.org

Yemen, *Embassy of the Republic of Yemen,* http://www.yemenembassy.org

Zambia, *World Factbook, Zambia,* https://www.cia.gov/library/publications/the-world-factbook/geos/za.html

Zimbabwe, *University of Pennsylvania African Studies Center, Zimbabwe Page,* http://www.africa.upenn.edu/Country_Specific/Zimbabwe.html

ACKNOWLEDGMENTS

Maps
Kartographisches Institut Bertelsmann, Gütersloh, Swanston Graphics, Euromap Ltd., Het Spectrum B.V., World Book Encyclopedia

Diagrams
Eugene Fleury, Ted McCausland, Jean Jottrand

Illustrations
John Davies, Bill Donahoe, Michael Gillah, Tom McArthur, Michael Saunders, John Francis, R. Lewis, Leslie D. Smith, Ed Stuart, George Thompson

Photographic acknowledgments
Abbreviations:
AL/Alamy
APA/Apa Photo Agency
APWW/AP Wide World
BPK/Bildarchiv Preussischer Kulturbesitz
BOV/Britain on View
B&U/B&U International Pictures
CP/Camera Press
CS/Colorsport
DP/Das Photo
DRE/Dreamstime
FS/Frank Spooner
GAM/Gamma-Liaison
GET/Getty
GFS/Gamma/Frank Spooner
HL/Hutchison Library
HPC/Hulton Picture Company
IS/istockphoto
JEP/Jürgens Ost und Europa Photo
LAN/Landov
MC/Mansell Collection
MPL/Magnum
MTX/Matrix International
NOV/Novosti
PCP/Paul C. Pet
PH/Photo Researchers
PEP/Planet Earth Pictures
PP/Popperfoto
RF/Rex Features
RHL/Robert Hunt Library
RHPL/Robert Harding Picture Library
SA/Survival Anglia
SAP/South American Pictures
SC/Scala
SCL/Spectrum Colour Library
SG/Susan Griggs
SHU/Shutterstock
SLMG/Soviet Life Magazine
SOV/Sovfoto-Eastfoto
SP/Sipa Press
SYG/Sygma
TAS/Tass
TIB/The Image Bank
TSM/The Stock Market
WC/Woodfin Camp, Inc.
Z/Zefa

VOLUME 1
Cover © Robert Harding Picture Library/SuperStock
2-3 SHU
10-11 Lonely Planet/SuperStock, George Hunter/APA
12-13 SHU, The Flag Research Center
14-15 Disappearing World/HL, Thomas Ives/SG, HL, APA
16-17 Robert Harding Picture Library/SuperStock, Victor Englebert/SG, MC
18-19 F. Jack Jackson/AL, Andre Singer/HL, J Hatt/HL
20-21 SHU, SCL
22-23 Robert Harding Picture Library/SuperStock, Th. Foto/Z, SCL, Th. Foto/Z
24-25 Frans Lemmens/SuperStock, Adam Woolfitt/SG
26-27 SHU, Adam Woolfitt/ SG
28-29 B&U, Sarah Errington/HL, Victor Englebert/SG, HL

30-31 Photononstop/SuperStock, Julian Calder/SG, HL, Cyril Isy-Schwartz/ TIB
32-33 Jocelyn Carlin/Panos Pictures, Michael MacIntyre/HL, SCL
34-35 SHU, Tor Eigeland/SG
36-37 SHU, Eric Miller/Panos Pictures
38-39 HL
40-41 SHU
42-43 Bryan & Cherry Alexander Photography/AL, National Geographic/SuperStock
44-45 Rick Price/SA, Edwin Mickleburgh/ SA
46-47 NASA, Prof Dr Franz, Greg Shirah, GSFC Science Visualization Studio, TOMS
48-49 SHU, Richard Laird/SG
50-51 Prisma/SuperStock
52-53 SHU, Mecky Fögeling
54-55 Mecky Fögeling, Mike Andrews/SG
56-57 Robert Frerck/SG, Goebel/Z, Nancy Durrell McKenna/HL, © David R. Frazier Photolibrary
58-59 Mecky Fögeling, N Russell/Liaison/GFS, Stephen Pern/ HL
60-61 Eduardo Rivero/SHU, age fotostock/SuperStock, Reflejo/SG, Mecky Fögeling
62-63 Diego Giudice/Bloomberg/GET, Reflejo/SG, Tony Morrison/SAP
64-65 Diego Giudice/Bloomberg/GET, AFP/GET, Magnum/PP, PP, International News Photo/PP
66-67 SHU, TAS from SOV
68-69 Stock Connection/SuperStock
70-71 SHU
72-73 Neale Cousland/SHU, Alastair Scott/SG, Guido Alberto Rossi/TIB, David Austen/SG
74-75 John Carnemolla/SHU, Dallas & John Heaton/APA, Jen & Des Bartlett/SA, Robin Morrison, Dallas & John Heaton/APA, L Kuipers/B&U, Jen & Des Bartlett/SA
76-77 William West/AFP/GET, Pete Turner/TIB, Dallas & John Heaton/APA, Robin Smith/Z
78-79 Royal Flying Doctor Service (www.flyingdoctor.org.au), Philip Little/APA, R Woldendorp/ SG, D Baglin/Z
80-81 Phillip Minnis/SHU, Otto Rogge/ TSM, Paul Steel/APA, Bill Bachman/PH, Paul Steel/APA
82-83 Jacky Ghossein/AFP/GET, APA, Paul Steel/APA
84-85 DRE, Joe Ng/SHU, Lee Torrens/SHU
86-87 Bruce Glikas/FilmMagic/GET, Australian Overseas Information Service, Art Seitz/GFS
88-89 SHU, Z, Peter Hendrie/TIB, Guido Alberto Rossi/TIB
90-91 Alastair Scott/SG, Z, Philip Little/APA, Christine Pemberton/HL
92-93 Gary Bell/GET, 88 Carl Roessler/PEP, APL/Z, Carl Roessler/PEP
94-95 Scholz/Z
96-97 SHU, Damm/Z
98-99 Imagno/GET, Heeresgeschichtlichtes Museum, Vienna, HPC, MC
100-101 M Thonig/Z, Adam Woolfitt/SG
102-103 Adam Woolfitt/SG,
104-105 Adam Woolfitt/SG,
106-107 E. J. Baumeister Jr./AL, Dieter Nagl/AFP/GET, W L Berssenbrugge/Z, Adam Woolfitt/SG, John Lewis Stage/TIB
108-109 Gala/SuperStock, Damm/Z, Adam Woolfitt/SG
110-111 SHU, Serguei Fedorov/WC
112-113 J Richardson/SCL, Sunak/Z, PCP
114-115 SHU, Walter Bibikow/Jon Arnold Images/AL
116-117 SHU, John Lewis Stage/TIB
118-119 DRE
120-121 DRE, Jehangir Gazdar/SG
122-123 Nicholas Cohen/SG, Sarah Errington/HL, Jehangir Gazdar/SG
124-125 Nicholas Cohen/SG, Sarah Errington/HL, Jehangir Gazdar/SG

126-127 Bartholomew/Liaison/GFS, Naray Angants/HL, Nicholas Cohen/SG
128-129 SHU, Reflejo/SG, Ted Spiegel/ SG
130-131 SHU, TAS from SOV
132-133 Mirenska Olga/SHU
134-135 DRE, Charles W Friend/SG
136-137 Z, Agence Belga, PCP, Weinberg Clark/TIB
138-139 Don Di Sante, James Davies/HL, PCP, SCL
140-141 Don Di Sante, Photononstop/SuperStock, HL, Charles W Friend/SG
142-143 PCP, Melanie Friend/SG, Louis De Poortere
144-145 DRE, age fotostock/SuperStock
146-147 DRE, Victor Englebert/SG
148-149 David Biagi/DRE, Bob Krist, Corbis Stock Market, Larry Dale Gordon/TIB
150-151 SHU, Sarah Errington/HL
152-153 Niv Koren/SHU
154-155 DRE, Jacques Jangoux/AL, Jan Csernoch/AL
156-157 SHU
158-159 SHU, Rob Cousins/SGS
160-161 SHU
162-163 SHU
164-165 APWW, Mary Evans Picture Library, RHL, Sipa/Colorsport, Wodder/TIB, Mary Evans Picture Library, Michael MacIntyre/HL
166-167 H J Anders/C&J Images/ TIB, S Jorge/TIB, Barbosa/TIB, J L Taillade/Colorsport
168-169 Frontpage/SHU, B&U, HL, Vautier-De Nanxe
170-171 Morten Elm/DRE, Reflejo/SG, J L Taillade/Colorsport
172-173 David Davis/SHU, HL, Giuliano Colliva/Colorsport
174-175 Paulo Whitaker/Reuters/LAN, Dr Nigel Smith/HL, Peter Frey/TIB, Sarah Errington/ HL
176-177 Caetano Barreira/ZUMA Press, Embraer/Brazil, Jesco von Puttkamer/HL
178-179 SHU, S Tucci/GFS, Michael Coyne/GFS, S Tucci/ Liaison/GFS
180-181 SHU, Dimitar Petarchev/DRE
182-183 Jürgens Ost und Europa Photo, Adam Woolfitt/SG, Jürgens Ost und Europa Photo
184-185 Jürgens Ost und Europa Photo, Adam Woolfitt/SG, Jean Michel Nossant/GFS
186-187 SHU, Chris Johnson/HL
188-189 Robert Harding Picture Library/ SuperStock, HL, Dr. Nigel Smith/HL

VOLUME 2
Cover © SIME/eStock Photo
2-3 Barnaby Chambers/SHU
8-9 Luciano Mortula/SHU
10-11 Nestor Noci/SHU, Leon Schadeberg/SG
12-13 Leon Schadeberg/SG, John Bulmer/SG, Leon Schadeberg/SG, Terry Madison/TIB
14-15 John Bulmer/SG, Kushch Dmitry/SHU, Luciano Mortula/SHU
16-17 Victor Englebert/SG, HL, Bernard Regent/HL, Victor Englebert/SG
18-19 Pozzo di Borgo Thomas/SHU, Art Directors & TRIP/AL
20-21 Victor Englebert/SG, Heinz Stucke/FS, V & A Wilkinson/HL
22-23 Andrew McDonough/SHU
24-25 SHU
26-27 Jordan Tan/SHU, Lorraine Swanson/SHU, Chris Lorenz/DRE, DRE
28-29 Christie's Images/Corbis
30-31 MC, MC, Guido Alberto Rossi/TIB, Chris Jackson/GET
32-33 R Vroom/Canadian High Commission, D Wilkins/APA, Greenpeace/Baptist/GFS, APWW
34-35 Bernard Regent/HL, Udo Weitz/GET, First Light/AL, David Boily, AFP/GET, SHU
36-37 Marc Romanelli/ TIB, K Kummels/Z, Damm/Z, Alexis Boichard/GET
40-41 The Flag Research Center, Juhana Lampinen/SHU, Juhana Lampinen/SHU
42-43 Jay Hocking/SHU, Victor Englebert/SG

44-45 Ecoimages/SHU, DEA/C. SAPPA, De Agostini Picture Library/GET

46-47 Wendy Watriss/SG, Chris Johnson/HL, HL

48-49 Mike Andrews/ SG

50-51 Alan Kraft/SHU, Edgard Garrido, Reuters/LAN

52-53 Mike Andrews/SG, Norwegian Archipelago Expedition/SG, Rob Cousins/SG

54-55 B&U, Victor Englebert/SG Brian Moser/HL, Marc C. Johnson/SHU

56-57 Brian Moser/HL, Mireille Vautier, SHU

58-59 Louis-Laurent Grandadam, The Image Bank/GET

60-61 Barnaby Chambers/SHU

62-63 MC, Robert Hunt Library, Hans Verkroost, SHU, From a 19th century Korean album, British Library, London (HIP/Art Resource)

64-65 Robert Hunt Library, John Dominis, Time Life Pictures/GET, APWW, Ed Nachtrieb/LAN

66-67 Gulbenkian Collection, Durham, Freer Gallery of Art/ET Archive Hans Verkroost

68-69 Mikhail Nekrasov/SHU, Hans Verkroost, NASA/JPL

70-71 Thiele/Z, Sarah Errington/HL, Dave Brinicombe/HL

72-73 Mike Andrews/SG, Harald Sund/TIB, Christina Dodwell/HL

74-75 GFS

76-77 Janeart Ltd/TIB, Yadid Levy, AL

78-79 Melanie Friend/HL, Leon Schadeberg/SG, Sarah Errington/HL, Don Klumpp/TIB, Peter Parks, AFP/GET

80-81 Melanie Friend/HL, Melanie Friend/HL, Guo Yifu, Xinhua/LAN, Geraint Tellem/AL

82-83 Felix Greene/HL, Harald Sund/TIB, Hans Verkroost, APWW

84-85 Hans Verkroost, Superstock, Jean Kugler/APA, Ho Philip/SHU

86-87 Adina Tovy/APA, RCA Nichols/APA, Sipa/RF, Julien Nieman, APA

88-89 Chi Hong/SHU

90-91 HL, Adam Woolfitt/SG HL

92-93 Reuters/LAN, Reuters/LAN, Reuters/LAN

94-95 Nevada Wier/TIB, Leon Schadelberg/SG, APA

96-97 Marcus Brooke, APA, Jean Kugler, Felix Green/HL APA,

98-99 José Antonio Sánchez Poy/DRE

100-101 SHU, Patrick Rouillard/TIB

102-103 Peter R Dickerson/SG, Brian Moser, HL, Victor HL, Daniel Munoz, Reuters/LAN

104-105 Brian Moser/HL, Victor Englebert/SG, Borderlands/AL

106-107 Tony Morrison/SAP, Adam Woolfitt/SG, Tony Morrison/SAP

108-109 J. S. Callahan, tropicalpix/AL, Al Venter/GFS

110-111 DRE, Z

112-113 IS

114-115 Andrew Hill/HL, Bruce Davidson/SA, Alissa Everett, Reuters/LAN

116-117 David Beatty/SG, Lee Lyon/SA

118-119 Kal Muller/SG

120-121 Gilles Merment/GFS, Peter Riley/South American Pictures, Gilles Merment/GFS

122-123 Jim Lipschutz/SHU, Timur Kulgarin/SHU

124-125 Gilles Merment/GFS, Peter Riley/South American Pictures, Gilles Merment/GFS

126-127 David Lyons/AL, M Isy-Schwartz/TIB

128-129 Nick Nickolson/TIB, Rentmeester/TIB, Nick Nickolson/TIB

130-131 SHU, T E Clark/HL, SCL

132-133 Damm/Z, Roxana Gonzalez/SHU, Jose Miguel Hernandez Leon/SHU

134-135 Joel Blit/SHU, Felix Greene/HL

136-137 Brian Moser/HL, Brian Moser/HL, John Hatt/HL

138-139 John Bulmer/SG, John Griffith, Adalberto Roque, AFP/GET

140-141 SHU, Brown/FS

142-143 A Gasson/RF, Brown/FS, Thompson/SCL

144-145 SHU

146-147 Roman Sigaev/SHU, Jürgens Ost und Europa Photo

148-149 Mansell Collection Jürgens Ost und Europa Photo, Robert Hunt Library, Nathan Benn/SG, Michal Krumphanzl, CTK from Eastfoto

150-151 Liba Taylor/HL, Jürgens Ost und Europa Photo, Z, Liba Taylor/HL

152-153 Mary Evans Picture Library, Jürgens Ost und Europa Photo, British Film Institute, Jürgens Ost und Europa Photo, B&U International Pictures, Kobal Collection, Bouvet-Merillon/GFS

154-155 Jürgens Ost und Europa Photo, Jaroslav Kubec/TSM, Peter Sellers/Impact

156-157 SHU, W Saller/Z

158-159 Irina Korshunova/SHU, Paul C Pet

160-161 Aleksi Markku/SHU, T Brettmann/Z, Paul C Pet, Andrei Nekrassov/SHU

162-163 VUSLimited/IS, Victor Englebert/SG

164-165 FloridaStock/SHU, Bernard Regent/HL, Adam Woolfitt/SG

166-167 SHU, Martin Rogers/SG

168-169 IS, Darren Whiteside, Reuters/Getty News, Adamini/ GFS

170-171 Michael Folco/GFS

172-173 Pete Oxford, Steve Bloom Images/AL

174-175 Vladimir Melnik/SHU, Mireille Vautier, Mike Andrews/AL

176-177 Mireille Vautier, Victor Englebert/SG, Mireille Vautier

178-179 Alan Root/SA, D & M Plage/SA, Jason Shenai/SG

180-181 SHU

182-183 SHU

184-185 PCP, NASA/Science Photo Library

186-187 Liba Taylor/HL, Sarah Errington/HL, Hutt/Mepha/FS, Dick Doughty/Impact Visuals, Robert Holmes

188-189 Charles W Friend/SG, Pasieka/Z, C Osbourne/Mepha/FS

190-191 PCP, Dave G. Houser, Z, A Hutt/Mepha/FS

192-193 PCP, Mitchell Beazley, PCP, Melanie Friend/HL, PCP, Sahm Doherty, Time Life Pictures/GET, SHU

194-195 Sacha Rocos/GFS, P F Bentley/Photoreporters, Luis Galdamez, Reuters/LAN

196-197 Andre Nantel/SHU, Reflejo/SG

198-199 Jeremy Bigwood/GFS, Reflejo/SG, HL, M C Kiernan/GFS, Boccon/Sipa/RF

200-201 DRE, Heinz Stucke/FS

202-203 IS, The Flag Research Center, Pascal Maitre/ MTX, MARKA/AL

204-205 Igor Sokolov/SHU, Betty

206-207 Marc C. Johnson/SHU, John Bulmer/SG

208-209 John Bulmer/SG, Kal Muller/APA, Photonews/FS, Kal Muller/APA

210-211 Adam Woolfitt/SG, Zeidl/Z, Alastair Scott/SG

212-213 Ros Drinkwater/RF, Annie Price/SA, Ros Drinkwater/RF

214-215 R. W. Brooks/SHU, The Flag Research Center, HL, Michael MacIntyre/HL

216-217 Courtney Keating/SHU, F Lochon/GFS, RF

218-219 Maurice Harvey/ HL, SCL, Charles Zuber/RF, Maurice Harvey/HL

220-221 F Lochon/GFS, Dallas & John Heaton/SCL, SCL, Michael MacIntyre/HL

222-223 Kotoh/Z

224-225 G Sirena/Z

226-227 Kotoh/Z, Mosler/Z, Michael St Maur Sheil/SG

228-229 Scanrail/DRE, Andy Caulfield/TIB, Bernard Gerard/HL, Lr Villota/Stockmarket/Z, Das Photo

230-231 Bernard Regent/HL, Yoray Liberman/GET

232-233 Angelina Dimitrova/SHU

234-235 SHU

236-237 Kerth/Z, Knight/ APA, Chad Ehlers/AL

238-239 Adam Woolfitt/SG, Molyneux/TIB, Picture Contact/AL

240-241 MC, MC, MC, Jeremy Horner/HL, Victor Englebert/SG

242-243 Adam Woolfitt/SG, Alexandra Allison/SG

244-245 MC, Robert Hunt Library, MC, Robert Hunt Library

246-247 Tom Craig/AL, Lazar Mihai-Bogdan/SHU, SHU

248-249 Bernard Regent/HL, Michel Braud/APA, B&U

250-251 HS/ B&U, DRE

252-253 APWW, Oster/Z, Adam Woolfitt/SG, PCP, *The Dance Class* (1874), oil on canvas by Edgar Degas; Musee d'Orsay, Paris (Image Asset Management Ltd./SuperStock)

254-255 John Lewis Stage/TIB, W-E Gudemann

256-257 John Bulmer/SG, Heinz Stucke/FS

258-259 Guido Alberto Rossi/TIB, Erwin Christian/Z

VOLUME 3

Cover © Margo Silver, Stone/Getty Images

2-3 Adrian Zenz/SHU

8-9 Manuel Velasco/IS, Greenshoots Communications/AL

10-11 Bernard Regent/HL, HL, HL, Bernard Regent/HL, Robert J. Ross from Peter Arnold, Inc.

12-13 Alan Kraft, SHU, HL

14-15 Genadi Yakovlev/DRE, Steenmans/Z

16-17 San Zarenber/TIB

18-19 SilverflyFoto/SHU

20-21 HPC, MC, Friedrichsruh/Bismarck Museum/BPK, Hulton Archive/GET

22-23 Vioujard/GFS, HPC, Alex Wong, GET, Wolfgang Rattay/LAN, Chris Niedenthal, Time Life Pictures/GET

24-25 Damm/Z, D Woog/Z, Helbing/Z

26-27 Damm/Z, K Goebel/Z, Damm/Z

28-29 Kolhas/Z, Damm/Z, Robert Hunt Library, Paul Van Reil/TIB, Alex Timaios Photography/SHU

30-31 JEP, Waldkirch/Z, Rhise/Z, Damm/Z

32-33 Iain Masterton/AL, Jan Kranendonk/SHU, Matthew Dixon/IS

34-35 Michael Pasdzior/TIB, Horst Munzig/SG, T Schneiders/Z, N Bahnsen/Z, JEP

36-37 SCL, JEP, Mecky Fögeling

38-39 JEP, Damm/Z, Adam Woolfitt/SG, Ine Beerten/SHU

40-41 Anna Tully/HL, Victor Englebert/SG, Anna Tully/HL, Michael St Maur Sheil/SG

42-43 Wessel Cirkel/DRE, Steve Heap/SHU

44-45 K F Scholz/Z, John Bulmer/SG, HL, Tom Smith/Robert Harding Library

46-47 Nathan Benn/SG, Robert Frerck/SG

48-49 Jim Lopes/SHU

50-51 Elpis Ioannidis/SHU

52-53 Lorraine Kourafas/SHU, Gordon Bell/SHU

54-55 David Beatty/SG, Rick Strang/APA, B&U, Michael St Maur Sheil/SG

56-57 Phil Robinson, PjrFoto/AL, Michael St Maur Sheil/SG, David Beatty/SG, PCP

58-59 PCP Key Color/Z, Peter Eastland/AL

60-61 David Beatty/SG, PCP, Bernard Regent/ HL, G P Reichelt/APA

62-63 MC, MC, K Kerth/Z, Bernard Regent/HL, AFP/GET

64-65 MC, Adam Woolfitt/SG

66-67 John Egan/HL, Adam Woolfitt/SG, Robert Knight/APA

68-69 PCP, Z, Karel Gallas/SHU

70-71 Konrad Helbig/Z, John Marmaras/SG, John Marmaras/SG

72-73 Ferchland/Z, blickwinkel/AL

74-75 Richard Semik/SHU, SHU

76-77 Nicolas Sainte Luce/GFS, Robert Fried/AL, Bruno Morandi, Robert Harding Picture Library/AL

78-79 Allen Brown, dbimages/AL, J Marshall-Tribaleye Images/AL

80-81 PCP, PCP, Mike Cohen/SHU

82-83 Henryk Sadura/SHU, Reflejo/SG

84-85 Robert Frerck/ SG, PCP, J G Fuller/HL

86-87 Raido Valjamaa/IS, Z

88-89 Heinz Stucke/FS, J C Francolon/GFS, M P Revelon/GFS, Joe Penney, APWW

90-91 Monika Wisniewska/DRE, Boundin/GFS, Luc Gnago, Reuters/LAN

92-93 Vladimir Meinik/SHU, Heinz Stucke/FS

94-95 Nick Gordon/SA, John Bulmer/ SG

96-97 Kassia Halteman/SHU, John Hatt/HL, Eric L Wheater/ TIB, Hulton Archive/GET, APWW

98-99 Sarah Errington/ HL, Patrick Farrell, MCT/LAN, Dick Rowan/SG, Sarah Errington/HL

100-101 Shawn Talbot/SHU, Grigory Kubatyan/SHU, Nicholas Gill/AL

102-103 Henryk Sadura/SHU, Adams/GFS, Jon Arnold Images/AL

104-105 Reflejo/SG, Reflejo/SG, Simon Fisher/RF, Elmer Martinez, AFP/GET

106-107 Mikhail Markovskiy/SHU

108-109 SHU, Dr H Kramarz/Z

110-111 Mansell Collection, Robert Hunt Library, INTERFOTO/AL

112-113 HL, Jürgens Ost und Europa Photo, Laszlo Balogh, Reuters/LAN, APWW

189

114-115 Jürgens Ost und Europa Photo, John Bryson/TIB, Zsolt Nyulaszi/SHU
116-117 Heidi Bradner, Panos Pictures, Jürgens Ost und Europa Photo, Barry Lewis Corbis, Jürgens Ost und Europa Photo, Horst Munzig/ SG
118-119 Bjartur Snorrason/SHU, Bob Strong, Reuters/LAN, International Pictures
120-121 J Behnke/Z, Sam Hall/GFS, Robert Francis/HL, Horst Munzig/SG, Konstantin Komarov/SHU
122-123 Paul C Pet, Horst Munzig/SG
124-125 Pep Roig/AL
126-127 Oksana Perkins/SHU
128-129 MC, PP, Adams/Liaison/GFS, Hulton Archive/GET
130-131 PCP
132-133 Stephanie Arnal/GFS, Punit Paranjpe, Reuters/LAN, Punit Paranjpe, Reuters/LAN
134-135 APA, Adam Woolfitt/SG, PCP
136-137 Ben Pipe/AL, Adam Woolfitt/SG, Jehangir Gazdar/SG, Punit Paranjpe, Reuters/LAN
138-139 David Pearson/AL, Lisl Dennis/TIB, Andre Singer/HL, C Maurice Harvey/HL
140-141 Luca Invernizzi/APA, Carlos Freire/HL, J Zittenzieher/Z, Michael MacIntyre/HL
142-143 David Beatty/SG, Nigel Hicks/AL,
144-145 PCP, Simon Hughes/APA, Jenny Pate/HL, PCP
146-147 Craig Lovell/APA, HL, PCP, Jehangir Gazdar/SG
148-149 PCP, David Beatty/SG, IS
150-151 © Tom Brakefield, © Tom Brakefield, Dieter & Mary Plage/SA
152-153 Lyle Lawson/APA
154-155 Pern/HL, Jehangir Gazdar/SG, Brett Froomer/TIB, Jehangir Gazdar/SG
156-157 MC, APA, PCP
158-159 Starfoto/Z, Charles Foale, Leon Schadeberg/SG, Mukesh Gupta, Reuters/LAN
160-161 Robert Francis/HL
162-163 SHU
164-165 R Waldendorp/SG, Robert Francis/HL, PCP
166-167 Peter Treanor/AL, R Ian Lloyd/HL, Guido Alberto Rossi/TIB, R Ian Lloyd/SG
168-169 Larry Dale Gordon/TIB, PCP, Victor Englebert/SG
170-171 Don King/TIB, Victor Englebert/SG, Michael MacIntyre/HL, SHU
172-173 Nigel Westwood/SA, Heinz Stucke/FS, Dieter & Mary Plage/SA
174-175 Valery Shanin/SHU
176-177 Morteza Safataj/SHU
178-179 dbimages/AL, APWW, GET
180-181 Globe Photos, Maroon/Z, Adam Woolfitt/SG, Anthony Howarth/SG, APWW, Hulton Archive/GET
182-183 Metropolitan Museum of Art, NY, Hughes Vassal/GFS, Radesslami/GFS, HL, Raheb Homavandi, Reuters/LAN
184-185 National Gallery of Art, Washington, Cleveland Museum of Art, Hutt/FS
186-187 APWW
188-189 DRE, APWW
190-191 Christine Osborne/FS, Southwell/HL
192-193 J Brown/FS, Ali Yussef, AFP/GET, APWW
194-195 Christine Osborne/FS, Ako Rasheed, Reuters/LAN, APWW
196-197 V Southwell/HL, J Simson/DP, Mario Tama, GET
198-199 BM/HL, Robert Aberman/HL, Christine Osborne/FS
200-201 Thierry Maffeis/SHU
202-203 SHU, Brian Lynch/Bord Fáilte
204-205 MC, Michael St Maur Sheil/SG, Robert Hunt Library
206-207 Adam Woolfitt/SG, Athlone Jerunladen/APA, GP Reichelf/APA, Roger Hutchings/AL
208-209 SHU
210-211 Bernard Regent/HL, Mikhail Levit/SHU
212-213 The Wiener Library, A Sharon & Massada Publishing/Wiener Library, Israel Sun/LAN
214-215 Ina Block/TIB, Sara Binovic/GFS, Uriel Sinai, GET
216-217 Z, SHU, Birute Vijeikiene/SHU
218-219 Nancy Durrell McKenna/HL, Marvin E Newman/TIB, Nathan Benn/SG
220-221 Danilo Ascione/SHU
222-223 Adrian Zenz/SHU
224-225 Starfoto/Z, Studio Benser/Z, PCP
226-227 MC, MC, Hulton Archive/GET, Andrea Seemann/SHU, Mirek Hejnicki/SHU
228-229 MC, MC, MC, PCP, Robin Lawrence/SG

230-231 SCL, Francesco Venturi/KEA/SG, Ted Spiegel/SG, APWW
232-233 Gert von Bassewitz/SG, John G Ross/SG, Adam Woolfitt/SG, Santi Visalli/TIB
234-235 Dennis/TIB, HL, Adam Woolfitt/SG, John Bulmer/SG
236-237 Galleria degli Uffizi, Florence/Bridgeman Art Library, Santa Maria della Victor, Rome/Bridgeman Art Library, Venice, *Annunciation* (late 1470's), by Leonardo da Vinci, Uffizi Gallery, Florence, Italy (SCALA/Art Resource)
238-239 RF, K Kerth/Z, Heinz Stucke/FS, Heinz Stucke/FS, Sonja Bullaty/TIB
240-241 APWW, APWW, Mike Kipling Photography/Alamy Images, Hans Wolf/TIB
242-243 John Heseltine/ SG, PCP, Adam Woolfitt/SG
244-245 John Heseltine/ SG, John Heseltine/SG, Studio Benser/Z
246-247 David W. Hamilton/TIB, John Marmaras/SG, Hans Hofer/APA, Nick Nickolson/TIB
248-249 Victor Istratov/SHU, Carp/Z
250-251 Jack Vartoogian, Lee Abel, Tim Beddow/HL, Giacomoni/GFS, John Hatt/HL
252-253 SHU
254-255 SHU
256-257 Orion Press/Z, G Coliva/TIB, Fred J Maroon/SG
258-259 Tony Martorano/APA, Spectrum Colour Library, Ippei Naoi/IS
260-261 Paul C Pet, Donald Smetzer, TSW/Click Chicago, Orion Press/Kuroda/Z, Paul Chesley
262-263 Grant V Faint/TIB, Neil Beer, Kaku Karita/GFS, Bloomberg/GET, Bloomberg/GET
264-265 Mansell Collection, ET Archive, Kurita-Wada/GFS, Junji Kurokawa, Reuters/LAN
266-267 Michael Holford/Victoria & Albert Museum, London, Alain Eyrard/SG, Michael MacIntyre/HL
268-269 Martin Mette/SHU, Spectrum Colour Library, Michael MacIntyre/HL, Spectrum Colour Library Z
270-271 SHU, Z, Adina Tovy/APA
272-273 Kazuyoshi Nomachi/RF, Paul C Pet, B Edelhajt/GFS, Robin Laurance/SG, David Ball/Spectrum Colour Library, Mark Bassett/AL
274-275 D Schmidt/Z, Ben Simmons/Sipa/RF, RF
276-277 A S I/Colorsport, Colorsport, John Bulmer/SG, Resource Foto/SG, Jim Anderson/SG
278-279 Paul C Pet, Z, Resource Foto/SG, Z
280-281 Paul C Pet, Spectrum Colour Library, Jean Kugler/APA, Spectrum Colour Library
282-283 Gerard Champlong/TIB
284-285 Paul Cowan/SHU, Michael Hardy/SG, Adam Woolfitt/SG
286-287 Kay Chernush/ TIB, Adam Woolfitt/SG, Z, Robert Paul Van Beets/SHU

VOLUME 4
Cover © Frans Lemmens/SuperStock
2-3 IS
8-9 SHU
10-11 SHU
12-13 Damm/Z, Arnold John Labrentz/SHU
14-15 HL, Timothy Beddow/HL, Victor Englebert/SG, Hoffman-Burchard/Z
16-17 David L. Moore/AL
18-19 IS
20-21 Frilet/Sipa/RF
22-23 Ball Miwako/AL
24-25 SCphotos/AL, DRE
26-27 J Reditt/HL, Alain Evrard/ SG, Nathan Benn/SG, Alain Evrard/ SG
28-29 Michael MacIntyre/HL, Alain Evrard/SG, Pat Behnke/AL
30-31 APA, Alain Evrard/SG
32-33 IS, Tony P. Eveling/AL
34-35 IS, SHU
36-37 Tor Eigeland/GFS, Christine Osborne Pictures/AL
38-39 Mike Andrews/SG, DRE
40-41 Jean-Leo Dugast/APA
42-43 Mark Maclean/IS, SHU
44-45 Jean-Leo Dugast/APA, Jean-Leo Dugast/APA, Brian Moser/HL

46-47 Heinz Stucke/FS, Blair Seitz/APA, Jean-Leo Dugast/APA
48-49 Alvis Upitis/TIB, Jürgens Ost und Europa Photo, Steven Burr Williams/TIB, DRE
50-51 B&U
52-53 Ramzi Hachicho/SHU, DRE
54-55 John Bulmer/SG, AFP/GET, Alan Gignoux/AL
56-57 DRE
58-59 Paul C Pet, Ted Spiegel/SG, William Strode/SG, DRE
60-61 James Davis World-wide Photographic Travel Library, Philip Wollmuth/HL, Dr Muller-Seeberg/Z, Dr Muller-Seeberg/Z
62-63 William Strode/SG, Austin J Brown/Z, DRE
64-65 IS, INTERFOTO/AL
66-67 DRE, blickwinkel/AL
68-69 G Noel/Figaro/GFS, John Bryson/TIB, B Regent/HL, HL, Art Directors & TRIP/AL
70-71 Charles W Friend/SG Mepha/FS, JTB Photo Communications, Inc./AL
72-73 B&U, parasola.net/AL
74-75 IS, JTB Photo Communications, Inc./AL
76-77 IS, SHU
78-79 A Davies/Z, Ying Geng/SHU
80-81 Bazin Scorceletti/GFS, Bazin Scorceletti/GFS, Peter Horee/AL, imagebroker/AL
82-83 Nikola Spasenoski/DRE, Bill Bachmann/AL
84-85 Timothy Beddow/HL
86-87 William Wang/IS, DRE
88-89 Timothy Beddow/HL, A & E Bomford/Ardea
90-91 Brian Moser/HL, DRE
92-93 SHU
94-95 P & G Bowater/TIB, DRE,
96-97 Frank Devlin/SG, Victor Englebert/SG, Alain Evrard/SG, HL, R Ian Lloyd/SG
98-99 David Alan Harvey/SG, R Ian Lloyd/HL, Alain Evrard/SG, Peter Horree/AL
100-101 Victor Englebert/SG, Iam Buswell/AL, Geof Kirby/AL
102-103 Michael Salas/TIB, SCPhotos/AL
104-105 Adam Woolfitt/SG, DRE
106-107 Wolfgang Freihen
108-109 IS
110-111 Timothy Beddow/HL, Horst Munzig/SG, HL
112-113 Mick Csaky/HL, Horst Munzig/SG, Bernard Regent/HL
114-115 B&U, IS
116-117 Bryan Lever/IS, Travel Pix/AL
118-119 L Kuipers/B&U, Mike Andrews/SG, Das Photo
120-121 Marcin Kirsz/IS, dbimages/AL
122-123 HL, Sarah Errington/HL
124-125 Derek Berwin/TIB, Tony Johansson/DRE, DRE
126-127 David Hiser/TIB
128-129 DRE
130-131 Horst Munzig/SG, Paolo Gori/TIB, Kal Muller/ APA, Kal Muller/SG
132-133 Robert Valladares/TIB
134-135 Liba Taylor/HL, DP, Liba Taylor/HL, Danita Delimont/AL
136-137 Liba Taylor/HL, PCP, Kal Muller/APA
138-139 Juergen Schmitt/TIB, Reflejo/SG, PCP, Tony Morrison/South American Pictures
140-141 MC, Mary Evans Picture Library, PP, Kal Muller/APA
142-143 Liba Taylor/HL, Al Satterwhite/TIB, Kal Muller/APA, Brett Froomer/TIB
144-145 E Gormsen
146-147 A Makarov/NOV, Maxim Bulat/IS
148-149 B&U, APA, Eric Brownstone/IS
150-151 Brian Moser/Granada TV/HL
152-153 The Flag Research Center, DRE, Frederic J. Brown/AFP/Getty Images
154-155 Brian Moser/HL, Xinhua-News/Z, Brian Moser/HL
156-157 Nikola Miljkovic/IS
158-159 DRE
160-161 B&U, SHU
162-163 HL, B&U, Tor Eigeland/SG, Tordai/HL
164-165 Don Klumpp/TIB, Victor Englebert/SG, Dave Brinicombe/HL
166-167 Adam Woolfitt/SG, Leon Schadeberg/SG, PCP
168-169 Adam Woolfitt/SG, Julian Calder/SG, John Hatt/HL
170-171 V Wentzel/Z, Nico Smit/IS
172-173 Reardon/RF, A Aarhus/Sipa/RF, Reardon/RF, Sarah Errington/HL
174-175 Joe Lynch/APA
176-177 Michael MacIntyre/HL, Xavi Arnau/IS,

190

178-179 Z, HL, S. Giles/HL, Tibor Bognar/AL, AFP/GET
180-181 Rob Cousins/SG, Lucyna Koch/IS
182-183 Michael Friedel/ RF, Frilet/Sipa/RF, Tim Graham/AL
184-185 Charles Foale, APA
186-187 SHU, Caro/AL
188-189 Joan Klatchko/HL, Charles Foale, Jeremy Horner/HL, Mick Csaky/HL
190-191 DRE
192-193 Jorgen Bitsch/Z, Sarah Errington/HL, Neville Presho/Z
194-195 Sarah Errington/HL Neville Presho/Z, Christine Osborne/APA
196-197 Dieter & Mary Plage/SA, APA, Craig Lovell/Eagle Visions Photography/AL
198-199 Shuji Kotoh/Z
200-201 PB/Rex Features, IS
202-203 Adam Woolfitt/SG, Rijksmuseum, Amsterdam, TIB, Bloomberg/GET
204-205 D & J Heaton/APA, T Schneiders/ Z, Puck-Kornetzki/Z, Adam Woolfitt/SG
206-207 Streichan/Z, PCP, DRE, Travelshots.com/AL
210-211 Paul C Pet, Charles Weckler/TIB, Paul C Pet
212-213 Michael MacIntyre/HL Pascule Sorgues/GFS
214-215 Kevin Ebi/AL
216-217 Robin Morrison, John Nguyen/SHU
218-219 RF, RHPL/AL, Nick Servian/AL
220-221 Adam Woolfitt/SG, Phil Walter/GET, Howard Sayer/AL
222-223 RF, RF, Adam Woolfitt/SG, Colorsport
224-225 R Ian Lloyd/SG, Danita Delimont/AL, Tom McHugh/PH
226-227 DRE, Peter Horree/AL
228-229 Enrico/Z, DRE
230-231 John Griffith, Gary Willis/South American Pictures, John Griffith
232-233 DRE
234-235 Dave Brinicombe/HL, Victor Englebert/SG, frans lemmens/AL
236-237 IS
238-239 IS
240-241 HL, Z, Andrew Holt/AL
242-243 HL, Wendy V Watriss/SG, Friedrich Stark/AL
244-245 HL, Werner Forman Archive, HL
246-247 Paul C Pet
248-249 Paul C Pet, IS
250-251 John Kegan/SG, Paul C Pet
252-253 Paul C Pet, B&U International Pictures, DRE
254-255 Jack Lentfer/Survival Anglia, B&U International Pictures, DRE
256-257 Das Photo, Paul C Pet, Tor Eigeland/SG

VOLUME 5
Cover © Pascal Deloche, Godong/Corbis
2-3 SHU
8-9 Marc C. Johnson/SHU, Robert Azzi/SG, Danita Delimont/AL
14-15 Globuss Images/AL
16-17 SHU
18-19 David Fleming/HL, Christine Osborne/APA
20-21 Leon Schadeberg/SG, HL, Lyle Lawson/APA, MARKA/AL
22-23 Christine Osborne/ APA, Christine Osborne/APA, Alain Evrard/APA, Christine Osborne/ APA
24-25 Hanif Raza/Z, Jehangir Gazdar/SG
26-27 R. W. Brooks/SHU, The Flag Research Center, HL, T Bryant/Liaison/GFS
28-29 SHU, Matt Ragen/SHU
30-31 Adam Woolfitt/SG, Adam Woolfitt/SG, Adam Woolfitt/SG, Robert Frerck/SG, JJM Stock Photography/Travel/AL
32-33 Robert Frerck/SG
34-35 Tyler Olson/SHU, Michael MacIntyre/HL, Jose Reis/GFS
36-37 Jose Reis/GFS, Michael MacIntyre/HL, Charlie Nairn/Disappearing World/HL, Michael Coyne/TIB
38-39 Larry Dale Gordon/TIB, Norberto Duarte, AFP/GET
40-41 Rafael Martin-Gaitero/SHU, Rob Cousins/SG, Eric Sander/GFS, Reflejo/SG, Melanie Friend/HL
42-43 H Sunak/Z, Noel/Figaro/GFS, John Bulmer/SG, P Aventurier/GFS
44-45 B&U
46-47 Neale Cousland/SHU, Victor Englebert/SG

48-49 Victor Englebert/SG, Mike Andrews/SG, Victor Englebert/SG, Mireille Vautier
50-51 B Hysebergh/GFS, Robert Frerck/SG, B Hysebergh/GFS
52-53 Mireille Vautier, H R Dörig/HL, Robert Frerck/SG, Cara Moore/TIB, Mireille Vautier
56-57 E Streichan/Z
58-59 SHU, Thomas Cockrem, AL
60-61 Terry Madison/TIB, Walter Ioos Jr/TIB, Michael Friedel/TIB
62-63 Ted Spiegel/ SG, Alain Evrard/SG, Michael MacIntyre/HL
64-65 Frilet/Sipa Press/RF, Alain Evrard/SG, Terry Madison/TIB, Alain Evrard/SG
66-67 Terry Madison/TIB, Jon Arnold Images/AL, Jay Directo, AFP/GET
68-69 Maurice Harvey/HL, Harald Sund/TIB, Walter Ioos Jr/TIB, Joel Nito, AFP/GET, RHPL/SuperStock
70-71 MC, Mary Evans Picture Library, Michael MacIntyre/HL, Robert Hunt Library, Time Life Pictures/GET
72-73 Bartlomiej K. Kwieciszewski/SHU
74-75 Jürgens Ost und Europa Photo, SHU
76-77 Mansell Collection, Jürgens Ost und Europa Photo, Attila Kisbenedek, AFP/GET
78-79 Jürgens Ost und Europa Photo, Morek/ Jürgens Ost und Europa Photo
80-81 Julian Nieman/SG, B&U International Pictures, Stanislaw Tokarski/SHU, APWW
82-83 Jürgens Ost und Europa Photo, CAF Warsaw/Z, Marc Deville/GFS, David Grossman/AL, Ints Kalnins, Reuters/LAN
84-85 Ekaterina Pokrovsky/SHU
86-87 SHU, ©Tony Arruza, Corbis
88-89 Robert Frerck/SG, K Kerth/Z, Jean Krugler/APA, Anthony Howarth/SG
90-91 imagebroker/AL, Reflejo/SG, Bill Wassman/ APA, parasola/AL
92-93 HL/B&U, PCP, Robert Frerck/SG, Tor Eigeland/Blackstar/SG
94-95 MC, MC, F Lochon/GFS
96-97 Cesar Lucas/TIB, Robert Valladares/TIB, Fulvio Roiter/TIB
98-99 SHU, J Brown/Mepha/FS, Jacky Naegelen, Reuters/LAN
100-101 Skinnet-Vernon/Fig. Mag/GFS, HL, Bernard Regent/HL, Heinz Stucke/ FS, Bernard Regent/HL
102-103 Ardelean Dan/SHU, PCL/AL
104-105 Jürgens Ost und Europa Photo, Adam Woolfitt/SG, Charles W Friend/SG
106-107 Adam Woolfitt/SG, Jürgens Ost und Europa Photo, Adam Woolfitt/SG, David Sutherland/AL
108-109 Mecky Fögeling
110-111 Steenmans/ Z, Denis Ukolov/SHU
112-113 Mansell Collection, Dave Brinicombe/HL
114-115 Robert Hunt Library, Jürgens Ost und Europa Photo, Jürgens Ost und Europa Photo, Robert Hunt Library, AFP/GET, AFP/GET
116-117 Jürgens Ost und Europa Photo, B Meyer/HL, Robert Francis/HL, Heinz Stucke/FS, Novosti/GFS, Alexandr Volozhanin, Reuters/LAN
118-119 Havlicek/Z, Jürgens Ost und Europa Photo
120-121 Leslie Woodhead/HL, Jürgens Ost und Europa Photo, Nik Wheeler/SG, Victoria Juleva/Z
122-123 Robert Francis/HL, SHU
124-125 Z, HL, G Ricatto/Z
126-127 Andreas Heumann/SG, Mecky Fögeling
128-129 Andreas Heumann/SG, Mecky Fögeling, HL
130-131 Guenter Guni/IS, Bernard Gerard/ HL, HL
132-133 Kippy Lanker/SHU, Heinz Stucke/FS, John Hatt/HL, CuboImages/AL
134-135 Fanny Reno/SHU, Z, Heinz Stucke/FS
136-137 L Kuipers/B&U, Ian Bradshaw/ SG, Maura Reap/DRE
138-139 SHU, Michael MacIntyre/HL, HL, Michael MacIntyre/HL
140-141 SHU, SCL, SCL
142-143 DRE, HL, Marion Kaplan/AL
144-145 Aidar Ayazbayev/DRE
146-147 Salem Alforaih/SHU
148-149 Anthony Howarth/SG, Peter Ryan/FS, Bernard Gerard/HL, Tor Eigeland/SG
150-151 Lister/ HL, PCP, R Azzi/SG

152-153 Juha Sompinmäki/SHU, Lisl Dennis/TIB
156-157 The Flag Research Center, Nikola Lucic/SHU
158-159 Popperfoto, Robert Hunt Library, Popperfoto, Tessore/Bavaria, SOV, SSPL/GET
160-161 AP/Wide World
162-163 Giaccone/Marka/Picture Group, Diomedia/Alamy Nenad Tomic/Diomedia
164-165 Steve McCurry/MPL, RHPL, allOver photography/Alamy
166-167 Adam Woolfitt/WC, Aleksandar Milosavljevic/Diomedia, Ingvar Bjork/SHU
168-169 Slobodan Ljubisic/Shutterstock, allOver photography/Alamy, Donøevic/RHPL
170-171 SHU, M Friede/G&J Images/TIB, Damm/Z
172-173 Ines Gesell/IS, David Clilverd/HL, Anna Tully/HL
174-175 Marc St Gil/TIB, SHU, SHU
176-177 Banagan/TIB, R Ian Lloyd/HL, P & G Bowater/TIB, R Ian Lloyd/HL, Michael Chua/TIB
178-179 SHU, John Eastcott & Yva Momatiuk/WC, Liba Taylor/HL
180-181 Tomas Sereda/SHU, Lubomir Grejtak/DRE
182-183 Adam Woolfitt/SG, Standa Riha/SHU
184-185 Leslie Woodhead/HL, Michael MacIntyre/HL, Giancarlo Liguori/SHU, Philip Game, AL
186-187 HL, Anthon Jackson/SHU, Omar Faruk, Reuters/LAN
188-189 Harlicek/Z
190-191 The Flag Research Center, SHU
192-193 Mansell Collection, Reuters/Archive Photos
194-195 Magubane/Liaison/GFS, Guy Hobbs/GFS, De Keerle/UK Press/GF, Caroline Penn, Panos Pictures Eric Bouvet/GFS
196-197 Robert Aberman/HL, Laurence Hughes/TIB, Heinz Stucke/HL, age fotostock/SuperStock
198-199 Javier Gil/SHU
200-201 Regien Paassen/SHU
202-203 Rob Cousins/SG, Jean Krugler/APA, Robert Frerck/SG, APA
204-205 Adam Woolfitt/SG, Yadid Levy/AL, Tor Elgeland/AL
206-207 HL, SCL, Tor Eigeland/SG, Vincent West, Reuters/LAN
208-209 Croxford/Z, Robert Frerck/SG, P Goycolea/HL
210-211 F R Damm/Z, F R Damm/Z, Robert Frerck/SG, BL Images/AL
212-213 MC, MC, MC, Adam Woolfitt/SG
214-215 MC, MC, Metropolitan Museum of Art, New York, C Vioujard/GFS, Vioujard Christian, LAN
216-217 PCP, K Kerth/Z, Dr Mueller/Z, PCP
218-219 MRE/B&U, APA, APA, MRE/B&U
220-221 Keresztes/Z, Robert Frerck/SG, B&U
222-223 SHU
224-225 Alain Evrard/SG, Dmitry Rukhlenko/SHU
226-227 Adam Woolfitt/SG, Charles Friend/SG, David Beatty/SG, Adam Woolfitt/SG
228-229 Eric Wheater/TIB
230-231 Gezira/HL, Galyna Andrushko/SHU
232-233 HL, Christine Osborne/Mepha/FS, HL, Christine Osborne/Mepha/FS
234-235 Zohra Bensemra/Reuters/LAN, Hemis/SuperStock, Sarah Errington/HL
236-237 DRE, Eye Ubiquitous/SuperStock
238-239 Adam Woolfitt/ SG, Claudia van Dijk/SHU
240-241 Z, Paul Slaughter/TIB, Elzbieta Sekowska/SHU
242-243 Leslie Garland Picture Library/AL
244-245 Paul C Pet, Antoine Beyeler/SHU
246-247 Popperfoto, United Nations, NY, Popperfoto, Robert Hunt Library, J Claude Francolon/GFS, Pascal Le Segretain/GET, Media24/Gallo Images/GET, Chip East, Reuters/LAN
248-249 Paul C Pet, Mohn/Z, Paul C Pet, Mosler/Z
250-251 Paul C Pet, D H Teuffen/Z, Colin Molyneux/TIB
252-253 Paul C Pet, Paul C Pet
254-255 V Phillips/Z
256-257 Michael St Maur Sheil/SG, Avner Richard/SHU
258-259 Horst Munzig/SG
260-261 Monique Jacot/SG, Ruth L Aebi, Gstaad Aelper/ APA, Horst Munzig/SG, Christian Hartmann, Reuters/LAN
262-263 Ciba-Geigy, Nestlé SA, Monique Jacot/SG, SHU

264-265 Andrew Hill/HL, International Red Cross, Credit Suisse, © CERN
266-267 Adam Woolfitt/SG
268-269 Jamie Simson/DP, SHU
270-271 Fetzer/Z, Kester J Eddy/FS, Jamie Simson/DP, Dr Nigel Smith/HL
272-273 Gunter Heil/Z, Jamie Simson/DP, Dr Nigel Smith/HL, Michele Falzone/AL
274-275 G L Scarfiotti/SG, K J Eddy/FS, G L Scarfiotti/SG

VOLUME 6
Cover © Peter Beck, Corbis
2-3 Liga Alksne/SHU
8-9 Guido Mangold/ TIB, Dick Rowan/SG, Bernard Regent/HL, Dick Rowan/SG
10-11 Charles Weckler/TIB, R Ian Lloyd/HL, Alexis Duclos/GFS, © Cai Yumao, IMAGECHINA
12-13 Jeff Hunter/TIB, SHU
14-15 Robert Hunt Library, Bob Croxford/Z, Alain Evrard/SG, Borderlands/AL
16-17 The Flag Research Center, Pascal Le Segretain/SYG, Arkady Chubykin/IS
18-19 Bruce Davidson/SA
20-21 Bullaty Lomeo/TIB, Liga Alksne/SHU, Graeme Shannon/SHU
22-23 Bullaty Lomeo/TIB, HL, Bullaty Lomeo/TIB, Timothy Beddow/HL
24-25 HL, Alan Root/SA, Paul C Pet, RF, Matthews Purdy/SA
26-27 SHU
28-29 Alain Evrard/APA, Chusak/SG, Nikolay Okhitin/SHU, Sukree Sukplang, Reuters/LAN
30-31 Sarah Errington/HL, Alain Evrard/SG, Pitiphotos/AL
32-33 Chiaroscuro Kessler/APA, Paul C Pet, APA, Guido Alberto Rossi/TIB, IS
34-35 APA, Guido Alberto Rossi/TIB, Adina Tory/APA, Michael MacIntyre/HL
36-37 Liba Taylor/HL, Leon Schadeberg/SG, Paul C Pet
38-39 Paul C Pet, John Lewis Stage/TIB, Alain Evrard/APA, Dr H Ginsburg
40-41 Victor Englebert/SG, Alan Tobey/IS
42-43 Esbobeldijk/DRE, dpa/LAN
44-45 HL, SCL, Michael MacIntyre/HL, SCL
46-47 Adam Woolfitt/SG, Richard Semik/SHU
48-49 Fulvio Roiter/TIB
50-51 © Kurt Scholz/Superstock, Igor Grochev/SHU
52-53 PCP, Bernard Regent/HL, Bulmer/SG, Ben Partner/Z, B&U, J Bulmer/SG, Fethi Belaid, AFP/GET
54-55 Brown/SS, Mahaux/TIB
56-57 Hugh Sitton/TSW, Click/Chicago, SHU
58-59 Jamie Simson/DP, Kaspars Grinvalds/SHU
60-61 H J Burkard/TIB, John Bulmer/SG, Maroon/Z
62-63 Stukke/FS, Charles Friend/SG, Morozova Oxana/SHU
64-65 Timothy Beddow/HL, Richard T Nowitz/APA, Dario Mitidieri, The Image Bank/GET
66-67 Brown/FS, Robert Aberman/HL, © Wolfgang Kaehler, Corbis, Imagestate Media Partners Limited - Impact Photos/AL
68-69 MC, John Lewis Stage/TIB, Finbarr O'Reilly, Reuters/LAN
70-71 Adam Woolfitt/ SG
72-73 K Benser/Z, Dallas & John Heaton/APA, Brett Froomer/TIB
74-75 Christopher Rennie/RHPL, SHU
76-77 G Ricatto/Z, SHU
78-79 Kal Muller/APA, SHU
80-81 Kal Muller/APA, Michael St Maur Sheil/SG, Dave Brinicombe/AL
82-83 Jürgens Ost und Europa Photo, Valery Bareta/SHU
84-85 Vladimir Melnik/SHU, Patrik Dietrich/SHU
86-87 HL, Barry Schofield/AL
88-89 RHPL/SuperStock
90-91 O Henson/BOV, J. Helgason/SHU
92-93 BOV, Verity Johnson/IS, Jon Arnold Images/AL, Patricia Hofmeester/SHU
94-95 BOV, Lukáš Hejtma/SHU, Jeremy Voisey/IS
96-97 Michael St Maur Sheil/SG, T Wood/Stockphotos/TIB, Julian Nieman/SG, SHU
98-99 Northern Ireland Tourist Board, Shorts Brothers Plc, Ailsa Kilpatrick/AL
100-101 Isle of Man Tourism, Imagegallery/AL
102-103 Adam Woolfitt/SG, SHU, DRE, wronaphoto/AL

104-105 Adam Woolfitt/SG, Paul Doyle, AL
106-107 G de Keerle/GFS, Leon Neal/GET, Jason Kench, DRE, Alessia Pierdomenico, Reuters/LAN
108-109 Adam Woolfitt/SG, Jonathan Tennant/AL, Lewis Whyld, PA Photos/LAN, Clive Chilvers/SHU
110-111 MC, FPA, Adam Woolfitt/SG, BOV
112-113 Oscar & Peter Johnson Ltd, London/Bridgeman Art, APA, MC, Alain Nogues, SYG, MC, HPC
114-115 SHU
116-117 Adam Woolfitt/SG, Mike Norton/SHU, Alan Freed/SHU
120-121 John Downman/HL, Peter Miller/TIB, Ian Macpherson/AL, Michael Ventura/AL
122-123 Harald Sund/TIB, Nathan Benn/SG, Michael Salas/TIB, S. Borisov/SHU
124-125 Nathan Benn/SG, Guido Alberto Rossi/TIB, David Davis/SHU, Katherine Welles/SHU
126-127 Reflejo/SG, A Boccaccio/TIB, Dallas and John Heaton/APA, Helen & Vlad Filatov/SHU, Tim Mosenfelder, GET
128-129 Bernard Regent/HL, Richard Rowan's Collection/SG, Gunter Marx, WA/AL, Greg Vaughn/AL
130-131 Paul Dix/SG, Tibor Hirsch/SG, Bob Hallinen, MCT/LAN
132-133 Adam Woolfitt/SG, D Nicol/Geoscience Features, Anton Foltin/SHU
134-135 MC, Virginia Museum of Arts, Richmond
136-137 Library of Congress, Alex Wong/GET, Sean Adair, Reuters/LAN, Library of Congress, NASA
138-139 David W Hamilton/TIB, Leslie Woodhead/HL, Alexis C. Glenn, UPI/LAN
140-141 Richard Erdos/APA, C Leroy/ Sipa–Press/RF, Adam Woolfitt/SG, Insight/Liaison/GFS
142-143 © Lawrence Migdale, Mark Wilson/GET
144-145 Brent Jones/APWW, Bill Bachmann, PhotoEdit
146-147 Adam Woolfitt/SG, Stock Connection Blue/AL, Marc Fischer/IS, David R. Frazier Photolibrary/AL
148-149 Adam Woolfitt/SG, Chad Palmer/SHU, Phil Degginger, NASA/Landsat/AL
150-151 Arthur D'Arazien/TIB, Mark Godfrey, The Image Works, P Guis-Figaro-Magaz/GFS, John Gress, Reuters/LAN
152-153 Adam Woolfitt/SG, Robert Galbraith, Reuters/LAN, David Banks, UPI/LAN, Paul J. Richards, AFP/GET
154-155 HL, Adam Woolfitt/SG, Ted Spiegel/SG, Tim Eagen/SG, SHU
156-157 John Egan/HL, Sean Locke/IS, Jim McIsaac/GET, Andre Jenny/AL
158-159 David J Maenza/TIB, FPI/AL, AfriPics/AL, Glow Images/SuperStock
160-161 B&U
162-163 Bernard Gerard/GFS, Kobby Dagan/SHU
164-165 Bernard Gerard/GFS, Alexandre Fagundes De Fagundes/DRE
166-167 Kramarz/Z, SHU
168-169 Larry Dale Gordon/TIB, Holger Mette/SHU
170-171 Franco Origlia, Getty Images, B&U, Ted Spiegel/SG, SHU
172-173 Mireille Vautier
174-175 Mireille Vautier, Ulrich Mueller/SHU
176-177 Andi Spicer, HL, Mireille Vautier
178-179 Victor Englebert/SG, Mireille Vautier, Andre Jenny/AL
180-181 Guido Alberto Rossi/TIB
182-183 Jean-Leo Dugast/APA, Kang Khoon Seang/SHU
184-185 Francoise Demulber/GFS, HL, Gilles Caron/GFS, Claude Johner/GFS
186-187 Guido Alberto Rossi/ TIB, J C Labbe/GFS, Leon Schadeberg/SG, HL
188-189 Austin J Brown/Z, Austin J Brown/Z
190-191 Damm/Z, Austin J. Brown/Z
192-193 HL
194-195 Magra/B&U
196-197 PCP, Vladimir Melnik/SHU
198-199 Charles W. Friend/SG, Peter Ryan/FS, PCP
200-201 B&U, PCP, B&U, B&U, Peter Ryan/FS
202-203 HL, Marc & Evelyne Bernheim/SG, Jan Zoetekouw/SHU
204-205 Marc & Evelyne Bernheim/SG, Campbell/SA, Obremski/TIB, Ulrich Doering/AL

206-207 Ian Murphy/SG, APWW
208-209 SHU, Desmond Kwande, AFP/GET, Chad Ehlers/AL
210-211 Heinz Stucke/HL, HL, Heinz Stucke FS

VOLUME 7
Cover © Roger De La Harpe, Gallo Images/Corbis
2-3 SHU
4-5 Joe Smoljan/SuperStock
8-9 John Reader/SPL/PH
10-11 Kenneth Garrett
12-13 Michael Haslam/Michael Petraglia, Danita Delimont Stock/awl-images
14-15 Geoff Bailey
16-17 APWW
20-21 Hitendra Sinkar Photography/AL, Borromeo/Art Resource
22-23 Douglas Anderson
24-25 Equinox Graphics/PH, Mauricio Anton/PH
26-27 APWW
28-29 Corbis, Peter Schouten
32-33 awl-images/Masterfile Corporation, Mission archéologique de Qafzeh/PH
34-35 Johannes Krause/Atelier Daynes, Paris, France/Museum of the Krapina Nean-derthals, Krapina, Croatia/Zeljko Kovacic/Jakov Radovcic/Max Planck Institute for Evolutionary Anthropology
36-37 Deutsches Archäologisches Institut
40-41 Balage Balogh/Art Resource, Steve Vidler/SuperStock
44-45 National Academy of Sciences
46-47 Hilde Jensen/University of Tübingen
48-49 Mark N. Boulton/PH
50-51 Eugene Ch'ng/The Open Virtual Reality Journal, Matthew Jacques/SHU
52-53 Gilles Tosello/PH, Kenneth Garrett
56-57 American Museum of Natural History, APWW
58-59 Reuters
60-61 MIXA/AL, The Art Gallery Collection/AL
62-63 The State Hermitage Museum
64-65 Mauricio Anton/PH
68-69 Warren Morgan/Corbis, The Art Archive (National Anthropological Museum, Mexico)
70-71 National Geographic/GET, Kenneth Garrett
72-73 Chase Studio/PH
74-75 Nuka Godfredtsen, University of Copenhagen
76-77 North Wind Picture Archives/AL, Cahokia Mounds State Historic Site, Granger Collection
80-81 Jesús Eloy Ramos Lara/DRE, SEF/Art Resource, The Trustees of the British Museum/Art Resource
82-83 SHU, The Art Archive, George Steinmetz/Corbis
84-85 Diego Goldberg/Sygma/Corbis, AAAS
86-87 Tom Dillehay/National Science Foundation
88-89 "The Fuegians and their Wigwams" (1890) from *Journal of Researches by Charles Darwin* (Ken Welsh, The Bridgeman Art Library)
92-93 Stephen Alvarez/National Geographic Society/GET
94-95 Herbert Kane/National Geographic Society/GET, Bettmann/Corbis, Ivonne Wierink/SHU
96-97 *Tamati Waka Nene* (1890), oil on canvas by Gottfried Lindauer; Auckland City Art Gallery, New Zealand (The Bridgeman Art Library)
98-99 NASA Goddard Space Flight Center
100-101 Aspect Picture Library
102-103 Erich Lessing/Art Resource, Jon Arnold Images/Masterfile
104-105 North Wind Picture Archives/AL, Huntington Library/SuperStock, Granger Collection
106-107 Granger Collection, Pictorial Press/AL

192